Global Environmental Change

AF181320

Springer-Verlag Berlin Heidelberg GmbH

Kirill Ya. Kondratyev · Vladimir F. Krapivin
Gary W. Phillips

Global Environmental Change

Modelling and Monitoring

With 104 Figures and 75 Tables

Springer

Professor Dr. Kirill Ya. Kondratyev
Russian Academy of Sciences
Research Centre for Ecological Safety
Korpusnaya Str. 118
197110 St.-Petersburg, Russia
E-mail: *kondratyev@kk10221.spb.edu*

Professor Dr. Vladimir F. Krapivin
Russian Academy of Sciences
Institute of Radioengineering and Electronics
1 Vvedensky Square
Fryazino, Moscow Region, 141120 Russia
E-mail: *vfk@ms.ire.rssi.ru*

Professor Dr. Gary W. Phillips
Georgetown University Medical Center
Radiation Medicine Department
3970 Reservoir Road NW, Room E202A
Washington, DC 2007, USA
E-mail: *phillipg@georgetown.edu*

ISBN 978-3-642-07773-9 ISBN 978-3-662-04941-9 (eBook)
DOI 10.1007/978-3-662-04941-9

Library of Congress Cataloging-in-Publication Data

Kondrat'ev, K. IA. (Kirill IAkovlevich)
Global environmental change : modelling and monitoring / Kirill Ya. Kondratyev, Vladimir F. Krapivin, Gary W. Phillips.
p. cm.
Includes bibliographical references and index.

1. Global environmental change. I. Krapivin, V.F. (Vladimir Fedorovich) II. Phillips, Gary W., 1940-III. Title.

This work is subject to copyright. All rights are reserved, whether the whole or part of the material is concerned, specifically the rights of translation, reprinting, reuse of illustrations, recitations, broadcasting, reproduction on microfilm or in any other way, and storage in data banks. Duplication of this publication or parts thereof is permitted only under the provisions of the German Copyright Law of September 9, 1965, in its current version, and permission for use must always be obtained from Springer-Verlag. Violations are liable for prosecution under the German Copyright Law.

http:/www.springer.de
© Springer-Verlag Berlin Heidelberg 2002
Originally published by Springer-Verlag Berlin Heidelberg New York in 2002.
Softcover reprint of the hardcover 1st edition 2002
The use of general descriptive names, registered names, trademarks, etc. in this publication does not imply, even in the absence of a specific statement, that such names are exempt from the relevant protective laws and regulations and therefore free for general use.

Cover Design: E. Kirchner, Heidelberg
Typesetting: Fotosatz-Service Köhler GmbH, 97084 Würzburg

SPIN: 10845012 32/3130 – 5 4 3 2 1 0 – Printed on acid free paper

Preface

The growing number of published works dedicated to global environmental change leads to the realization that protection of the natural environment has become an urgent problem. The question of working out principles of co-evolution of man and nature is being posed with ever-increasing persistence. Scientists in many countries are attempting to find ways of formulating laws governing human processes acting on the environment. Numerous national and international programs regarding biosphere and climate studies contribute to the quest for means of resolving the conflict between human society and nature. However, attempts to find efficient methods of regulating human activity on a global scale encounter principal difficulties. The major difficulty is the lack of an adequate knowledge base pertaining to climatic and biospheric processes as well as the largely incomplete state of the databases concerning global processes occurring in the atmosphere, in the ocean, and on land. Another difficulty is the inability of modern science to formulate the requirements which must be met by the global databases necessary for reliable evaluation of the state of the environment and forecasting its development for sufficiently long time intervals.

Many scientists are trying to find answers to the above questions. The majority suggest the creation of a unified planetary-scale adaptive GeoInformation Monitoring System (GIMS) as one of the efficient ways of resolving the conflicts between nature and man. This system should be based on regenerated knowledge bases and global data sets. The adaptive nature of such a system should be provided through correction of the data-acquisition mode and by varying the parameters and structure of the global model.

Attempts to create the major GIMS component, a global model, have not been implemented so far, despite the large capacities of modern computers. It is clear that the creation of a model adequately representing the real processes going on in the environment is not possible: on the one hand, taking into account the whole multitude of environmental parameters is sure to lead to boundless multidimensionality; on the other hand, simplified models taking into account a restricted number of parameters would be inadequate for the task. Moreover, creation of an adequate global model is impossible in principle, due to the practical impossibility of providing the exhaustive information needed. As to the simulation of socio-economic processes, more often than not, large uncertainties are inevitable. In attempting to surmount these difficulties, many researchers resort to methods of the theory of games, evolution modeling, and describing the behavior of systems in the form of sets of scenarios. Such approaches help to take expert opinions into account. However, without a concept of global changes, all such attempts are doomed to fail. A mechanism is necessary for singling out information-carrying elements and determining their interrelation so that redundancy in the global model is minimized.

The first steps in understanding the key problems of global ecology and climate have been made recently in a number of publications by K.Ya. Kondratyev. In addition, results and discussions in recent publications by V.F. Krapivin and G.W. Phillips permit the production of a constructive synthesis of a global model of natural processes on a set of spatial scales.

The present volume proposes some ideas directed toward the solution of the above problems and describes a simulation system based on sets of computer algorithms for processing data of global and regional monitoring.

A global simulation model describing the spatiotemporal interactions in the nature/human society system is synthesized. The model comprises units describing biogeochemical cycles of carbon, nitrogen, sulfur, phosphorus, oxygen, and ozone; the global hydrologic balance in liquid, gaseou,s and solid phases; productivity of soil-plant formations with 30 types defined; photosynthesis in ocean ecosystems taking into account depth and surface inhomogeneity; demographic processes and anthropogenic changes. The model is designed to be connected to a global climate model.

An informational interface between the global model, algorithms, and field experiments is created. It makes possible the use of global sets of vegetation-related parameters, data sets characterizing meteorological fields and hydrologic processes, and other global environmental and anthropogenic data to forecast global change. The informational interface supports the relationships between different components of the global model.

The theoretical part of the volume contains chapters which describe various algorithms and models. The applied part of the volume considers some specific problems of environmental dynamics. The purpose of the theoretical and applied parts is the development of a universal information technology to assess the state of environmental subsystems functioning under various climatic and anthropogenic conditions. The basic idea of the approach proposed in this volume is the combination of Geographical Information System (GIS) techniques with modeling technology to estimate the functioning of the nature/society system. This idea is implemented by the use of new methods for spatiotemporal reconstruction of incomplete data.

Algorithms, models, methods, criteria and software are suggested that are aimed at the synthesis of the GIS with the modeling functions for a complex estimation of the state of the nature/society subsystems. The newly developed GeoInformation Monitoring System (GIMS = GIS + Model) is focused on the use of systematic observation data and evaluation of changes attributable to human impacts on the environmental subsystems. One of the important functional aspects of the integrated system is the possibility of a forecasting the capability to warn of undesirable changes in the environment as a result of human activities.

The first four chapters describe a mathematical model of the nature/society system representing the interaction of the atmosphere with the land and ocean ecosystems under anthropogenic impacts. The model comprises units describing biogeochemical cycles of carbon, nitrogen, sulfur, phosphorus, oxygen, and ozone; the global hydrologic balance in liquid, gaseous, and solid phases; the productivity of soil–plant formations with specific types defined; photosynthesis in the ocean ecosystems taking into account depth and surface inhomogeneity; demographic processes and anthropogenic changes.

In the following six chapters, various applications of GIMS technology are considered. These chapters describe the application of GIMS technology to study the dynamics of radioactive pollutants, heavy metals, and oil hydrocarbons in the Angara/Yenisey River system and in the Arctic basin; to assess the Peruvian Current and Okhotsk Sea ecosystems dynamics; to estimate excess CO_2 distribution in the biosphere; to study the Aral/Caspian aquageosystem water regime; and to monitor environmental system changes in the oil and gas extraction zones.

These chapters characterize the subsystems of the global model and demonstrate their local structure and interrelationships with the climate system. The biogeochemical unit of the global model makes it possible to compute the dynamics of industrial CO_2 distribution between the ocean, terrestrial biota, and atmosphere. The ocean is described by a three-dimensional model with due regard for water chemistry. The terrestrial biota is represented by 30 types of ecosystems with their spatial distribution defined by a geographical grid.

Chapter 11 describes the components of the GIMS technology which relate to the decision-making and data collection procedures.

This volume brings together a wide spectrum of theoretical and applied techniques for assessments of global change. It presents systematically the theoretical and practical notions of simulation modeling associated with environmental and human systems. Applied mathematicians, hydrologists, geophysicists, ecologists, socioeconomists, and other researchers of global change will find this volume useful.

Contents

Abbreviations

ACS	Aral-Caspian System
ACES	Aral-Caspian Expert System
AD	Amsterdam Declaration
ANWAP	Arctic Nuclear Waste Assessment Program
AMAP	Arctic Monitoring and Assessment Program
APS	Atmosphere/Plant/Soil
APSO	Atmosphere/Plant/Soil/Ocean
ARCSS	Arctic System Science
AYRS	Angara-Yenisey River System
AYRSSM	Angara-Yenisey River System Simulation Model
BMDS	Ballistic Missile Defense System
BOD	Biological Oxygen Demand
BSS	Biosphere-Society System
CIESIN	Consortium for International Earth Science Information Networks
CLIMBER	CLIMate and BiosphERe
CM	Climate Model
CNTBT	Comprehensive Nuclear Test Ban Treaty
DOM	Dead Organic Matter
ESAIEP	Expert System for Adaptive Identification of the Environment Parameters
GCD	Gas Condensation Deposit
GDP	Global Gross Product
GEZ	Gas Extraction Zone
GIMS	GeoInformation Monitoring System
GIS	Geographical Information System
GWP	Gross World Product
GSM	Global Simulation Model
ICGGM	International Centre for Global Geoinformation Monitoring
IEA	International Energy Agency
IHDP	International Human Dimensions Programme
IGBP	International Geosphere-Biosphere Programme
IPCC	Intergovernmental Panel on Climate Change
MCBS	Magnetosphere/Climate/Biosphere/Society
NS	Nature/Society
OSE	Okhotsk Sea Ecosystem

PCE	Peruvian Current Ecosystem
POEM	POtsdam Earth system Modelling
SAT	Surface Air Temperature
SD	Sustainable Development
SEM	System of Ecological Monitoring
SIB	Simple Biosphere Model
SMOSE	Simulation Model of Okhotsk Sea Ecosystem
SSMAE	Simulation Model of the Arctic Ecosystem
START	Strategic Arms Reduction Treaty
TDS	Total Dissolved Solids
TRMM	Tropical Rainfall Measuring Mission
TSS	Total Suspended Solids
WCRP	World Climate Research Programme

Chapter 1

Introduction 1

1.1 Contemporary Stage of Civilization Development

The problem under discussion of modern civilization and its perspectives is broad-scale and multicomponent. This is why the objective of this article could be only a fragmentary consideration of the key aspects of this problem, taking into account that the essence of it is the interaction of society and nature. In this connection, it should be emphasized that "sustainable development" is perhaps the most discussed subject, though this subject has not yet acquired a clear enough definition. Nevertheless, the scheduled conference, which will take place in Johannesburg on 2–11 September 2002, is called "World Summit on *Sustainable Development*" ("Rio+10"), being a continuation of two similar conferences (Stockholm, 1972 and Rio de Janeiro, 1992). The Rio de Janeiro conference has become known as "The Second United Nations Conference on Sustainable Development" and after it in 1995 another global environmental conference "Special Session of the United Nations" ("Rio+5") took place. The documents adopted at the above-mentioned conferences have been widely discussed, not only in scientific publications, but also in the mass media as well (Kondratyev 1990, 1997, 1998a, b, 1999a–e, 2000b, 2002; Kondratyev et al. 1996b; Moiseyev 1998; Nakicenovic et al. 1998; Moiseyev 1999; Allen et al. 2001; Munasinghe et al. 2001; Krapivin and Kondratyev 2002). However, until now, the problems have not been clearly enough formulated, nor have priorities or the ways of further civilization development been defined. The first predominant reaction to the "Black Tuesday" tragedy (11 September 2001) of the acts of terrorism in New York and Washington, which is concentrated on military actions, has turned out to be the most expressive example of the inability of a thorough analysis of the contemporary stage of civilization. In this context, the Amsterdam Open Science Conference (June, 2001), and "The Amsterdam Declaration" (AD) acquire great importance. The preparation of this conference and the elaboration of the AD have been held within the framework of the four international nongovernmental programs: The International Geosphere-Biosphere Programme (IGBP), the International Human Dimensions Programme on Global Environmental Change (IHDP), the World Climate Research Programme (WCRP), and the International Biodiversity Programme (DIVERSITAS).

The main conclusions made at the Amsterdam Open Science Conference (*The Amsterdam Declaration on Global Change, 2000*) are the following:

- "The Earth System behaves as a single, self-regulating system comprised of physical, chemical, biological and human components. The interactions and feedbacks between the component parts are complex and exhibit multi-scale temporal and spatial variability. The understanding of the natural dynamics of the Earth System has advanced greatly in recent years and provides a sound basis for evaluating the effects and consequences of human-driven change.
- Human activities are significantly influencing Earth's environment in many ways in addition to greenhouse gas emissions and climate change. Anthropogenic changes to Earth's land surface, oceans, coasts and atmosphere and to biological diversity, the water cycle and biogeochemical cycles are clearly identifiable beyond natural variability. They are equal to some of the great forces of nature in their extent and impact. Many are accelerating. Global change is real and is happening *now.*
- Global change cannot be understood in terms of simple cause-effect paradigm. Human-driven changes cause multiple effects that cascade through the Earth System in complex ways. These effects interact with each other and with local- and regional-scale changes in multidimensional patterns that are difficult to understand and even more difficult to predict. Surprises abound.
- The Earth System dynamics are characterized by critical thresholds and abrupt changes. Human activities could inadvertently trigger such changes with severe consequences for Earth's environment and inhabitants. The Earth's system has operated in different states over the last half million years, with abrupt transitions (a decade or less) sometimes occurring between them. Human activities have the potential to switch the Earth System to alternative modes of operation that may prove irreversible and less hospitable to humans and other life. The probability of a human-driven abrupt change in Earth's environment has yet to be quantified but is not negligible.
- In terms of some key environmental parameters, the Earth system has moved well outside the range of natural variability exhibited over the last half million years at least. The nature of changes now occurring simultaneously in the Earth System, their magnitudes and rates of change are unprecedented. The Earth is currently operating in a no-analogue state.
- Based on the mentioned conceptual circumstances the AD authors reason that it is necessary to come to agreement about the following:
- An ethical framework for global stewardship and strategies for Earth System management are urgently needed. The accelerating human transformation of the Earth's environment is not sustainable. Therefore, the business-as-usual way of dealing with the Earth System is not an option. It has to be replaced – as soon as possible – by deliberate strategies of good management that sustain the Earth's environment while meeting social and economic development objectives.
- A new system of global environmental science is required. This is beginning to evolve from complementary approaches of the international global change research programs and needs strengthening and further development. It will draw strongly on the existing and expanding disciplinary base of global change science; integrate across disciplines, environment and development issues and the natural and social sciences; collaborate across national boundaries on the basis of shared and secure infrastructure; intensify efforts to

enable the full involvement of developing country scientists; and employ the complementary strengths of nations and regions to build an efficient international system of global environmental science.

- The global change programs are committed to working closely with other sectors of society and across all nations and cultures to meet the challenge of a changing Earth. New partnerships are forming among university, industrial and governmental research institutions. Dialogues are increasing between the scientific community and policy-makers at a number of levels. Action is required to formalize, consolidate and strengthen the initiatives being developed. The common goal must be to develop the essential knowledge base needed to respond effectively and quickly to the great challenge of global change."

No doubt the Amsterdam Conference is a new important step forward to the realization of ways of development of modern civilization with due regard to the interactive complexity of the problems to be solved and, most important, to the elaboration of a new paradigm of development instead of the continued movement on the basis of the "business as usual" principle. However, regretfully, it should be noted that during the AD preparation the proper conceptual researches, made in the former USSR since the 1960s and later in Russia, the results of which had been summarized already in the 1990s in monographs (Gorshkov 1990; Kondratyev 1990), and later on in the series published by Springer/PRAXIS (UK) (Kondratyev 1998b, 1999b; Gorshkov et al. 2000; Kondratyev and Varotsos 2000), had been ignored. During the last few years, quite a number of fundamental publications appeared in Russia in the field of natural sciences as well as in the social sciences (Abalkin 1999; Arbatov 2000; Bokov and Lushchik (1998); Schwartz et al. 2000; Andreev 2001; Barlybaev 2001; De Rivero 2001; Emery and Thomson 2001; Shutt 2001; Schrope 2001).

However, let us refer to the commentaries concerning the booklet *Global Change and the Earth System: A Planet Under Pressure* (Steffen and Tyson 2001), which contains the detailed substantiation of the Amsterdam Declaration, beginning with a statement reflecting nonrecognition of "somebody else's research":

"Somewhat more than a decade ago it was recognized that the Earth behaves as a system in which the oceans, atmosphere and land, and the living and non-living parts therein, were all connected... Over the last 10 years the understanding of how humans are bringing about global change has undergone a quantum jump. Attempts to separate natural and anthropogenically induced variability in the Earth System have proved to be successful in many respects."

We will further refer to the substantial commentary of the statement mentioned above, without paying too much attention to defend scientific priorities. First, we will discuss the main scientific achievements formulated in Steffen and Tyson (2001), where "Earth System" has been defined as a complex of interacting physical, chemical, biological components and connected with human activity components, which determine the processes of transportation and transformation of substances and energy, and thus keeps the planet habitable.

- "The Earth is a system that life itself helps to control. Biological processes interact strongly with physical and chemical processes to create the planetary environment, but biology plays a much stronger role than previously thought in keeping Earth's environment within habitable limits."

It should be mentioned that this thesis has undoubtedly a fundamental meaning, but in this context it would be appropriate to refer to the conception of the biotically regulated environment extensively elaborated by Gorshkov (1990) and Gorshkov et al. (2000).

- "Global change is much more than climate change. It is real, it is happening now and it is accelerating. Human activities are significantly influencing the functioning of the Earth System in many ways; anthropogenic changes are clearly identifiable beyond natural variability and are equal to some of the great forces of nature in their extent and impact."

We have to emphasize that in the contemporary world, when the scientific discussion on the extent of anthropogenic pressure on climate, has been transformed into a political discussion and reached the level of state leaders, it has been pointed out that greater attention should be paid to global climate changes in the context of the Kyoto Protocol, which has already been done in a number of publications (Kondratyev et al. 2000, 2001a,b; Kondratyev and Demirchian 2001; Palumbi 2001; Robertson and Kellow 2001; Kondratyev 2002; Krapivin and Kondratyev 2002). However, we agree with Steffen and Tyson (2001) in the sense that climate is not the first priority in the totality of global change problems.

Paying attention to the introduction of a new term "Anthropocene Era" (about which V.I. Vernadsky wrote long ago), *Global change and the Earth System: A Planet Under Pressure* (Steffen and Tyson 2001) contains a list of some of the most important manifestations of anthropogenic pressures on the Earth System:

- "In a few generations humankind is in the process of exhausting fossil fuel reserves that were generated over several hundred million years.
- Nearly 50% of the land surface has been transformed by direct human action, with significant consequences for biodiversity, nutrient cycling, soil structure and biology, and climate.
- More nitrogen is now fixed synthetically and applied as fertilizers in agriculture than is fixed naturally in all terrestrial ecosystems.
- More than half of all accessible freshwater is used directly or indirectly by humankind, and underground water resources are being depleted rapidly in many areas.
- The concentrations of several climatically important greenhouse gases, in addition to CO_2 and CH_4, have substantially increased in the atmosphere.
- Coastal and marine habitats are being dramatically altered; 50% of mangroves have been removed and wetlands have shrunk by one-half.
- About 22% of recognized marine fisheries are overexploited or already depleted, and 44% more are at their limit of exploitation.
- Extinction rates are increasing sharply in marine and terrestrial ecosystems around the world; the Earth is now in the midst of its first great extinction event caused by the activities of a single biological species (humankind).
- The human enterprise drives multiple, interacting effects that cascade through the Earth System in complex ways. Global change cannot be understood in terms of a simple cause-effect paradigm. Cascading effects of human

activities interact with each other and with local- and regional-scale changes in multidimensional ways.

- The Earth's dynamics are characterized by critical thresholds and abrupt changes. With catastrophic consequences for the Earth System. Indeed, it appears that such a change was narrowly avoided in the case of depletion of the stratospheric ozone layer. The Earth System has operated in different quasi-stable states, with abrupt changes occurring between them over the last half million years. Human activities clearly have the potential to switch the Earth System to alternative modes of operation that may prove irreversible."
- Though, the last two theses are very important, we must keep in mind that until now there are only preliminary quantitative assessments and only fragmentary data exist; there is not enough information for concrete judgements about the potential changes. This is true, for instance, in the case of the impact on the global ozone layer in the atmosphere. The question about the processes which affect the total ozone content (including the atmospheric dynamics of the processes and purely natural factors) is far from clear (Kondratyev and Varotsos 2000). In this context, serious attention should also be paid to the problem of the anthropogenic impact on the global biogeochemical cycles of carbon, sulfur, nitrogen, and other chemical compounds (Dimitroulopoulou and Marsh 1997; San Jose 1999; Kondratyev and Demirchian 2000; Kondratyev et al. 2000; Nesje and Dahl 2000; Swanson and Johuston 2000; Kondratyev and Krapivin 2001b; Malinetsky and Kurdumov 2001; Kondratyev 2002).
- "The Earth is currently operating in a no-analog state. In terms of key environmental parameters, the Earth System has recently moved well outside the range of the natural variability exhibited over at least the last half million years. The nature of changes now occurring simultaneously in the Earth System, their magnitudes and rates of change are unprecedented."

By all means, such general and exclusively important assessments have to be more concrete and, for example, in the case of climate change issues, they do not correspond to the facts (Kondratyev et al. 2001a), whereas for other phenomena they are, no doubt, topical problems (Kondratyev 2002). Contrary to the conclusion about the unique present-day global changes, the authors of *Global Change and the Earth System: A Planet Under Pressure* (Steffen and Tyson 2001) justly refer to the paleoclimatic data, which represent much more significant changes of mean global annual surface air temperature (SAT) which reached 10°C in 10 years or even less in some regions. The general opinion about the difficulties of the recognition of anthropogenic changes in the framework of the cause and effect paradigm is certainly correct in view of the fact that these changes interact with the natural ones. In the context of perspectives of the development of contemporary civilization a key factor is undoubtedly the continuing global population increase: if during the second half of the twentieth century the population doubled, then at the same time the production of grain crops tripled, the energy consumption increased fourfold and economic activity scaled up to fivefold. Gorshkov (1990) clearly demonstrated that *Homo sapiens* left their ecological niche and this gave rise to the first signs of an accelerating ecological catastrophe (later, we will discuss these problems in more detail).

Taking into account the above-mentioned conceptual circumstances, the authors of *Global change and the Earth System: A Planet under pressure* (Steffen and Tyson 2001) came to two important conclusions: (1) there was a need for a public response to the emerging ecological menace and (2) a requirement for relevant scientific studies, which should provide an adequate understanding of those processes that determine the variability of the Earth System.

- "Ethics of global stewardship and strategies for Earth System management are urgently needed. The inadvertent anthropogenic transformation of the planetary environment is, in effect, already a form of management, or rather mismanagement. It is not sustainable. Therefore, the business-as-usual way of dealing with the Earth has be replaced – as soon as possible – by deliberate strategies of good management.
- A novel system of global environmental science is emerging. The largely independent efforts of various international research programs and numerous national projects create the basis for an Earth System science that is capable of tackling the cognitive tasks suggested by the research findings above. This new science will employ innovative integration methodologies, organize itself into a global system with transnational infrastructures and embark on a continuing dialogue with stakeholders around the world."

Putting aside for a while these two quite rhetorical conclusions, we should stress the absence of a clear statement in the document *Global Change and the Earth System: A Planet Under Pressure* (Steffen and Tyson 2001) of the question of Earth System studies, which have to be focused on the interactions between nature and society as well as on the analysis of "sustainable development" (and this is more than strange in view of the forthcoming Earth Summit on *sustainable development*).

Keeping in mind the above-mentioned priorities of global change problems, attention should be paid to the absence of even a mention (in priorities) of such an important priority as global observation systems, as well as global changes connected with cosmogenic influence (this refers to, e.g., the influence of solar activity [Kondratyev 1998a, b] and ecological disasters [Grigoryev and Kondratyev 2001]).

1.2
Contemporary Global Ecodynamics

In order to avoid repetition (see References), we will confine ourselves to consideration of a new statistical data discussion published by the Worldwatch Institute (Brown et al. 2001). Table 1.1 is borrowed from Losev (2001) and contains general characteristics of the environmental changes for the period from 1972 to 1995 and the possible trends until 2030.

The data of Table 1.1 help to draw two main conclusions: (1) for the period under consideration, there is a continuing deterioration of the environmental characteristics; (2) relevant potential trends will be negative and become even worse in the future. That means that the tendencies of unsustainable development will increase. Now we will discuss the dynamics of the population growth

Table 1.1. Environmental changes from 1973 to 1995 and expected trends until 2030

Characteristic 1	Trend 1972–1995 2	Scenario until 2030 3
Reduction of the natural ecosystem's spatial extent	Reduction at a speed of 0.5–0.1% per year on land; about 37% of them preserved by the beginning of the 1990s	The trend retains, approaching the total disappearance of the terrestrial ecosystems
Consumption of the clean primary natural biological production by humans	Consumption increase: 40% in case of terrestrial ecosystems and 25% of global ecosystems (estimation of 1985)	Consumption increase: 80–85% for terrestrial ecosystems and 50–60% of global
Change of the greenhouse gases (GHGs) concentration in the atmosphere	Increase in the greenhouse gases from the decimals of 1% to the percents ranging from 1 to 10 annually	Increase in GHG concentration; accelerated growth of CO_2 and methane concentrations due to biota destruction
Ozone depletion, the Arctic ozone "hole" growth	Ozone layer depletion by 1–2% per year; the growth of ozone hole size	The same trend even if emissions of CFCs (chloro-fluoro-carbons) will be stopped by 2000
Decrease in the forest landscapes especially of tropical forest areas	Decrease at a speed of 10 million ha (1990–1995) per year; the ratio of forest regeneration to the forest reduction is 1:10	The same trend retains; decrease in the tropical forests areas from 18 (1990) to 9–11 million km^2, decrease in temperate forest cover
Desertification	Increase in arid areas (40% of land), increase in the technogenic desertification as well as toxicity of deserts	The same trend; there is a process due to water cycle changes on land as a result possibility of speeding the of deforestation and the accumulation of pollutants in soils
Soil degradation	Increase in soil erosion (24 billion t/year), soil fertility decrease, pollutant accumulation, acidification, salinisation	The same trend, increase in erosion and pollution, reduction of agricultural land per person
Rise of the ocean level	Rising of the World Ocean level of 1–2 mm/year	The same trend, speeding up of ocean level is possible by 7 mm/year
Natural calamities, technological hazards	Natural calamities increase by 5–7% per year, damage increase by 5–10 per year	The same trends and their enhancement
The loss of biodiversity (biological species disappearance)	The fast exhaustion of species, the exhaustion speed is 100–1000 times higher than it has been observed on the Earth before	Enhancement of trends during the process of biosphere degradation

Table 1.1 (continued)

Characteristic 1	Trend 1972–1995 2	Scenario until 2030 3
The freshwater resources depletion	Increase of the sewage water volume, enhancement of local and distributed pollution as well as number and their concentration	The same trends and their enhancement, two from three people in the world will survive shortage of fresh water
Accumulation of the pollutant in the environment and organisms, migration in trophical chains	Increase in the mass and quantity of pollutants, accumulated in the environment and organisms, increase in radioactivity, "chemical bombs" phenomena	The same trends and their possible reinforcement
Quality of life deterioration, increase of the number of disease rate due to destruction of the humans' ecological niche and with the environmental contamination, including genetic illnesses increase, emergence of new diseases	Poverty increase, shortage of food, a high infant mortality, high level of illness rate, lack of fresh water in developing countries; increase of genetic disease number, high level of accident rate, increase of medicine consumption, increase of allergies in developed countries, panepidemic AIDS in the world, decrease of immunologic status	The same trends, increase of shortage of food supply, number of diseases connected with the environmental degradation, including genetic diseases, enlarge ment of infected areas, emergence of new diseases
Global propagation, including mankind, of super-toxicants through trophical chains	Problems of the mankind endocrine system, which cause reproduction problems, problems of cerebral activity and other vital human organs	Increase in trends, spreading of diseases connected with the endocrine system, increase of a number of childless couples
Artificial introduction and accidental invasion of strange species into ecosystems	Ecosystems destruction, migration of pests, diseases of plants, animals and humans, biodiversity decrease	Enhancement of invasion process
Changes in the World Ocean: reef destruction, mangrove ecosystems decrease, fish supply exhaustion as a result of the intensive fishing, whale's herds decrease, contamination of the inland and coastal waters and littoral zones, "red tides"	Fast changes of all these characteristics	Enhancement of these changes

Table 1.2. World population, total and annual addition, 1950–2000. Designations: G is the population size (billion), ΔG is the population addition (million)

Year	G	ΔG	Year	G	ΔG
1950	2.555	38	1984	4.774	81
1955	2.780	53	1985	4.855	83
1960	3.039	41	1986	4.938	86
1965	3.346	70	1987	5.024	87
1970	3.708	78	1988	5.110	86
1971	3.785	77	1989	5.196	87
1972	3.862	77	1990	5.284	83
1973	3.939	76	1991	5.367	83
1974	4.015	74	1992	5.450	81
1975	4.088	72	1993	5.531	80
1976	4.160	73	1994	5.611	80
1977	4.233	72	1995	5.691	78
1978	4.305	75	1996	5.769	78
1979	4.381	76	1997	5.847	78
1980	4.457	76	1998	5.925	78
1981	4.533	80	1999	6.003	78
1982	4.613	81	2000	≈ 6.080	77
1983	4.694	80			

rate and the life support systems (food, energy, environment) that prove the general conclusion.

Table 1.2 (Vital Signs 2001–2002, 2001) illustrates the dynamics of the global trends of population, main features of which are the continuing growth of population (this trend should remain, at least during the first half of the twenty-first century) and the speed of this tendency, which is slowing down. Of course, the mean global trends mask the existence of real huge regional variations. In general, there is a very slow increase in industrial-country populations or even its recession (this last process is very typical for the former Soviet Republics). The USA is an exception, where a third of the nearly 1% growth rate is fuelled by immigration. Of the global population growth in the year 2000, 95% has been observed mainly in the developing countries, including Asia (57%, some 45 million people), African (23%), Latin America (9%), and the Middle East (5%). About half of the global population growth in 2000 occurred in India, China, Pakistan, Nigeria, Bangladesh, and Indonesia. A number of positive trends account for most of the reduction in fertility rates and growth rates, such as improvement of the economic situation in many countries and the public health system, progress in women's education with the increase in their status, broad access to contraceptives, and others. In Iran, for example, the population growth rate was reduced from 3.2% in 1986 to 0.8% in 2000. The deceleration of the birth rate is caused by the spread of AIDS (about 3 million people died in 2000 of AIDS, bringing this disease's cumulative total to nearly 22 million).

It is natural that during the second half of the last century the global gross domestic product (GDP) was growing, though slowing down (Table 1.3). The global economy in 2000 expanded by 4.7% and exceeded the level of 1999

Table 1.3. Gross World Product (GWP), 1950–2000. Designation: U is the total GWP (trillion 1999 dollars), U_G is the GWP per person (1999 dollars)

Year	U	U_G	Year	U	U_G
1950	6.4	2502	1984	26.2	5485
1955	8.1	2921	1985	27.1	5582
1960	10.0	3306	1986	28.0	5673
1965	12.8	3822	1987	29.0	5778
1970	16.3	4407	1988	30.3	5938
1971	17.1	4505	1989	31.2	5996
1972	17.8	4599	1990	31.9	6031
1973	19.0	4819	1991	32.0	5957
1974	19.4	4829	1992	32.4	5941
1975	19.7	4816	1993	33.2	6000
1976	20.7	4977	1994	34.5	6150
1977	21.5	5083	1995	35.8	6295
1978	22.4	5210	1996	37.3	6475
1979	23.1	5282	1997	39.0	6666
1980	23.6	5306	1998	39.9	6732
1981	24.2	5329	1999	41.2	6871
1982	24.4	5280	2000	≈ 43.2	≈ 7102
1983	25.1	5341			

(3.4%). The global GDP of US$ 43 trillion provided an average GDP per person of US$ 7,102. This progress was achieved due to strong economic development in the US and western Europe, continuing recovery of the Asian economy from the 1997 financial crisis, and recovery of the economy of Latin America from the crisis of 1998, as well as a significant improvement of the economic situation in the transition economies. It is interesting to mention that the economic growth in China in 2000 was 7.5%, the most significant in Asia. India's economy grew by 6.7%, followed by Pakistan at 5.6%, and Bangladesh at 5%, this growth is especially important for this region with 1.2 billion people who live on US$ 1 a day and even less. The authors of Vital Signs 2001–2002 (2001) consider the significant economic growth of the transition economies as the biggest surprise. The economic growth in 2000 in Russia, for example, reached 7.2% in comparison with 3.2% in 1999 (of course, mainly due to the increase in oil prices), in Hungary the economy grew by 5.5% and in Poland by 5%. According to an IMF forecast, the global economic growth will continue in 2001, but at a somewhat slower rate.

An important indicator of the economic and technological development is the fossil fuels consumption (Table 1.4). The year 2000 was the second consecutive year of the decrease in global fossil fuel consumption after a long period of increasing, by more than 3.5 times since 1959. At present, the share of fossil fuels accounts for 90% of commercial energy use and coal provides 25% of world commercial use. The coal world average share is gradually decreasing, but in the USA, in contrast, coal use (about 25% of the world total) increased in 2000 by 1.6%, as growing electricity demand spurred coal-fired power generation. In contrast, coal use in China (also about 25% of the world total) has been

Table 1.4. World fossil fuel consumption, 1950–2000 (million tons of oil equivalent)

Year	Coal	Oil	Gas	Year	Coal	Oil	Gas
1950	1043	436	187	1983	1918	2632	1463
1955	1234	753	290	1984	2001	2670	1577
1960	1500	1020	444	1985	2100	2654	1640
1965	1533	1485	661	1986	2135	2743	1653
1970	1635	2189	1022	1987	2197	2789	1739
1971	1632	2313	1097	1988	2242	2872	1828
1972	1629	2487	1150	1989	2272	2921	1904
1973	1668	2690	1184	1990	2244	2968	1938
1974	1691	2650	1212	1991	2189	2967	1970
1975	1709	2616	1199	1992	2179	2998	1972
1976	1787	2781	1261	1993	2171	2969	2012
1977	1835	2870	1283	1994	2186	3027	2019
1978	1870	2962	1334	1995	2218	3069	2075
1979	1991	2998	1381	1996	2298	3150	2170
1980	2021	2873	1406	1997	2285	3224	2155
1981	1816	2781	1448	1998	2243	3241	2181
1982	1878	2656	1448	1999	2034	3332	2277
				2000	≈ 2034	≈ 3332	≈ 2277

reduced by 3.5% (since 1996 this reduction reached 27%). India is the third leading user; its coal consumption rose in 2000 by 5.4%, amounting to 7% of the global total.

Global oil use (its share of world commercial energy is 41%) increased by 1.1% and the USA, the leading petroleum user (with a 26% share), increased consumption by only 0.1%, while Asian countries of the Pacific Ocean region increased oil consumption by 2.6%. Western Europe's share is 22% of global oil use, with an increase of 0.2%. Natural gas consumption (whose share of world commercial energy use is 24%) expanded by 2.1%. The strongest growth of natural gas consumption was in the Baltic States of the former Soviet Republics, in Lithuania by 29%, Estonia by 30%, and Latvia by 45%. Spain in western Europe and South Korea in Asia became the leaders of natural gas use (each increased by 16%). The highest price for oil (US$ 35 per barrel) was in 1985. This was the reason for renewed interest in oil and natural gas exploration in the untapped Arctic fields of Russia and Alaska, though the reserves of Alaska are relatively small and drilling in Russia would require huge investments.

According to the IEA (the International Energy Agency), overall consumption of fossil fuels for the period 1997–2020 is expected to grow by 57% (2% annually), maintaining their 90% share of world energy use. The predicted world coal use is expected to increase by 1.7% annually, with two-thirds of this growth occurring in China and India. Petroleum will remain the dominant energy source (1.9% annually) and its share in primary energy will reach 40%, but natural gas consumption growth will be fastest (2.7% annually), mainly due to an increased need for power generation.

Table 1.5. World carbon emissions due to fossil fuel (1950–2000) and CO_2 concentration in the atmosphere (1960–2000). Designations: H_1 is the CO_2 emission to the atmosphere (million tons of carbon), C_A is the CO_2 concentration in the atmosphere (ppm)

Year	H_1	C_A	Year	H_1	C_A
1950	1612	–	1984	5098	344.2
1955	2013	–	1985	5271	345.7
1960	2535	316.7	1986	5453	34.0
1965	3087	319.9	1987	5574	348.7
1970	3997	325.5	1988	5789	351.3
1971	4143	326.2	1989	5892	352.7
1972	4305	327.3	1990	5931	354.0
1973	4538	329.5	1991	6020	355.5
1974	4545	330.1	1992	5879	356.4
1975	4518	331.0	1993	5861	357.0
1976	4776	332.0	1994	6013	358.9
1977	4910	333.7	1995	6190	360.9
1978	4950	335.3	1996	6315	362.6
1979	5229	336.7	1997	6395	363.8
1980	5155	338.5	1998	6381	366.6
1981	4984	339.8	1999	6340	368.3
1982	4947	341.0	2000	≈ 6299	≈ 369.4
1983	4933	342.6			

The trend of CO_2 emission into the atmosphere (Table 1.5) shows only a small reduction in the world emissions and an increase in CO_2 concentration during the last 3 years, despite all discussions about the necessity of CO_2 reduction and the recommendations of the Kyoto Protocol. Since 1959 some 217 billion tons of carbon have been released into the atmosphere, with annual emissions nearly quadrupling over this period. The important fact is that the amount of carbon emitted per unit of global economic output continued to drop, by 3.6% (to 148 tons per million dollars of the gross world product, GWP), and the carbon/GWP ratio declined by nearly 41% over the past half century. In contradiction to the Kyoto Protocol, the Western industrially developed countries have increased CO_2 emissions into the atmosphere since 1990 on average by 9.2% and the USA by 13%. The total share of the developed countries has reached 22.8%. According to the IPCC (Intergovernmental Panel on Climate Change) estimations, global carbon emissions are projected to reach 9–12.1 billion tons of carbon by 2020 and 11.2–23.1 billion tons by 2050. The problem of global climate changes and the Kyoto Protocol is reviewed in detail by (Kondratyev and Demirchian 2000; Kondratyev et al. 2001a; Soon et al. 2001).

The most important component of economic development is the grain production/harvest (Table 1.6), which has decreased during the last period from 1869 million tons in 1999 to 1840 million tons in 2000. The year 2000 harvest was down more than 2% from the historical high in 1997 (1881 million tons). The main reason for this decrease was a reduction of crop yield in China from 391 million tons in 1998 to 353 million tons in 2000, approximately 10%. The cause of this reduction was the combined influence of low prices that discour-

Table 1.6. World grain production, 1950–2000. Designations: P is the total production (million tons), P_G is the consumption (kg per person)

Year	P	P_G	Year	P	P_G
1950	631	247	1984	1632	342
1955	759	273	1985	1647	339
1960	824	271	1986	1665	337
1965	905	270	1987	1598	318
190	1079	291	1988	1549	303
1971	1177	311	1989	1671	322
1972	1141	295	1990	1769	335
1973	1253	318	1991	1708	318
1974	1204	300	1992	1790	328
1975	1237	303	1993	1713	310
1976	1342	323	1994	1760	314
1977	1319	312	1995	1713	301
1978	1445	336	1996	1872	324
1979	1411	322	1997	1881	322
1980	1430	321	1998	1872	316
1981	1482	327	1999	1869	311
1982	1533	332	2000	$\approx$ 1840	$\approx$ 303
1983	1469	313			

aged the farmers planting nationally and of drought and the tightening of water supplies in the northern half of the country. Grain production in the USA (the world's second ranking grain producer after China), climbed for the same period from 332 to 343 million tons, the increase almost all due to improvement in the corn yield.

The world grain yield of 2.75 t/ha in 2000 was down slightly from the 2.77 t/ha of 1999, the multiannual maximum. There were only slight fluctuations of the grain yield during the last few years from the average, which is 2.75 t/ha, but grain harvest per person was reduced to 303 kg in 2000 and that harvest was down 13% from the all time maximum in 1984. This decline was mainly due to Eastern Europe, the former Soviet Union, and Africa. In 2000 and the two previous years, the world corn harvest (of 588 million tons) exceeded the wheat (580 million tons) and rice (401 million tons) harvest. The use of corn, the main source of food for livestock, poultry, and farmed fish, was concentrated in the United States (43% of the world harvest). China, India, and the United States are the leaders in wheat export. China is also a leader for rice production. The United States accounts for over three-quarters of the world corn export. For wheat, the United States is also the leader, followed by France, Canada, and Australia. Rice exports, which are quite small compared with those of wheat and corn, are distributed among China, Thailand, the United States, and Vietnam. World wheat imports, until recently dominated by Japan, are now beginning to shift; thus Brazil has become the world's leading wheat importer, while Iran and Egypt have moved into second and third places, with Japan dropping to fourth place. In 2001, world carryover stocks of grain, the amount in the bin when the new harvest begins, are estimated at 60 days, which means a decrease to the

Table 1.7. Foreign debt of developing and former Eastern Bloc Nations, 1970–1999. Designation: V is the total debt (trillion 1999 dollars)

Year	V	Year	V
1970	0.26	1985	1.47
1971	0.29	1986	2.57
1972	0.32	1987	2.73
1973	0.37	1988	1.68
1974	0.42	1989	1.70
1975	0.51	1990	1.77
1976	0.59	1991	1.80
1977	0.74	1992	1.85
1978	0.86	1993	1.97
1979	0.98	1994	2.15
1980	1.07	1995	2.30
1981	1.18	1996	2.35
1982	1.23	1997	2.40
1983	1.33	1998	2.61
1984	1.35	1999	2.57

lowest level in comparison with previous years. This means if the 2001 harvest is below average, grain prices could become higher.

One of the key indicators of the level of socio-economic development is the foreign debt. Table 1.7 presents data for the developing and the former Eastern bloc nations. There is a continuing debt increase, which was largest in 1998, essentially unchanged in 1999, and which fell slightly after adjusting for inflation (US$ 2.57 trillion). Serious financial debt trends in such nations as Brazil, Indonesia, Russia, and South Korea are largely explained by the financial crises in 1997 and 1998.

A matter of exceptional importance is the data on human health, which give a dynamic pattern of some diseases (Vital Signs 2001–2002, 2001). We will limit ourselves to the data on HIV-infected people and AIDS deaths (see Table 1.8). Every 6 s in the year 2000, someone was infected with HIV. In total, 5.3 million people were HIV-infected. By the end of the year, one of every 100 adults worldwide between the ages of 15 and 49 was infected. Since the epidemic started 20 years ago, AIDS has killed almost 22 million people, more than the population of greater New York City.

Two-thirds of the world's HIV-infected people live in sub-Saharan Africa. In 2000, for the first time, the number of new infections in the region decreased, from 4 to 3.8 million. However, this region has already some of the highest infection rates (e.g., in Botswana, one in three adults is infected). The consequence of AIDS spreading during just two decades is that life expectancy in many African countries has dropped. It is important that AIDS strikes at young, sexually active people, the cornerstone of the work force. This is why by 2020 it could reduce the labor forces in Botswana, Mozambique, Namibia, South Africa, and Zimbabwe by less than three-quarters. In addition, as countries lose people, fewer children will have the chance to obtain a proper education. Students are already leaving school to support their families after the loss of other breadwinners.

Table 1.8. Cumulative HIV infections and AIDS deaths worldwide, 1980–2000. Designations: G_C is the HIV infections (million), M_G is the AIDS deaths (million)

Year	G_C	M_G	Year	G_C	M_G
1980	0.1	0.0	1990	10.0	1.7
1981	0.3	0.0	1991	12.8	2.4
1982	0.7	0.0	1992	16.1	3.3
1983	1.2	0.0	1993	20.1	4.7
1984	1.7	0.1	1994	24.5	6.2
1985	2.4	0.2	1995	29.8	8.2
1986	3.4	0.3	1996	35.3	10.6
1987	4.5	0.5	1997	40.9	13.2
1988	5.9	0.8	1998	46.6	15.9
1989	7.8	1.2	1999	52.6	18.8
			2000	≈ 57.9	≈ 21.8

The Caribbean region harbors the second highest rates of infection, with 2 adults in every 100 infected. In Asia, AIDS has spread from the heroin-producing Golden Triangle to remote corners of China and India, infecting 6.4 million people. The number of people in Eastern Europe living with HIV and AIDS jumped nearly 70 % (from 420,000 to 700,000). Despite all warnings, the number of infected people in Western Europe and North America in 2000 was 70,000 people.

The data considered concern only one of the critical aspects of human health and mainly relate to the industrially developed countries. Generally, the critical situation may be characterized by data on money spent on health care, which illustrates existing and even growing social contrasts. In total, some US$ 3 trillion is spent on health care around the world annually, making this sector one of the largest in the global economy. However, the distribution of these expenses is very uneven. The 84 % of the world population living in low- and middle-income countries obtain just 11 % of the global health expenses, but bear 93 % of the world's disease burden (disease burden is measured by years of healthy life lost from illness combined with those lost from premature death). While on the average, countries earmark 5.5 % of their gross domestic product (GDP) for health care, the United States dedicates 13.7 % to health care and Somali only 1.5 %. The contrasts of annual expenses per person are also very huge; the world's poorest countries each spend US$ 50 or less per person on health, while the United States figure was more than US$ 4,100.

We conclude this survey of new statistical data (Vital Signs 2001-2002, 2001) on ecological and socio-economic trends by consideration of the dynamics of the world nuclear arsenal (Table 1.9). The data show an increase until 1986 and then a significant decline in the number of nuclear warheads. According to the data, the United States retains 10,500 strategic and tactical warheads, while Russia has 20,000, after a twofold reduction (however, these data are not reliable enough). The three other original nuclear powers – France, China, and the United Kingdom – have roughly 1,000 nuclear warheads. Israel, India, and Pakistan possess an unknown, but clearly smaller, number of warheads.

Table 1.9. World nuclear arsenal, 1945–2000. Designation: Y is the number of nuclear warheads

Year	Y	Year	Y
1945	2	1983	66,979
1950	303	1984	67,585
1955	2490	1986	69,478
1960	20,368	1987	68,835
1965	39,047	1988	67,041
1970	39,691	1989	63,645
1971	41,365	1990	60,236
1972	44,020	1991	55,772
1973	47,741	1992	52,972
1974	50,840	1993	50,008
1975	52,323	1994	46,542
1976	53,252	1995	43,200
1977	54,978	1996	40,100
1978	56,805	1997	37,535
1979	59,120	1998	34,535
1980	61,480	1999	31,960
1981	63,054	2000	≈ 31,535
1982	64,769		

Since 1945, 128,00 warheads have been built; the United States produced 70,000 and the Soviet Union/Russia about 55,000. There are some 260 tons of weapons-grade plutonium and that is enough to manufacture about 85,000 warheads. The size of the global nuclear stockpile has declined as a result of arms control efforts by Russia and the United States. However, the remaining arsenals still contain 5,000 megatons of explosive power, including the 4,600 United States and Russian warheads that are still on the alert state.

An important aspect of this problem are nuclear tests; more than 2,050 tests have been conducted since 1945 including the recent series of test explosions conducted in May 1998 by India and Pakistan. By the end of 2000, the Comprehensive Nuclear Test Ban Treaty (CNTBT) had been signed by 160 countries and ratified by 69. This treaty will go into effect only if the 44 countries with nuclear reactors ratify it, but until now only 41 of these countries have signed (the nonsignatories are India, North Korea, and Pakistan) and 30 have ratified it. Among the 11 nonratifiers is the United States (the Senate of the US rejected ratification in October 1999).

Although the nuclear powers have ceased their test explosions, they are pursuing computer test simulations and so-called subcritical tests to develop nuclear warhead technology further. The United States, for example, conducted 8 subcritical tests in 1997–2000 and is planning at least another 14. A number of countries oppose such experiments as a violation of the spirit of the CNTBT.

An important fact is that the United States aims to develop a ballistic missile defense system (BMDS) that is considered to be in violation of the 1972 Anti-Ballistic Missile Treaty between the Soviet Union and the United States, which led Russia to warn against inadmissibility of the common agreement violation. China announced in May 2000 that it might significantly expand its nuclear arsenal in response to any US antimissile defense. This intention of the United States would likely increase pressure on India and Pakistan to step up their own programs and jeopardize the progress in the United State's and Russia's negotiation development on the planned third strategic arms reduction treaty (START III). Nevertheless, the President of the United States, G.W. Bush, called upon Russia to denounce the 1972 Anti-Ballistic Missile Treaty as an out-of-date document (after the event of "Black Tuesday"), and thus open the way to BMDS. All this indicates serious contradictions in solving the problem of strategic nuclear weapons.

To summarize the results of global civilization development during the second half of the twentieth century it is necessary, first of all, to underline that this was a period of unprecedented fast changes in the global population level, biosphere, economy and society as a whole (see Table 1.1). The world has become more economically rich (with the increase in socio-economic contrasts between the countries), but poorer from the environmental point of view to such an extent that had not been possible to predict. The following trends, which were distinguished during the second half of the twentieth century, are especially interesting (Lonergan 1999; Lee and Kirkpatrick 2000; Vital Signs 2001–2002, 2001; Martin and Shumann 2001):

- The global population has exceeded 6 million and increased by 3.5 million during the last half century; that means it has more than doubled. Most of the growth has come in the developing countries, which are already overpopulated. The growth was especially fast in the urban areas where population increased by about fourfold. A remarkable feature of the dynamics of the industrially developed countries is an increase in the proportion of old people, due to improved health care.
- The world economy has grown (approximately sevenfold over 50 years) even more dramatically and has exceeded the rate of the population growth; this provided a significant improvement in living standards, but at the same time 1.2 billion people still live in severe poverty and 1.1 billion do not have access to clean drinking water.
- The global grain yield has nearly tripled since 1950; this allowed people to enrich their diets, but the abundance of food has come with dangerous ecological consequences: depletion of freshwater aquifers and intensification of natural water pollution from the massive use of fertilizers and pesticides. Despite the increase in production of food, more than 1 billion people in the world are still undernourished, while another 1 billion is actually overnourished, which has caused a global epidemic of obesity that is now spreading to the developing world.
- Only a small part of the primary forest (boreal and tropical) has been preserved; more than half of the wetlands and over one-quarter of the coral reefs are lost. This was accompanied by a considerable decline in biodiversity and, what is even more important, the mechanisms of the biotic regulation of the environment have been destroyed (Gorshkov 1995; Gorshkov et al. 2000). The anthropogenic impacts on the global climate and ozone layer are a cause of great concern, although there are many uncertainties in the relevant assessments (Kondratyev 1998b, 1999c; Kondratyev and Varotsos 2000; Kondratyev and Demirchian 2001; Reilly et al. 2001; Krapivin and Kondratyev 2002).

It is especially important that the socio-economic and ecological dynamics during the second half of the last century were, as a rule, unpredictable, because environmental changes very often happened suddenly and unexpectedly. It is not possible to predict where and when a new economic crisis or ecological catastrophe will come from, but it is obvious that the projections of smooth, gradual change that computer models indicate are far from perfect. Until the 1970s, for example, oil forecasts were projecting exponential growth in demand and steady, low prices through the end of the century, until severe oil shocks forced a wholesale revision in this sanguine outlook. Then later the forecasts of the specialists moved to the other conclusion that a period of permanent shortages would drive oil prices over $100 per barrel by 2000. However, just at that time, there was a collapse of oil prices to $10 per barrel (in the mid-1980s).

It is becoming more and more obvious that as the world becomes more complex, predicting the future is becoming more difficult, although the value of forecasting has increased, because only adequate planning can minimize risks and maximize advantages given by the fast changing world. The main challenge for this century is to provide conditions for a continuation of the economic progress that has taken place since the middle of the twentieth century and to restrain ecological decline and social degradation. There is no doubt that the consequences of the ecological recession will finally be destructive for the economy as well. Though this relates to the social contrasts to a greater degree. Under these conditions, for instance, none of the new medical achievements are able to stop new epidemics. The more difficult problem is resistance to the developing global terrorism and no way has been found to solve it. However, it is obvious that military counteraction alone is not an acceptable long-term perspective.

In the Introduction to Vital Signs 2001–2002 (2001) the President of the Worldwatch Institute, Ch. Flavin, and the Executive Director of UNEP, K. Topfer, correctly pointed out: "The new century has begun with many surprises, most of them unwelcome. One thing is virtually certain: the next half-century will not see a repeat of the trends of the one just past. Earth simply will not support it. The question is whether humanity will forge a healthier, sustainable future or risk the downward spiral that would be the result of failing to understand the ecological and economic threshold on which we now stand."

1.3 Sustainable Development

The paradox of the contemporary situation is that despite all the unprecedented scale discussions on sustainable development (SD), the world is continuing an unsustainable way of development. This last fact was confirmed at the session of the UN General Assembly "Rio+5", and certainly, the same conclusion will be the main result of the discussion of the World Summit on sustainable development in Johannesburg, which should be correctly called Summit on Problems of Unsustainable Development. The definition of sustainable development proposed by the International Commission on Environment and Sustainable Development ("Bruntland Commission"), as "a development that provides the needs of the present generations of mankind, and at the same time, gives the opportunity to the coming generations to satisfy their needs as well", of course, is too general and vague. As mentioned above, only recently has the understanding of the key role of the biosphere been recognized in international documents and widely presented in Russian scientific publications in the form of a concept of the biotic regulation of the environment (Gorshkov 1990, 1995; Kondratyev 1998a, 1999b; Gorshkov et al. 2000) and even earlier in the publications by N.V. Timofeev-Resovsky and others (see Losev 2001).

The ideas developed in the N.V. Timofeev-Resovsky et al. studies to some extent have something in common with a Gaia hypothesis suggested by Lovelock (1995). Firstly, they assert that communities of organisms within their biogeocenoses and the whole biosphere regulate the chemical composition of two important media – the atmosphere and hydrosphere; secondly, it was pointed out that the biosphere and biota are not one super-organism, but they are biogeocenotic populations which form landscapes or ecosystems. The question was also raised about the necessity to understand specific features of the dynamic equilibrium of these systems, in other words, about the certain borders of their permissible disturbance. The problems of the landscape of science and ecology, in the context of the concept of biotic regulation, have been discussed in a recent article by Kondratyev et al. (2001a).

Since the last quarter of the twentieth century the scientific community has started to pay more attention to problems such as the ecosystem destruction, its role in the biosphere, an understanding of the necessity of preservation of biodiversity, including ecosystem diversity, problems of the limits of growth, and sustainable development and studying the biosphere as a whole system. Another important element, which becomes apparent in ecological problems and in sustainable development, is the necessity to change the point of view on development problems, to show respect to nature and to consider technologies as only one of the elements of solving problems of the environment and sustainable development.

The detailed discussion of the notion "sustainable development", including the SD indicators problem can be found in a number of monographs and articles (Kondratyev et al. 1996b; Kondratyev 1998a, 1999e; Kondratyev and Galindo 2001), in the References below. As has been mentioned by Losev (2001), there are four problems that should be resolved for the transfer to sustainable development.

The first problem is a clash of interests of our civilization with nature. Modern civilization is the result of 10,000 years of spontaneous human development, when people could build their history without any restrictions and the passage given them by the biosphere. However, by the beginning of the 20th century, as a result of the continuing expansion, humanity had gone beyond the limits and now the global environmental changes indicate such an evolution. *As a result of this evolution the twentieth century has become the century of severe environmental crisis development*; which has not yet been realized by all people and politicians.

The environmental crisis is a signal that humanity cannot further continue the random spontaneous development of its history and mankind must coordinate its history with the biosphere from which humanity is not separable. First, the development should be coordinated with the energy flow distribution law in the biosphere, which determines the safe limits of ecological development (Gorshkov 1990; Gorshkov et al. 2000; Losev 2001; Munasinghe et al. 2001; Raven 2001).

The signals from the biosphere destroyed by humanity do not yet affect the majority of the people enough, nor do they connect them with the fast developing environmental crisis. At the same time, virgin ecosystems still exist on our planet and with them there is a hope that irreversible changes have not yet started, and that this process can be stopped and reversed. However, disturbing signals should be adequately perceived as a guide for action. *None of the national programs or strategies on sustainable development contain an adequate response to such signals. Such a response cannot exist, however, because none of the relevant documents is based on adequate theoretical considerations.* Humanity has collided with the environment and that means that the laws of nature which determine the limitations of development have been violated. Such laws must be taken into account and understood as a guide for action. Some of these laws are already known, though they have been discovered too late. However, it is important that their knowledge gives ideas as to how the sustainable development can be provided.

The second problem under discussion belongs to the field of economics. This is economic development, which has led to the destruction of the natural mechanisms of the environmental stabilization and the development of ecological crises. The environmental expenses of civilization have begun to convert into economical, social, and demographic expenditures. *They produced the tendencies of decreased profitability of the world economy, decreased investment activity, slowing down of the pace of growth, and stimulated an increase in poverty, destruction of genetic programs and the detriment of human health, etc.* Humanity still uses all natural resources for free and destroys the biosphere, which is the basis of its life. To save the environment, humanity must radically change its ways of production. *It is necessary to replace the production cycles that use natural resources for free and give wastes unrestrictedly back into closed production cycles, which take into account the limitation of the pressure on the biosphere.* Thus, progress will be much less efficient, but much better controlled. Meanwhile, a latent economic crisis is developing, the manifestations of which can be seen from time to time in the form of regional and local financial crises, trade conflicts, recession and stagnation of the economy, increase in economic migration and unemployment.

The third problem of sustainable development, which needs resolution, is of a social nature. The present socio-economic system has not only led civilization in the twentieth century to a collision with nature, but has not even solved social problems, as has been mentioned earlier. Poverty is increasing and hunger spreading. During the last decade of the twentieth century, the number of poor people increased by more than 500 million, which is more than half of the population growth for the same period. Further, the gap between poor and rich people in all countries and in the world between poor and rich countries keeps growing. All this means that the global socio-economic system, no matter whether market or not, does not satisfy the requirements for the sustainable development of mankind. This system is robbing the people and destroying nature.

Thus, a social crisis is growing with all its features, such as the growth of a number of poor and starving people, the increase in the gap between rich and poor, growing unemployment, which is a constant factor in the developed countries, etc.

The fourth problem is a demographic crisis. It is known that wealthy people are getting wealthier and the poor families are growing larger and larger. The majority of new generations are doomed to live in poverty and hunger in the twenty-first century. This question is not being solved. There are a lot of obstacles, beginning with economics, up to traditions of the past, because people live with certain stereotypes and myths. Intelligent people concerned about these problems are few.

When the President of Botswana says that it is necessary to increase the population of the country from 6 to 12 million people, as if this would help his country to be taken into account, shows that his thinking reflects the stereotypes of tribal society organization. However, in the twenty-first century absolutely different criteria determine the significance of a country. The population of Switzerland is also 6 million people, but this country is of world importance.

The increasing demographic crisis provokes a social one. This leads to the decrease in arable land and the volume of food per person, aggravates ecological crisis, which in turn deteriorates such an important criterion as human health. An enormous number of young people in the developing countries emerge into life, demanding their own share of social benefits, which are not enough for everybody today. This destabilizes the situation in the developing countries, provokes conflicts, gives rise to migration, extremism and terrorism, and finally could result in global conflict.

Thus, modern civilization is in a system crisis, which was remarked on in the Bruntland Commission Report and subsequent documents. The alternative for this crisis is seen in the transition to sustainable development; not in the mythical one that is described in many international and national documents, but in the real change of the vector of the development of civilization.

At present, a certain gap exists in reaching the aims of sustainable development. The United Nations are going to reveal in the year 2002 what has been done for the accomplishment of the Rio Conference recommendations. For this purpose a *system of so-called indicators of sustainable development* has been elaborated and it is supposed to give qualitative and quantitative assessments of further progress in this direction. However, until now, a reliable scientific basis for this purpose does not exist except for the experience of the previous assessments, which are not fit for the new conditions of development under conditions of collision with nature. There is still no clear definition of what economic, ecological, social, and demographic sustainability is. *It is important to start from substantiation of values and goals of sustainable development*, that can be formulated under conditions of collusion between humanity and nature and the potential perspective of socio-economic catastrophe, with support from natural sciences and an understanding of the limits of development, stipulated by natural laws of the biosphere.

A special case is the problem of sustainable development in Russia. Generally speaking, the idea of sustainable development was born in the industrially developed countries. The world has supported this idea and taken it to be right. However, all countries are different: some of them still live in medieval times and some in the twenty-first century. Each country is very specific. If developed countries, which proposed the idea, have said to themselves: "We do not live the right way, we need something new", other countries can say: "We have not even reached your starting conditions which are needed in order to just to think about sustainable development." *This is why Russia, before thinking about sustainable development, should overcome the economic and structural post-Soviet crisis, just because it is more important for Russia.* However, the restrictions which are defined by the aim of sustainable development should be taken into account. First, sustainable development is a system of restrictions which, if they are accepted, could bring economic and social benefits.

For Russia, ecological restrictions for the period of overcoming the economic and structural crisis are profitable both economically and socially. With due regard to the perspectives of transition to sustainable development, the most important point is to preserve the land expanses of undisturbed (virgin) or hardly ever disturbed natural ecosystems (especially forests), which occupy 65% of the country (Bazilevich and Rodin 1967; Doos 2000; Schrope 2001). These are the sparsely populated territories of the North, Siberia and the Far East, where economic development is unprofitable, except for natural resources extraction, but at a certain level of world prices the value of such ecosystems will grow.

The principal economic activities should be concentrated in the regions which have been populated for a long time, in the middle and southern parts of the European part of Russia, and in the southern regions of Siberia as well as the Far East. The economic and infrastructure conditions and labor force available in these territories will allow reduced cargo transportation and decreased energy consumption, and besides, these territories are more acceptable habitats for people. The second requirement is the maintenance of a certain stability of population rate, which is ecologically optimum for Russia at present. The decrease in the population, which is predicted for the coming years, should be compensated by a reasonable supporting policy of immigration of Russian-speaking people from the neighboring territories. The last requirement is a complete refusal of extensive economic development, including a refusal of gigantic projects, and then a transition to an intensive way of development.

The realization of the restrictions mentioned for the period of transition from the economic and structural crisis, in the future will provide Russia with the starting conditions for sustainable development, especially ecologically sustainable development. It might become clear in time what sustainable economic and social development is which is not equivalent to sustainable economic growth, as many economists believe, and also clear what justice is. Anyway, the modern socio-economic system, as has been mentioned above, leads only to a deeper global system crisis and the danger of a rapid ecological catastrophe.

1.4
Conclusion. Unsolved Problems

The forthcoming World Summit on Sustainable Development "Rio+10" in 2002 in Johannesburg determines the necessity of substantiation of priorities, the main content of which is connected with the interaction of society and nature.

Despite the fact that much attention has been paid to the problem of sustainable development by some states, governments, unions and the United Nations, the contemporary human evolution is following the way of unsustainable development, which leads to the inevitable global ecological catastrophe and finally, to the downfall of modern civilization.

The obvious feature of such an evolution is a continuing growth of the population (especially in the poor developing countries) and the growth of social contrasts between a "golden billion" and the rest of the world population (there are different important indicators of these contrasts, such as unacceptable differences in the standard of life, poverty growth and enormous debts in the developing countries). Another important indicator of unsustainable development is the permanent degradation of the environment on the global scale.

The manifestations of such a situation are the anthropogenic impacts on the global climate and ozone layer as well as an increase in toxic emissions into the environment. The bothersome symptoms concern the depletion of natural, renewable and nonrenewable resources and the decrease in biodiversity.

It has become clear that the modern market economic system does not provide for an adequate transition from unsustainable development to a sustainable one. The efforts of the specialized institutions of the United Nations and the UN as a whole, in the spheres of education, legislation, and management improvement are not sufficient. NGOs and religious organizations could play a more constructive role.

The transition to sustainable development requires a number of measures. The measures of key importance are the following:

1. There is a necessity to realize that further development of the consumption society will lead to a global ecological catastrophe and the collapse of civilization. The solution of this problem can only be a refusal of the traditional paradigm of the consumption society and a radical change of life style towards spiritual values.
2. To remove the present social contrasts between "the south" and "the north", there is an urgent need to help developing countries in many ways, according to the recommendations which have been accepted by the UN.
3. Since the level of understanding of the problem of social and environmental dynamics remains low, the support of interdisciplinary studies has particular importance. It would be important to organize the international scientific discussions on sustainable development problems in the context of "Rio+10", concerning first of all the problem of priorities and quantitative assessments of various threats for modern civilization.

In this connection, the unsolved problems have been formulated in *Global Change and the Earth System: A Planet Under Pressure* (Steffen and Tyson 2001).

Analytical Questions

- What are the critical thresholds, bottlenecks and switches in the Earth System? What are the major dynamic patterns, teleconnections and feedback loops?
- What are the characteristic regimes and time scales of Earth's natural variability?
- What are the regimes of anthropogenic disturbance that matter at the Earth System level, and how do they interact with abrupt and extreme events?
- Which are the most vulnerable regions under global change?

Methodological Questions

- What are the principles for constructing representations of the Earth System that aggregate unnecessary details and retain important system-level features? What levels of complexity and resolution are required?
- What is the optimal global strategy for generating, processing, and integrating relevant Earth System data sets?
- What are the best techniques for analyzing and possibly predicting irregular events?
- What are the most appropriate methodologies for integrating natural science and social science paradigms, research approaches, and knowledge?

Normative Questions

- What are the general criteria and principles for distinguishing nonsustainable futures?
- What are the human carrying capacities of Earth under various assumptions and values? What processes, both human and natural, are most likely to constrain or compromise these goals?
- What are the potential changes in the state of the Earth System that can be triggered by human activities, but should be avoided?
- What are the equity principles that should govern global environmental response strategies?

Strategic Questions

- What is the optimal mix of adaptation and mitigation measures to respond to global change?
- What is the best distribution of nature reserves and managed areas on the planetary surface?
- What are the knowledge levels, value/culture bases, options, and caveats for technological fixes like geoengineering and genetic modifications?
- What is the structure of an effective and efficient system of global environment and development institutions?

Needless to say, this list of unsolved problems is not comprehensive. Many of the above-mentioned problems have been widely discussed in scientific publications (Diatlov 1988; Roden 1988; Maguire et al. 1991; Prykin 1998; Ernst 1999;

Kondratyev and Filatov 1999; Lvov 1999; Moller 1999; Gaffin et al. 2000; Boehmer-Christiansen 2000; Chistobaev 2001; Houghton et al. 2001; Hurell and Woods 2000; Pokazeev and Medvedev 2000; Kates et al. 2001; Pimm 2001). There is no doubt that the world becomes more and more complicated, multicomponent and contradictory. Its main features are a great number of feedbacks, chaotic dynamics, and the possibility of strong and sudden changes.

The document, *Global change and the Earth System: A Planet under pressure* (Steffen and Tyson 2001) justly declares: "Humanity is already managing the planet, but in an unconnected and hazardous way driven ultimately by individual and group needs and desires. As a result of the innumerable human activities that perturb and transform the global environment, the Earth System is being pushed beyond its operating domain. This excursion into *planetary terra incognita* is accelerating as the consumption-based Western way of life becomes more widely adopted by a rapidly growing world population. The management challenges to achieve a sustainable future are unprecedented. As a business-as-usual approach to the future could trigger abrupt changes with potentially catastrophic consequences, alternative strategies of planetary management are clearly needed."

In conclusion, one more reference to the Vail Regional Roundtable Report (2002) is proper: "The present generation may be among the last that could correct the current course of the world development before it reaches a point of no return, due to depletion of the natural resource base and degradation of the environment. It has the knowledge and technological ability to achieve this. What it still lacks is political will and individual commitment for action and broad public awareness of the consequences of inaction. There is a need for a new level of commitment, responsibility and partnerships. There is also a need for new ethics that are based on the recognition that growth is limited by the health and carrying capacity of the natural processes, their connectedness with the natural world and their capacity for positive action."

"The ways in which the industrial countries have developed in the past are not sustainable in the future, either for themselves or for others. Taking forward the switch to sustainability, this region must take special responsibility for the poorer countries in the world by directly addressing environmental, social and economic problems. A new degree of global solidarity and partnership in the world – a world that is increasingly inter-connected and inter-dependent – will be key to the health and quality of life and the sustainable future of all citizens of our planet. Quality of life in one part of the world should not be at the expense of quality of life in other parts."

Chapter 2

The Basic Principles of Global Ecoinformatics 2

2.1 The Main Idea of Global Ecoinformatics

The accelerating accumulation of knowledge, vast scientific and engineering progress and unprecedented growth of human influences on the environment have already been felt since the 1970s posing the problem of global evaluation of the state of the environment and suggesting the possibility of its long-term forecasting. The scientific research in this field has led scientists throughout the world to the conclusion that the solution of the problem of objective control of the environmental quality is possible only through creation of a unified international monitoring system in combination with a global Nature/Society (NS) modeling system. Many international and national environmental programs are dedicated to the realization of such a system (Abalkin 1999; Drake 2000; French 2000; Ivashov 2000; Barlybaev 2001; Demirchian 2001). In the framework of these programs, sufficiently voluminous databases of environmental parameters are created, information about the dynamics of natural and anthropogenic processes on various scales is accumulated and models of biogeochemical, biogeocenotic, climatic and demographic processes are created. The technical base of global geoinformation monitoring is combined with efficient means of data acquisition, recording, accumulation and processing of measurement data obtained from onboard spacecraft, aircraft, ground platforms, and floating laboratories.

However, despite significant progress in many fields of natural monitoring, the main problems remain unsolved, consisting of designing the optimum combination of all technical means of data acquisition, creating efficient and economical monitoring structures, and creating reliable evaluation methods for forecasting environmental dynamics under the influence of anthropogenic consequences. The experience of recent years shows that the possibility exists of creating a global model capable of use in adaptive mode for providing recommendations on the optimum monitoring structure and the requirements for formation of the databases (Wefford 1997; Young 2001). However, the solution to the problem of creating this model is lagging. The lack of such a model leads to unjustified expenses in conducting new expeditions for environmental investigations and in building new systems of observation (Kondratyev 1990, 1999b; Krapivin and Mkrtchyan 1991; Aota et al. 1993; Krapivin 1993; Kozoderov 1998; Kozoderov and Kosolapov 1999).

Recently, many investigators (Krapivin and Shutko 1989; Maguire 1991; Sellers et al. 1995; Kondratyev et al. 1996a) have put forth the problem of the synthesis of complex systems for collection of environmental information and for uniting GIS, remote and local measurements with models. Such systems are called geo-information monitoring systems (GIMS) and they are aimed at the systematic observation and evaluation of the environment and its changes under the effects of people's economic activities. One of the important aspects of the functioning of these systems is the possibility of forecasting the state of the surrounding medium and warning about undesirable changes in its characteristics (Martchuk and Kondratyev 1992). The realization of this monitoring function is possible while applying mathematical modeling methods ensuring simulation of the functioning of natural complexes (Armand et al. 1987, 1997; Hidayat 1994; Bui and Krapivin 1997; Phillips et al. 1997; Tikunov 1997; Demirchian and Kondratyev 1998, 1999).

The development of models of biogeochemical, biocenotic, demographic, socio-economical and other biospherical and climatic processes on the whole requires the formation of the necessary requirements to the GIMS structure and its database. According to the proposed GIMS structure in the paper of Kelley et al. (1992b), the simulation of the biosphere dynamics is one of the important functions of the GIMS. As a result, the necessity arises of a new approach to the estimation of the biosphere state. Eventually, the basic aim of all investigations in the direction of development of the GIMS technology comes down to the following tasks:

- acceleration of the optimization of expenditures on the reconstruction of environmental survey systems (Krapivin et al. 1996a);
- creation of conditions for optimal planning of the organizational structure of human society as it affects the environment (Kondratyev 1999b; Krapivin et al. 1996b);
- ensuring the purposeful direction of global environmental processes so that they are for the good of mankind and do not cause damage to nature (Kondratyev 1998a; Krapivin et al. 1994).

As was shown by these investigations, balanced criteria of information selection covering the hierarchy of causal-investigative constraints in the biosphere exist. These include the coordination of measurement tolerances, the depth of spatial quantitation in the course of describing the atmosphere, land and oceans, the degree of detailing of biomes, etc. At an empirical level, as expressed in the evaluations of the results of computing experiments by experts, these criteria allow the selection of the informational structure of the geoinformational monitoring system representing the hierarchical subordination of the models at various levels (Krapivin et al. 1991; Nitu et al. 2000a; Krapivin and Kondratyev 2002).

At the same time, the development of an adequate global system for monitoring the environmental dynamics is supposed to be based on requirements imposed by the regional industrial and socio-economic structure of the society. This is expressed by the nonuniform development of ecologically impure industrial production, the distribution of its concentration by regions, differences of regional facilities for control and collection of information, the state of their technical equipment, etc. Such differences inevitably influence the choice of the GIMS structure and its informational technical base. Thus the hierarchic structure of the combination of mathematical models, which determine the GIMS structure, is determined by nature dynamics and socio-economic factors, as well as by technical possibilities. The degree of detail in the models depends mainly on the spatiotemporal resolution reached at the given level of simulating both natural processes and socio-economic dynamics.

Global ecoinformatics suggests the development of a set of models for various processes in the biosphere, taking into account their spatial inhomogeneity, and the combination of existing global databases with already functioning systems for environmental observations. Relevant solutions should be based on the cooperation between experts who have developed climatic, biospheric and socio-economic models with the eventual aim of creation of a global model of the Nature/Society (NS) system. As a subsequent improvement of the model, we may study the interaction of the NS system with processes in the near-earth space (first of all in the magnetosphere) to further develop the global simulation model (GSM). As a result, a system will be created that will be capable of forecasting the development of natural processes and of evaluating the long-term consequences of large-scale impacts on the environment. Applications of this system will encompass the problems of environmental protection on the global, continental, regional, and local scales by the accomplishing projects for expert ecological examinations of topsoil, hydrological regimes and the changing composition of the atmosphere etc.

The realization of the GSM allows integration within a complex structure of all international and national means of environmental monitoring and provides a tool for objective evaluation of the environmental quality. The furnishing of the system with modern efficient processing techniques allows it to solve a wide spectrum of problems regarding the identification of pollution sources and contributes to the elimination of conflicts due to transboundary transport of pollution, etc. (Krapivin et al. 1997a; Kondratyev 1998b, 1999c–e; Bratimov et al. 2000; Kondratyev et al. 2000; Gerard et al. 2001).

New global modeling information technologies in the framework of the International Centre for Global Geoinformation Monitoring (ICGGM; Fig. 2.1) will be used to create a principally new structure of monitoring which depends on a database of information of various quality and many mathematical and physical models of various types. Evolutionary technology will solve many contradictions which arise from incompleteness and an indeterminate information base, fragmentary knowledge about natural laws, and the absence or under-development of observational systems and of simulation techniques (Bukatova et al. 1991; Krapivin et al. 1997b; Kozoderov 1998; Kalinkevitch and Kutuza 2001).

The ICGGM functional structure is based on realization of the idea of evolutionary neurocomputer technology. This will have the architecture of modern computer system software including a network of tutorial servers of an evolutionary type. This will lead to an ICGGM structure which will simulate the NS system with forecasting capabilities for optimum evaluation of nature protection measures and other structural decisions concerning the interactions of human society with nature (Kohlmaier et al. 1987; Kondratyev 1990; Demirchian and Kondratyev 1998).

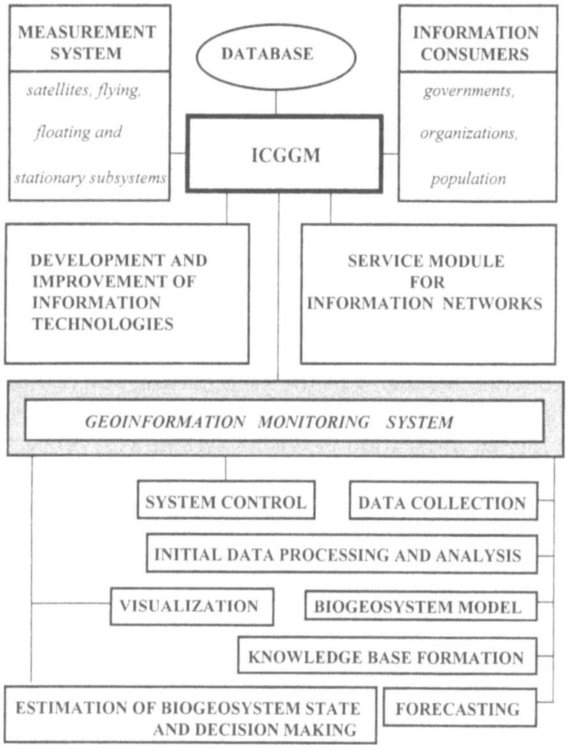

Fig. 2.1.
Structure of the International Centre for Global Geoinformation Monitoring

The practical embodiment of the idea of NS model creation requires the realization of purposefully complex studies, among which the following questions are the most important:

- Systematization of the study of global changes and formation of complex ideas about biospheric processes and the structure of the biosphere.
- Development of a conceptual model of the biosphere as an element of the global geoinformation monitoring system.
- Inventory and analysis of the existing environmental databases and choice of the global database structure.
- Classification of space–time characteristics and cause-effect connections in the biosphere with the aim of working out a coordination of space-time scales of ecological processes.
- Creation of typical model sets of ecological systems, biogeochemical, biogeocenotic, hydrological, and climatic processes.
- Study of processes of interaction of the biosphere and climate. Search for regularities in the influence of the sun on biospheric systems.
- Systematization of information about oceanic ecosystems. Description of geophysical and trophic structures with their regional division and coordination with space–time scales.
- Construction of models of biocenographic fields and development of algorithms of synthesis based on the oceanic unit of a global model.

- Formation of the scenarios set for coevolution development of the biosphere and human society.
- Creation of demographic models.
- Parameterization of scientific and technical processes in the utilization of land resources.
- Search for new information technologies of global modeling providing for the reduction of requirements on the databases and knowledge bases.
- Development of architectural and algorithmic principles of the functioning of computer systems consisting of neuron-like elements for evolutionary processing of information with high speed and efficiency and integration of these systems into the ICGGM structure.
- Development of the ecological monitoring concept and creation of the theoretical ecoinformatics base. Development of methods and criteria for evaluation of the stability of global natural processes. Analysis of biospheric and climatic structural stability.
- Synthesis of the NS model and development of the means for realization of numerical experiments in the framework of evaluation of the consequences of scenarios of various anthropogenic activities.

An analysis of recent investigations in these fields shows that for success in global modeling it is necessary to develop new methods for the system analysis of complicated natural processes and new data processing methods directed at the synthesis of balanced criteria of information selection taking into consideration the hierarchy of cause–effect connections in the NS system.

The accumulated knowledge and experience of global monitoring make it possible to separate the main units of the NS model: magnetosphere (Korgenevsky et al. 1989), climate (Demirchian and Kondratyev 1999), biogeochemical cycles (Nitu et al. 2000a), biogeocenotic processes (Kozoderov and Kosolapov 1999), socio-economic structure (Kondratyev 1999b) as well as scientific and engineering progress (Kondratyev et al. 1997). The development of methods of parameterization of these units has reached a level where the synthesis of the NS model is possible based on the principle of systematic coordination of the inputs and outputs of the individual units. The realization of this process requires the solution of the principal tasks of coordination of the space-time scales of the natural processes and the choice of the algorithms connecting the NS model with the databases. Two upper spatial levels, global and continental, are differentiated into the NS. Below these, the national and state levels encompass three spatial scales: national, regional, and local. There are also intermediate levels as required by the particular databases.

Creation of the NS model requires systematization of the models and databases at the national and state levels and connection of them to the global models and databases. One result of this systematization must be the creation of typical models of natural processes to serve as NS model base elements encompassing the national, regional, and local levels.

Realization of the hierarchical structure of the complex of mathematical models included in the NS model superimposes this structure on the modeling algorithms being used. Fragmentary, noisy, and incomplete space–time data necessitate research into new modeling methods which facilitate the parameterization processes of model phenomena under the conditions of irreducible information interdeterminacy and the possible time-dependent character of data measurements of environmental parameters. Among these new approaches to the simulation-evolution modeling technology, one has been developed recently (Nitu et al. 2000b). This technology allows the isolation, beside traditional natural phenomenon models, of new types of models providing the description of weakly parameterized processes. For example, units may appear in the NS model describing such processes as scientific and engineering progress, agricultural production, extraction and use of mineral resources, demography, etc. (Krapivin and Nazaryan 1995).

2.2
The Technology of Geoinformation Monitoring

There are many parameters describing the environmental conditions on the Earth. Among them are soil moisture and moisture-related parameters such as the depth of a shallow water table and contours of wetlands and marshy areas. The knowledge about these parameters and conditions is very important for agricultural needs, water management, and land reclamation, for measuring and forecasting trends in regional to global hydrological regimes and for obtaining reliable information about the water conservation estimates.

In principle, the required information may be obtained by using on-site measurements and remote sensing and by getting access to a prior knowledge-based data in the GIS databases (Maguire et al. 1991; Tikunov 1997). However, the following problems must solved:

- what kind of instruments are to be used for conducting the so-called ground-truth and remote sensing measurements?
- what is the cost to be paid for the on-site and remote sensing information?
- what kind of balance is to be taken under consideration between the information content of on-site and remote sensing and the cost of these types of observations?
- what kind of mathematical models may be used for both the interpolation of data and the extrapolation of them in terms of time and space with the goals to reduce the frequency and thus the cost of the observations and to increase the reliability of forecasting the environmental behavior of the observed items?

These and other problems are solved by using a monitoring system based on combining the functions of environmental data acquisition, control of the data archives, data analysis, and forecasting the characteristics of the most important processes in the environment. In other words, this unification forms the new information technology called the GIMS-technology. The term "GeoInformational monitoring system (GIMS)" is used to describe the formula: *GIMS = GIS + model*. The relationship between GIMS and GIS is shown in Fig. 2.2. Evidently, GIMS is a superset of the systems shown in Fig. 2.1. There are two views of the GIMS. In the first view, the term "GIMS" is synonymous with "GIS". In the second view, the definition of GIMS expands on the GIS. In keeping with the second view, the main units of the GIMS are considered below.

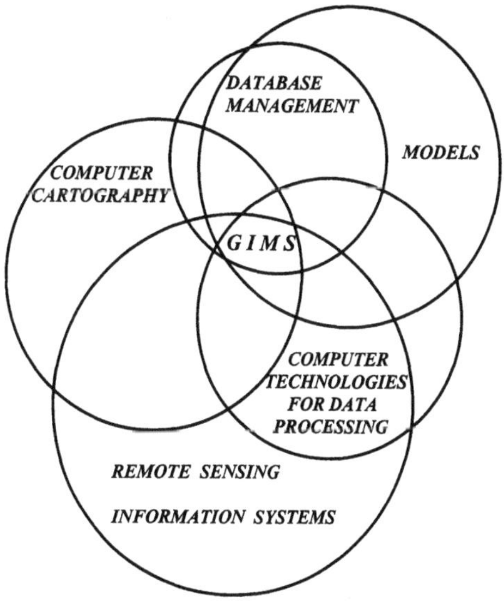

Fig. 2.2.
Conceptual diagram showing the definition of the GIMS-technology as the intersection of many scientific disciplines

The basic component of the GIMS is considered as a natural subsystem interacting through biospheric, climatic, and socio-economic connections with the global NS system. A model is created describing this interaction and the functioning of various levels of the space–time hierarchy of the whole combination of processes in the subsystem. The model encompasses characteristic features for typical elements of the natural and anthropogenic processes and the model development is based on the existing information base. The model structure is oriented to the adaptive regime of its use (Fig. 2.3).

The combination of the environmental information acquisition system, the model of the functioning of the typical geoecosystem, the computer cartography system, and the means of artificial intelligence will result in creation of the geoinformation monitoring system of a typical natural element capable of solving the following tasks:

- evaluation of global change effects on the environment of the typical element of the NS system;
- evaluation of the role of environmental change occurring in the typical element of climatic and biospheric changes on the Earth and in its territories;
- evaluation of the environmental state of the atmosphere, hydrosphere, and soil–plant formations;
- formation and renewal of information structures on ecological, climatic, demographic, and economic parameters;
- operative cartography of the situation of the landscape;
- forecasting the ecological consequences of the realization of anthropogenic scenarios;
- typifying land covers, natural phenomena, populated landscapes, surface contamination of landscapes, hydrological systems, and forests;
- evaluation of population security.

Fig. 2.3.
Conceptual block-diagram of geoecoinformation monitoring and use of the GSM model

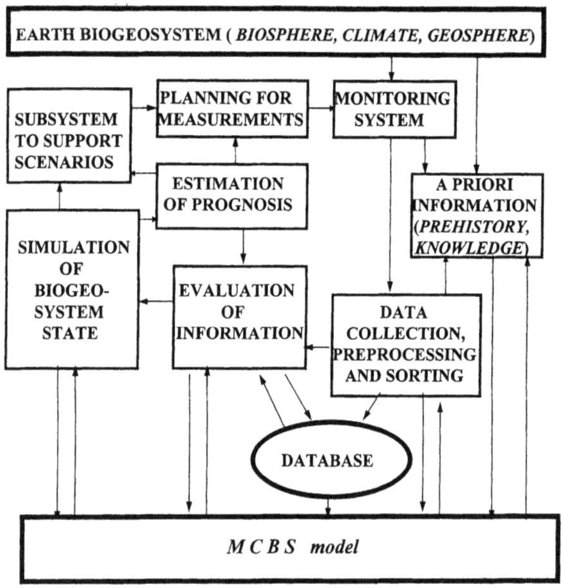

Construction of the GIMS is connected with consideration of the components of the biosphere, climate and social medium characterized by the given level of spatial hierarchy.

■ **Subsystem for Planning and Analysis of the Data Acquisition Systems.** This subsystem solves the task of experimental planning by analysis of the structure of the environmental data acquisition system, making use of satellite data, flying laboratories, and movable and stationary ground observation means. The laboratories are equipped with the necessary software and hardware tools to allow determination of the degree of environmental contamination, of the ecological situation, mapping of the characteristic geological formations, detection of soil subsurface centers of ecological injury, performing the all-weather land-cover typification and detection of permafrost disturbances, oil spills, forest states and pollution of bodies of water.

■ **Subsystem for Initial Data Processing and Data Acquisition.** Methods and algorithms for synchronous analysis of aerospace information and ground measurements are realized using space–time interpolation methods. Retrieval of the data and their reduction to the common time scale is performed. Model parameters are determined. Thematic classification of the data is carried out and space–time combination is performed of images in the optical, IR, and microwave ranges and of trace measurements obtained from devices of various types.

■ **Subsystem for Computer Mapping.** Algorithms are realized for creation of computer maps with characteristic markings for evaluating the ecological situation. Multilevel scaling and fragmentation of the territory are envisaged.

The overlaying of output maps with the information needed by the user is provided through the user interface.

Subsystem for Evaluation of the State of the Atmosphere. Models of atmospheric pollution spread due to evaporation and burning of oil products, natural gas and other outputs of industrial enterprises are suggested. The problem of evaluation of the atmosphere dust content is solved. The gas and aerosol composition of the near-earth atmospheric layer are provided and forecasting maps of their distribution over the Earth's surface are created.

Subsystem for Evaluation of the State of the Soil-Plant Cover. This subsystem solves the following tasks:

- typifying of the floristic background taking into account the microrelief, soil type and its salinity, humidification, and degree of soil brine mineralization;
- revealing of micro-and macrorelief peculiarities and subsurface anomalies;
- determination of the structural topology of the land cover;
- indication of forests, swamps, agricultural crops, and pastures.

Subsystem for Evaluation of the State of the Water Medium. A complex simulation model of the territory is developed taking into account seasonal changes of surface and river runoff, the influence of snow cover and permafrost and the regime of precipitation and evapotranspiration. A model is constructed of water quality dynamics for the hydrologic network of the territory.

Subsystem for Risk Evaluation of the Ecological Safety and the Health of the Population. Algorithms are developed for evaluation of the damage to nature, economic stability, and population health depending on changes in the environment connected with natural trends of meteorological, biogeochemical, biogeocenotic, microbiological, radiological, and other natural processes as well as the enhancement of environmental stress of anthropogenic origin.

Subsystem for Identification of Causes of Ecological and Sanitary Disturbances. The task of revealing the sources of environmental pollution is solved. This subsystem determines the source coordinates, the magnitude, and the possible time of nonplanned introductions of contaminate substances. The dynamic characteristics of the pollution sources are given. A priori unknown pollution sources are revealed and the directions of possible transborder transfer of pollutants are determined.

Subsystem for Intelligent Support. Software-mathematical algorithms are realized for providing the user with intelligent support in performing the complex analysis of objective information formed in the framework of the simulation experiment. The necessary information for the objective dialogue with the MCBS model is provided in a convenient form for the user. The introduction of data processing corrections is also provided. The knowledge base of anthropogenic, demographic and socio-economic processes on the territory is formed.

2.3 Elements of the Evolutionary Computer Technology

The synthesis of the intelligent support subsystem of the GIMS is based on evolutionary simulation methods (Bukatova et al. 1991; Nitu et al. 2000b). These methods were the source of a new scientific direction called evoinformatics. This science lies on the boundary between such sciences as neurocybernetics, cognitive psychology, artificial intelligence, theory of systems, theory of survivability, and systemology. Recent advances in informatics and information technology have promoted the extension of mathematical modeling and computer technology applications to such domains as ecology, biophysics, and medicine. The notion of the model experiment has already lost its novelty, and the term "numerical experiment" is used in many serious studies covering a vast range of subjects including biosphere dynamics (Marchuk and Kondratyev 1992; Krapivin 1993, 1996; Boysen 2000). All such studies a priori imply the availability of a more or less adequate model implemented as an array of tools in an algorithmic language. To manipulate such a model in order to perform a series of specific numerical experiments one needs a general-purpose computer. It is at this point that the researcher may be challenged with insurmountable difficulties caused by constraints on the computer's memory and speed of operation.

Many such endeavors indicate that modern hardware can handle relatively complex models. Yet, the same experience prompts the need for constant improvement of modeling techniques, because the researcher runs into a conflict between his desire to enhance the accuracy of a model and the limited capabilities of the computer. It is obvious that the building of a model completely adequate to a real-world entity is not feasible: on the one hand, the account of all parameters of the entity leads to the "evil of multidimensionality", and on the other hand, simple models allowing for a small number of parameters are inadequate to model the complex entities under consideration. Besides, in projects involving ocean physics, geophysics, global ecology, socio-economics, etc., the building of an adequate model is in principle impossible because of the unattainability of complete information. It is only with such systems that applied problems from the domains of global ecology, biophysics and medicine deal. Furthermore, difficulties in these subject areas arise at the early stages of research, i.e., at an attempt at formulation of a model. What is to be done when the currently available knowledge does not allow for synthesis of a mathematical model of an entity or a process? The answer to this question leads to a theory of learning computers of the evolutionary type.

We shall retain the term "model", although it is used here in a somewhat different sense. What is implied, is the description of entities changing over time in an unpredictable manner, and by virtue of this ensuring the irremovability of information uncertainty at any moment. Such are natural systems studied, e.g., in global ecology, geophysics, biophysics, and medicine. Consequently, a model treated in a broad sense must provide for continuous adaptation to the changing behavior and structure of the observed entity. It is clear that universal models can be built only through the synthesis of particular models. Models of this kind are implemented for problems of recognition and prediction.

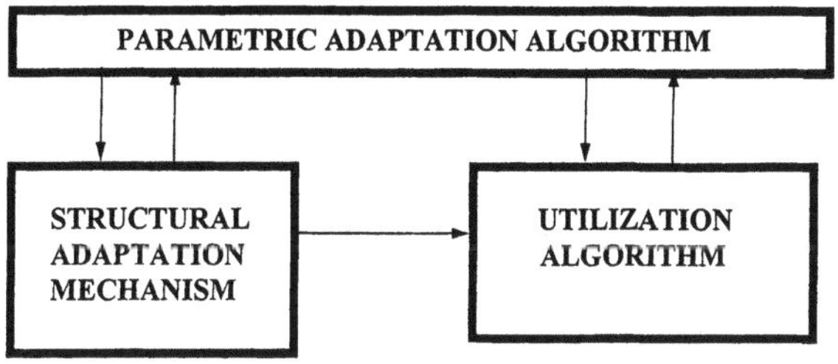

Fig. 2.4. Schematic representation of the evolutionary modeling concept

Thus let a real-world object A have some unknown algorithm of operation, with only some previous history of operation of finite length known. It is required to simulate efficiently the object A using models built in some system operating in the real-time scale. It is necessary to build a sequence of improving models $\{A_k\}$, where k = 1, 2,.... Cybernetics began to ponder the problem of how this could be attained virtually from the time of emergence of the first, primitive by today's standard, computing machines. It was then that the idea of creating artificial intelligence and designing "thinking machines" appeared, giving rise to the development of robotics. Being aware that a computer is an obedient executor of some program, cybernetics doggedly worked on the possibility of imparting to the machine a measure of unpredictable behavior with some facets of "innovation". Treating intelligence as the ability to correctly respond to a novel situation, scientists came to the conclusion that machines were feasible that would be capable of adaptation at the level of individual components and their structural organization.

One of the first to suggest the idea of model-free learning for computers was D. Fink, who in 1966 proposed to teach a computer by changing the structure of relations between its elements. This analogy with the neural operation of the brain established in neurophysiology yielded fruit and helped, by far, to advance the scale of artificial intelligence capabilities. The volume by Fogel et al. in 1966 gave impetus to an entirely new cybernetic trend.

Evolutionary modeling on the whole can be represented by a hierarchical two-tier procedure (Fig. 2.4). At the first tier, there are two constantly alternating processes conventionally termed the structural adaptation process and the utilization process. At the kth stage of adaptation during the structural adaptation algorithm, models of the sequence $\{A_{s,i}\}$, $i = 1,...,Ms$ are synthesized. Special rules create memory from the most effective models $< A_s^1,...,A_s^k >$. Following the adaptation stage, the system selects the most efficient model for the utilization stage. A schematic diagram of the ith step of kth stage of adaptation is represented in Fig. 2.5. The "object" unit here denotes that the real-world object is defined by some previous history.

The procedure of the evolutionary selection of models provides for a virtually time-unlimited operation of the system under irremovable information uncertainty. Apart from the previous history which, as a rule, does not meet the requirements of traditional statistical analysis, the researcher has no other available information. Under such conditions one obviously has to use the maximum available information, in particular, that on the operation of the adaptation and utilization stages. In the parametric adaptation algorithm this information is

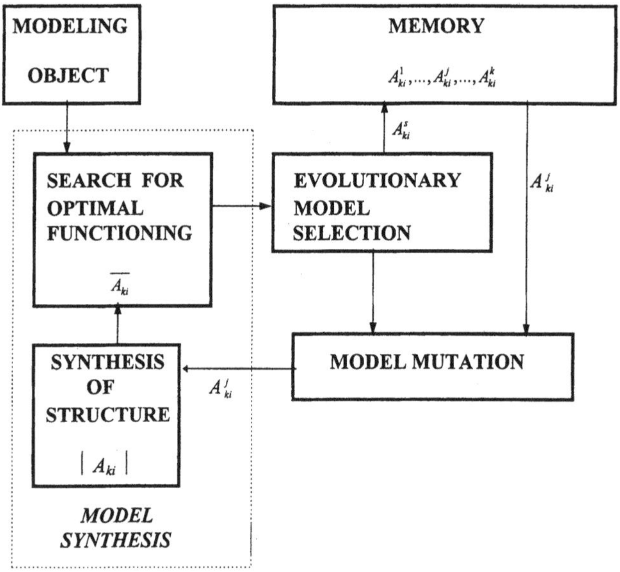

Fig. 2.5. Conceptual flow-diagram of the *i*th step on the *k*th stage of the adaptation process

used as early as the adaptation of the first-tier parameters, i.e., the characteristic number for each mode of changes; multiplicity of the change mode utilization; length of the change mode list; the distribution of change modes in the list; the distribution of *k*-models of *Aik*, memory size at the stage of adaptation, the prehistory length, etc.

Given some results, the modeling process endeavors to search for a building-block base to synthesize a network with a variable structure in an attempt to create special hardware with a flexible structure based on the new principles of information processing. Recent advances in microelectronics have helped solve the problem of selecting components for structures with variable fields of relations.

The unification and specialization are characteristic features of the evolutionary software as a consequence of the minimum of a priori information, effective mechanisms of adaptation, and the modular principle of realization. With an orientation to up-to-date personal computer engineering and a diverse range of application problems, these specific features have made it possible to work out evolutionary computation technologies in which, with the aid of an active dialogue with a user, a set of software modules is realized as well as the adjustment of the evolutionary facility to the specificity of the problem being solved (Bukatova et al. 1991). Regardless of the field of application, the evolutionary technology or artificial neural networks with software support is characterized by adaptability, flexibility, dynamism, and self-correction, while the main distinction consists in the high effectiveness and adjustability under the conditions of maximum informative uncertainty, including that of an irremovable character. This approach allows the realization of the adaptive procedure in Fig. 2.5. Actual evolutionary modeling technology gives a new structure of the GIMS (Fig. 2.6).

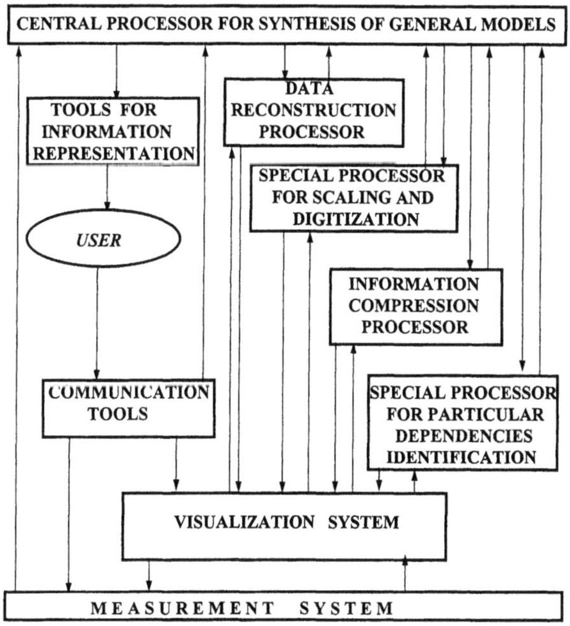

Fig. 2.6. Structure of the geoinformation monitoring system with evolutionary devices

2.4 A New Type of Global Model

Various approaches to the synthesis of a global model exist to describe all aspects of man's interactions with environmental bodies and with their physical, biological and chemical systems. One such way has its origin in the studies of the Computer Centre of the Russian Academy of Sciences in Moscow (Krapivin et al. 1982). A global model of this type is formed on the basis of the detailed description of the climate system with the consideration of a small set of biospheric components. This strategy of global modeling is adhered to in the Potsdam Institute for Climate Impact Research studies (Boysen 2000) where the Moscow Global Model prototypes are developed. More than 30 climate models are being developed in different countries as attempts to bring forth new trends in the science of global change (Claussen et al. 1999; Demirchian and Kondratyev 1999). Unfortunately, global- and regional-scale studies on the processes and impacts of global change using this approach have not produced satisfactory results. That is why another approach to the global modeling problem has been developed by many authors (Krapivin 1993; Sellers et al. 1996; Kondratyev et al. 2000). This approach is known as evolutionary modeling.

The traditional approaches to building a global model encounter some difficulties of the algorithmic description with respect to many socio-economic, ecological, and climatic processes (Kondratyev 1999a,d) so that one has to deal with information uncertainty. These approaches to global modeling simply ignore such uncertainty and, consequently, the structure of the resultant models does not adequately reflect the real processes. Evolutionary modeling makes it possible to remove this drawback by the synthesis of a combined model whose structure is subject to adaptation against the background of the history of a system of the biosphere and climate components. The implementation of such a model can also be combined in various classes of models using conventional software and hardware and special-purpose processors of the evolutionary type. The form of such a combination is diverse, depending on the spatiotemporal completeness of the databases (Rochon et al. 1996).

The experience in global modeling abounds in examples of unsolvable problems encountered when looking for ways to describe the scientific and technological advances and human activity in its diverse manifestations. No lesser difficulties arise in modeling climate described by a superimposition of processes with different temporal variability rates. As to completeness of description in the global model, it is impossible to clearly delineate the bounds of information availability and the extent of the required spatial and structural detail. Therefore, without going into natural-philosophical analysis of global problems, and skirting the issue of the ultimate solution to global modeling, we will confine ourselves to the discussion of only one of the possible approaches. This approach will demonstrate in which way evolutionary modeling implemented on special processors can help overcome computing, algorithmic, and other difficulties of global modeling. All of this implies that a search for effective models of the traditional type can aptly be considered. At present, the development of global biogeocenotic models is not seen as difficult. Many such models have been created, and the gathering of information to support them is under way. The history of the interaction of the biosphere with the climate system and human society is not known sufficiently, which is one of the hurdles, e.g., in the description of climatic cycles. To build a global model accounting for the interaction of the biosphere, climate system, magnetosphere, etc., it is necessary to apply the evolutionary approach which helps to overcome the uncertainties in the description of this interaction. As a result of the adjustment of such a model to the history of the prescribed cycle, we will obtain a model implicitly tracing various regularities of the dynamics of the biosphere in the past and allowing for forecast assessments to be made in the same temporal cycle. A special-processor version of this model completely removes all the existing algorithmic and computing hurdles arising from the large dimensionality of the global model and the conditions of irreducible nonparametric uncertainty.

Figure 2.7 shows the conceptual diagram of this new type of global model. The data archive is formed here as two structures. Data of the first type for the computer models of the biosphere processes are stored as climatic maps and as tables of the model equation coefficients. It is necessary to fill in all cells of the schematic maps. Data of the second type are represented as fragments recorded disparately (possibly irregularly) in time and space, i.e., CO_2 concentration, temperature, precipitation, pressure, population numbers, availability of resources, etc. Data of this type are used to adjust the evolutionary processor to the given class of models, e.g., finite automata. As a result of this procedure the model is adapted to the history of the prescribed time cycle. As has been shown by Bukatova et al. (1991) a stable forecast is produced with 75–95% reliability covering several temporal steps. The extent of a forecast is determined by the depth of the history against the background of the saturation of the effect. Given the need for a forecast under the conditions of change in the trends of human economic

Fig. 2.7.
Structure of the global model based on using the evolutionary units

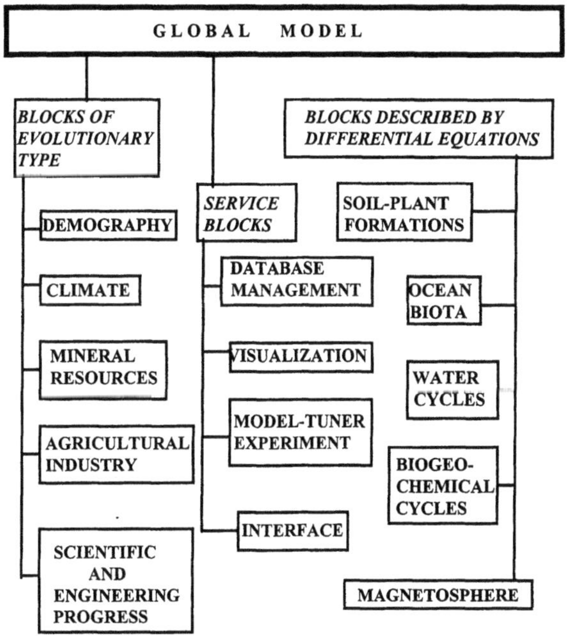

activities, an evolutionary processor is adjusted to the assigned scenario, thus automatically providing for simulation of the corresponding response of the biosphere to this change.

The suggested structure of the global model thus ensures a flexible combination of models of the traditional and evolutionary types. The proposed approach helps escape the need to model nonstationary processes (climatic, socio-economic, demographic, etc.) and provides for overcoming uncertainty. A model of this new type makes it possible to go from learning experiments to the assessment of the viability of the biosphere with regard to actual trends of anthropogenic stresses in all regions of the globe.

A departure from the established global modeling techniques based on the new information technology makes it possible to proceed to creating a global system of monitoring with the global model as a portion of the support for the system. The structure of such a system is represented in Fig. 2.8. Application of the evolutionary computer technology provides for categorization of the whole system by a class of subsystems with variable structures and for making it adaptable to changes in the natural process or entity under observation. Furthermore, it becomes possible to detail heterogeneously the natural systems under study in the space of phase variables, and to select nonuniform geographical grids in a sampling analysis of the planetary surface, i.e., arbitrary insertion of significant regularities at the regional level becomes possible.

The automatic system for processing global information is aimed at the acquisition of combined models reflected in the real-time scale of the climatic and anthropogenic changes in the biosphere and based on the known history (or, rather, its simulation). The system relies on the set of models of biosphere

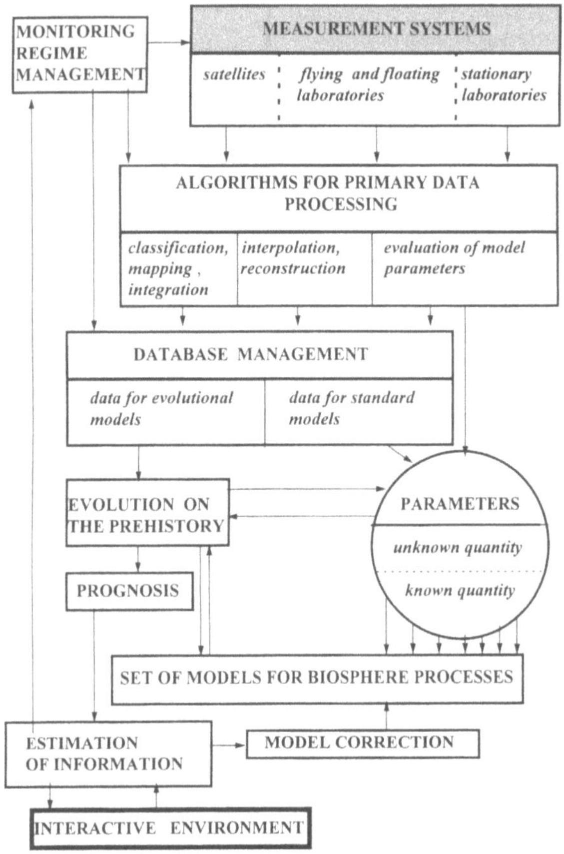

Fig. 2.8. Structure of the learning monitoring system of the environment implementing evolutionary computer technology and relying on the set of models of biosphere processes

processes and, using software of other units with the help of the scenario of anthropogenic behavior formulated at input, provides for prompt assessment of the environmental state and for forecast assessments within the framework of this scenario. The automatic system for processing global information also consists of the advantage in that it formulates the entry and bound conditions for particular built-in models in the study of regional systems and virtually substitutes for field measurements of those conditions.

The first version of the global model (Krapivin et al. 1982) was oriented towards rigid spatiotemporal detailing and, therefore required a large quantity of information. The subsequent development of an automatic system for processing global information has made it possible, owing to the evolutionary technology, to discard the generally accepted regular geophysical grid in archive development (Rosen 2000) and to solve this problem using algorithms for the recovery of spatiotemporal information.

2.5 Global Modeling and Theory of Complex Systems Survivability

One of the main problems in the framework of modern science is the estimation of the biosphere survivability under conditions of increasing anthropogenic impact. Such estimations can be realized by using the biosphere model. Well-known attempts in the direction of the synthesis of a global model give unsatisfactory results. Such global models have an educational character but give no possibility for real assessments. For the last few years many investigators put forward the problem of creating reliable and effective systems for control of the environmental state on a global scale. Generally, this problem includes elaboration of technical means for collection, storage, and transfer of the data about the state of the natural media and (on the other side) development of methods for processing these data (Kondratyev 1999a,b; Gorshkov et al. 2000; Nitu at al. 2000a,b; Svirezhev 2000).

The existing means for collection of information about natural objects and processes makes possible the formation of the data set, covering large domains up to the whole biosphere. Remote sensing means of environmental monitoring become especially effective here. The aim of this chapter is to formulate the basic model of the biosphere survivability and to propose a new viewpoint on global modeling.

The behavior of any system is determined by the values characterizing the state of the system, which can take on different terms. Upon interaction with an external medium and in particular with other systems, the values of these terms can vary in one way or another. For any technological or biological system, it is always possible to show the field of change of the characteristic parameters, wherein the system can be considered to be functioning. Outside this field the system does not exist.

Thus one can substitute the complex behavior of a system by a description depicting this system by a point in the phase space of the characteristic parameters. If the change of any coordinate leads to the disappearance of the depicting point from the allowable field, the system is demolished (the organism as a whole perishes).

The significant variables are not identical with respect to the degree of threat to the system. Such variables as the oxygen content of blood or the structural integrity of the medulla oblongata do not tolerate any significant changes, since such changes would almost certainly lead to immediate death. However, there are changes, such as changes in the temperature of the individual areas of the skin, the sharp fluctuations of which do not necessarily lead to death. Such a separation of all variables characterizing the state of the system into significant variables makes possible the simplification of the problem of the behavior strategy of the system upon its interaction with an external medium or other systems. The most constructive approach to the description of the global environment was proposed by Gorshkov et al. (2000). These authors consider that a different path of development compatible with long-term environmental safety lies through conservation and restoration of a substantial part of the Earth's biosphere in its natural nonperturbed state, bearing in mind the stabilizing potential of the natural biota of Earth with respect to the global environment.

Really, the problem of biosphere survivability correlates with mechanisms of biotic regulation, physical and biological stability, sensitivity of the biota, ecological limitations, and other basic principles of biology. In general, it is evident that each living organism plays its role in global change. The problem is in the description of this role to estimate the significance of interactions between the hierarchy of the biospheric elements having various spatial scales and different influences on the levels of biological organization. Thus, the global model described in this volume is based on the parametric description of the spatiotemporal hierarchy of biosphere elements. The precision of the discretization of this hierarchy is defined by the existing global data sets and knowledge base. The main sense of the ecoinformatics technique entitled in Section 2.2 as GIMS-technology is in the global simulation model (GSM) which is focused on the systematic observation and evaluation of the natural environment as related to changes attributable to human impacts. Questions to be addressed in this volume are related to:

- Synthesis of the present GSM to describe interactions across multiple spatial and temporal scales.
- Development of a simulation approach to effectively join the GSM with existing databases.
- Addressing criteria for incorporating human activities and effects.
- Sensitivity of the global *Biosphere/Society* subsystems to alterations in its components.

The methodology involves the synthesis of a hierarchical structure of models describing the interactions between environmental phenomena and processes and elements having various physical, biological, chemical, and socio-economic origin. Another objective of the research is to investigate requirements that must be met by global databases necessary for the evaluation of the environmental state and leading to reliable forecasting for sufficiently long time intervals. Some results concerning the criteria to estimate these interactions are given in Chapter 3 (see Sect. 3.6).

This chapter considers the *Nature/Society* system as the complex of systems having inputs and outputs. A system is defined by its structure and behavior. The behavior of such a system is aimed at providing uninterrupted functioning by means of a correspondingly organized structure and behavior. This characteristic of the complex system to actively withstand the hostile action of an external medium, we shall refer to as survivability.

In the present study, an analysis is made of a system, the elements of which are subdivided into working, defending, and active external agents of the system, which suppress or neutralize the hostile actions of the external medium. By taking into account the possibility of suppressing the hostile action of an external medium and the vulnerability of all the elements of the system, it becomes possible to reach a theoretical-game definition of the problems. The use of game theory in the investigation of the survivability of complex systems makes possible the classification of the most unfavorable action of an external medium on the system and the working out of the optimal strategy of behavior for the system. This study of antagonistic situations between systems enables us to understand the mechanisms of the adaptability of living systems to the varying conditions of an external medium.

2.5.1
Basic Definitions

The biosphere is a complex unique system. From a historical viewpoint, Man was an element of the biosphere. However, at the present time, the problem of co-evolution between *Human Society* (H) and *Nature* (N) has arisen. The influence of human activity on natural systems has reached the global scale, but it is possible to make the conditional division between anthropogenic and natural processes (Kondratyev 1997; Khor 2001; Tickel 2001a,b; Vital Sings 2001–2002, 2001). The use of system analysis permits a more formal description of this division to be made. Commonly, there are two interacting systems: *H* with the technologies, sciences, economics, sociology, agriculture, industry, etc. and *N* with climatic, biogeocenotic, biogeochemical, hydrological, geophysical, and other natural processes.

The solution of practical problems in investigating complex systems often involves evaluating their effectiveness and, in particular, their stability under indeterminate conditions of functioning. The theory of potential effectiveness of complex systems deals with the solution of these problems. A constructive mathematical apparatus has been developed within the framework of this theory by which it is possible to solve different problems of optimizing the structure and behavior of the functioning of *H* and *N* in certain situations.

The systems *H* and *N* are determined by their structures (number of elements and relations among them) and behavior (responses to impacts). The internal behavior of such a system is aimed at maintaining its uninterrupted functioning. The external behavior of the system is aimed at achieving a certain outside goal. Temporal stability of a complex system is a necessary property without which all its other properties become meaningless. It is connected with the structural stability of the material composition and energy balance of the complex system as well as with the regularity of its responses to the same external stimuli.

A breach in stability of a system may result from internal causes (the aging of its elements) or external causes associated with the unfavorable influence of the environment (an ill-intentioned enemy in particular). Survivability of biological systems is determined by environmental conditions where man's interference with nature is one of the important factors. In connection with this and the prospects of constructing artificial biological systems, optimization problems also arise; an increase in productivity of a biological system being the main criterion of optimality.

For the global scale the problem of survivability of the interacting systems is complicated at the expense of the hierarchy of interaction levels. For a complete explanation of *H* and *N* systems their openness has to be taken into account. Commonly, let us consider the interaction of two open complex systems *H* and *N*, defined by their goals H_G and N_G, structures H_S and N_S and behavior H_B and N_B, respectively. It is suggested that the functioning of such systems should be described by equations of (*V*, *W*) exchange. Namely, the interaction of an open system with the environment (or other system) is represented as a process whereby the system exchanges a certain quantity *V* of resources spent in exchange for a certain quantity *W* of resources consumed. The aim of the systems is the most advantageous (*V*, *W*) exchange, i.e., maximum *W* in exchange for minimum V. The V is a complex function of the structure and behavior of both systems:

$$V = V(W, H_S, N_S, H_G, N_G) = V(W, H, N) \tag{2.1}$$

As a result of their interaction, the systems H and N obtain the following (V, W) exchanges:

$$\begin{aligned} V_{H,\max} &= V_{H,\max}\left(W_H, H^*, N^*\right) = \max_{\{H_B, H_S\}} \min_{\{N_B, N_S\}} V_H(W_H, H, N) \\ V_{N,\max} &= V_{N,\max}\left(W_N, H^*, N^*\right) = \max_{\{N_B, N_S\}} \min_{\{H_B, H_S\}} V_N(W_N, H, N) \end{aligned} \tag{2.2}$$

where H^* and N^* are the optimal systems H and N, respectively.

From (2.1) and (2.2) one can see that the value of (V, W) exchange depends on the goal of the system and may vary within certain limits: $V_{1,\min} \leq V_H \leq V_{1,\max}$, $V_{2,\min} \leq V_N \leq V_{2,<\max}$, where $V_{i,\min}$ $(i = 1,2)$ corresponds to the case when both systems are most aggressive, and $V_{i,\max}$ $(i = 1,2)$ to the case, when they are most cautious. In one word, there is some *spectrum* of the interactions between H and N. For a formal description of these interactions we shall divide all the elements of both systems into three classes: the working (working), protective (defensive) and active elements, the latter designed to act on the environment. For short, we shall refer to the working elements of systems H and N as a and b elements, to the protective elements as R_a and R_b elements and to the active elements as C_a and C_b elements, respectively.

Let us assume that before interaction the systems H and N have certain limited energy resources (vital "substrates") V_a and V_b, where $V_a = \{V_{aj}, j = 1,..., m_a\}$, $V_b = \{V_{bi}, i = 1,...,m_b\}$. These substrates generate working elements in such a way, that the substrate V_{aj} (V_{bi}) can generate H_j (N_i) $a(b)$ elements of the $j(i)$-th type of values $a_j(b_i)$.

The protective and active elements of each system are generated by the working elements. First, the protective $E^a_{Rm}(E^b_{Rm})$ and active $E^a_{Cm}(E^b_{Cm})$ substrates are created which, in their turn, generate R and Celements of the mth type. These processes are described by the following dependences:

$$E^a_{Cm} = E^a_{Cm}(V_a, H_1, ..., H_{m_a}) = \sum_{j=1}^{m_a} w'^a_{mj} f^a_{jC}(V_{aj}, H_j),$$

$$E^b_{Cm} = E^b_{Cm}(V_b, N_1, ..., N_{m_b}) = \sum_{j=1}^{m_b} w^b_{mj} f^b_{jC}(V_{bj}, N_j),$$

where $w^{a(b)}_{mj}, w'^{a(b)}_{mj}$, and $f^{a(b)}$ are certain present weights and functions, respectively.

We shall assume that as a result of such hierarchical synthesis, elements in the systems H and N have at the beginning of the interaction $(t = 0)$:

1. m_j and n_j working elements of the jth type with values a_j and b_j, respectively, where

$$\sum_{j=1}^{m_a} a_j H_j = M_a(0), \sum_{j=1}^{m_b} b_j N_j = M_b(0); \tag{2.3}$$

2. r_a and r_b types of protective elements, the mth type having α_m and β_m elements, and:

$$\sum_{m=1}^{r_a} \alpha_m = M_{R_a}(0),\ \sum_{m=1}^{r_b} \beta_m = M_{R_b}(0); \tag{2.4}$$

3. s_a and s_b types of active elements the mth type having v^a_m and v^b_m and:

$$\sum_{m=1}^{s_a} v_m^a = D_a(0),\ \sum_{m=1}^{s_b} v_m^b = D_b(0) \tag{2.5}$$

respectively.

In the discrete case, the change of the average number of system elements that have survived till the moment t_{i+1} will be described by the following relations:

$$H_s(t_{i+1}) = \max\left\{0, H_s(t_i) - \sigma_{hs}^n(t_i) p_{hs}^n(t_i)\right\}, \quad s = 1,\ldots,m_h; \tag{2.6}$$

$$\alpha_j(t_{i+1}) = \max\left\{0, \alpha_j(t_i) - \sigma_{Rj}^n(t_i) p_{Rj}^n(t_i)\right\}, \quad j = 1,\ldots,r_h \tag{2.7}$$

$$v_m^h(t_{i+1}) = \max\left\{0, v_m^h(t_i) - \sigma_{Cm}^n(t_i) p_{Cm}^n(t_i)\right\}, \quad m = 1,\ldots,s_h \tag{2.8}$$

$$N_l(t_{i+1}) = \max\left\{0, N_l(t_i) - \sigma_{nl}^h(t_i) p_{nl}^h(t_i)\right\}, \quad l = 1,\ldots,m_n \tag{2.9}$$

$$\beta_s(t_{i+1}) = \max\left\{0, \beta_s(t_i) - \sigma_{Rs}^h(t_i) p_{Rs}^h(t_i)\right\}, \quad s = 1,\ldots,r_n \tag{2.10}$$

$$v_m^n(t_{i+1}) = \max\left\{0, v_m^n(t_i) - \sigma_{Cm}^h(t_i) p_{Cm}^h(t_i)\right\}, \quad m = 1,\ldots,s_h \tag{2.11}$$

where the $\sigma_{\omega i}^{a(b)}(t)$ values characterize the external behavior of both these systems:

$$\overline{H}_e^{(i)} = \{\|\sigma_{bl}^a\|, \|\sigma_{Rs}^a\|\},\ \overline{N}_e^{(i)} = \{\|\sigma_{as}^b\|, \|\sigma_{Rj}^b\|\};$$

and $p_{\omega i}^{a(b)}(t)$ are the respective probabilities of death of the elements as a result of their interaction.

The following limiting conditions should be taken into account here:

$$\sum_{i=0}^{T} \{\sigma_{Cs}^b(t_i) p_{Cs}^b(t_i) + \sum_{j=1}^{m_b} \sigma_{bj}^a(t_i) + \sum_{j=1}^{r_b} \sigma_{Rj}^a(t_i) + \sum_{j=1}^{s_b} \sigma_{Cj}^a(t_i)\} = v_l^h(0), \tag{2.12}$$

$$\sum_{i=0}^{T} \{\sigma_{Cs}^a(t_i) p_{Cs}^a(t_i) + \sum_{j=1}^{m_a} \sigma_{aj}^b(t_i) + \sum_{j=!}^{r_a} \sigma_{Rj}^b(t_i) + \sum_{j=1}^{s_a} \sigma_{Cj}^b(t_i)\} = v_l^n(0) \tag{2.13}$$

The stochastic solution of Eqs. (2.1)–(2.13) can be in the unrealized form in practice. There are many real situations when the realization of H^* or N^* system is impossible. Some tasks and algorithms were described by Krapivin and Klimov (1995, 1997).

2.5.2
Study of the Simple Survivability Model

Let us consider the interaction of the two systems in the framework of the diagram shown in Fig. 2.9. At the start, systems *H* and *N* have, respectively, $N_a(0)$ and $N_b(0)$ working elements, $N_{Ra}(0)$ and $N_{Rb}(0)$ protective elements, and $M_a(0)$ and $M_b(0)$ active agents for undertaking action against an external medium. In this case, we shall assume that the initial structures H_S and N_S of the systems are uniformly filled with elements. This means that at the time $t = 0$ in any sphere with a fixed radius ε, which is completely confined within system *H*, there are a constant numbers of elements.

We shall consider that all elements of systems *H* and *N*, independent of their spatial location, are accessible to the same degree to the active agents of the external medium. The interaction of the systems consists in the situation that each system in a fixed interval of time [0,T] at discrete moments $t_i = ih$, $i = 0,1,2,\ldots, k$ ($k = [T/h]$), determines its behavior by a set of numbers:

$$H_B = \{m_b(t_i), m_{Rb}(t_i), \rho_i\}, N_B = \{m_a(t_i), m_{Ra}(t_i), r_i\}.$$

Here, m_a and m_b are parts of C_b and C_a elements aimed at destroying *a* and *b* elements, respectively; analogously, the parts $(1-r_i)m_{Ra}$ and $(1-\rho_i)m_{Rb}$ of C_b and C_a elements are directed towards the destruction of the corresponding protective elements. In the course of time, portions of C_a and C_b elements uniformly fill the opposite system, and in such a manner the elements of the systems become with time *uniformly depleted.*

It is assumed that system *H(N)* is put out of operation, if at the time $t_i \leq T$, $N_a(t_i) \leq \theta_a N_a(0)$ $[N_b(t_i) \leq \theta_b N_b(0)]$; that is, if more than $(1-\theta_a)$-th $((1-\theta_b)$-th) portion of the initial amount of its working elements is out of operation, where $0 \leq \theta_a, \theta_b \leq 1$. On the other hand, when $N_a(T) > \theta_a N_a(0)$ $[N_b(T) > \theta_b N_b(0)]$, then system *H(N)* is considered to have survived. The assignment of the values θ_a and θ_b is determined by the peculiarities of the system under consideration. It is clear that the smaller the number θ, the more *survivable* is the system.

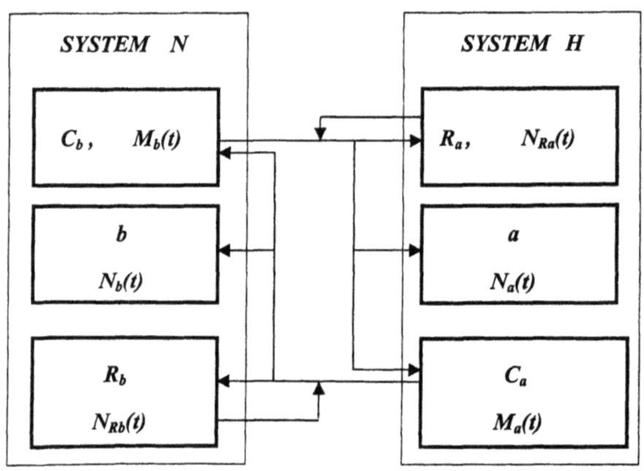

Fig. 2.9. Schematic diagram of the interaction of two systems in the survivability problem

Further, we shall assume that one of the elements of system H or N is put off operation by the action of one of the elements C_a or C_b with a probability $p_1[N_{Ra}(t)]$ or $p_2[N_{Rb}(t)]$, respectively. Consequently, in the interval of time $[t_i,$ $t_i + h]$ with step length, h, on average the following changes in the structure of the system occur:

$$
\begin{aligned}
&N_a(t_i) - N_a(t_i + h) = m_a(t_i)p_1[N_{Ra}(t_i)],\\
&M_a(t_i) - M_a(t_i + h) = r_i m_{Ra}(t_i)[N_{Ra}(t_i) + m_b(t_i) + m_{Rb}(t_i),\\
&N_{Ra}(t_i) - N_{Ra}(t_i + h) = (1 - r_i)m_{Ra}(t_i)p_1[N_{Ra}(t_i)],\\
&N_b(t_i) - N_b(t_i + h) = m_b(t_i)p_2[N_{Rb}(t_i)],\\
&M_b(t_i) - M_b(t_i + h) = \rho_i m_{Rb}(t_i)p_2[N_{Rb}(t_i)] + m_a(t_i) + m_{Ra}(t_i),\\
&N_{Rb}(t_i) - N_{Rb}(t_i + h) = (1 - \rho_i)m_{Rb}(t_i)p_2[N_{Rb}(t_i)]
\end{aligned}
\tag{2.14}
$$

The above-discussed interaction scheme of the two systems can be readily implemented with the aid of game-theory methods. Indeed, from Eqs. (2.14) it follows that in the interval of time [0,T] (T = kh, here, one can take $h = 1$) both systems lose working a and b elements in the following amounts:

$$
Q_1 = \sum_{n=0}^{k-1} m_a(n)p_1[N_{Ra}(n)], Q_2 = \sum_{n=0}^{k-1} m_b(n)p_2[N_{Rb}(n)] \tag{2.15}
$$

In this case, system H with its behavior of

$$H_B(t) = \{m_b(t), m_{Rb}(t), \rho(t)\}$$

tends to minimize the function Q_1, which characterizes its losses, and to maximize Q_2, the losses of system N. On the other hand, system N with its behavior $N_B(t) = \{m_a(t), m_{Ra}(t), r(t)\}$ tends to maximize function Q_1 and minimize function Q_2.

In practical problems, the win function is taken as such as a characteristic of the antagonistic situation which most fully describes a given conflict. In this case such a function is $Q = Q_1 - Q_2$. The maximizing participant in this case will be system N, the minimizing opponent, system H. In this manner, the solution of the set problem on the optimal behavior of the two systems in an antagonistic situation becomes reduced to the solution of a game with a win function:

$$
Q(H_B, N_B) = \sum_{n=0}^{i-1} \left\{m_a(n)p_1[N_{Ra}(n)] - m_b(n)p_2 N_{Rb}(n)]\right\} \tag{2.16}
$$

where, according to Eq. (2.14),

$$
\begin{aligned}
&N_{Ra}(n) = N_{Ra}(0) - \sum_{i=0}^{n-1} \left(1 - r_i\right) m_{Ra}(i)p_1[N_{Ra}(i)],\\
&N_{Rb}(n) = N_{Rb}(0) - \sum_{i=0}^{n-1} \left(1 - \rho_i\right) m_{Rb}(i)p_{21}[N_{Rb}(i)]
\end{aligned}
\tag{2.17}
$$

and conditions are imposed on the behavior of the system related to the limitations of C_a and C_b elements:

$$\sum_{i=0}^{k-1} \{r_i m_{Ra}(i) p_1[N_{Ra}(i)] + m_b(i) + m_{Rb}(i)\} = M_a(0),$$
$$\sum_{i=0}^{k-1} \{\rho_i m_{Rb}(i) p_2[N_{Rb}(i)] + m_a(i) + m_{Ra}(i)\} = M_b(0) \tag{2.18}$$

Thus we have a k-step survival game. At the start of each step, system H provides a certain amount of its resources $u = \{m_b, m_{Rb}, \rho\}$ and N provides a certain amount of its resources $v = \{m_a, m_{Ra}, r\}$, so that limitations (2.18) are maintained. As a result of this distribution of resources, system N gets the advantage:

$$R(u,v,M_a,M_b) = m_a p_1[N_{Ra}] \tag{2.19}$$

However, this win of N and the loses of H are not counted on the basis of their initial resources and cannot be added to the remaining amounts of C elements. After each step of the game [Eq. (2.14)], the change in the resources in the participants takes place, and as a result of the k-step process a total win of system N can be described by the equation:

$$Q_k = Q_k\left[u_0, u_1, ..., u_{k-1}; v_0, v_1, ..., v_{k-1}; M_a(0), M_b(0)\right] = \sum_{n=0}^{k-1} \left\{m_a(n) p_1[N_{Ra}(n)] - m_b(n) p_2[N_{Rb}(n)]\right\} \tag{2.20}$$

There are several methods with the aid of which an analysis can be made of this k-step process. One can consider this k-step game as a one-step game, in which case system H must select simultaneously a plurality of the vector $\{u_0, u_1, ..., u_{k-1}\}$, and system N, a plurality of the vector $\{v_0, v_1, ..., v_{k-1}\}$, where the selection of u_k and v_k depends on the previous values obtained according to Eq. (2.14).

For a solution of this quite complex problem, it is suggested to substitute for it a similar problem of multi-step optimization. The values of this game can be expressed as:

$$V_k = \max_F \min_G \{\int Q_k dF(v_0, v_1, ..., v_{k-1}) dG(u_0, u_1, ..., u_{k-1})\} = \min_G \max_F \{...\} \tag{2.21}$$

where, the distribution functions F and G are determined on the boundaries of the complex form:

$$\begin{array}{ll} 0 \le v_0 \le M_b(0) & 0 \le u_0 \le M_a(0) \\ 0 \le v_1 \le M_b(1) & 0 \le u_1 \le M_a(1) \\ \dots & \dots \\ 0 \le v_{k-1} \le M_b(k-1) & 0 \le u_{k-1} \le M_a(k-1) \end{array} \tag{2.22}$$

By utilizing the optimality principle and taking into account the dependence $V_k = V_k[M_a(0), M_b(0)]$, we obtain the following functional equation:

$$V_{n+1}[M_a(0), M_b(0)] = \max_F \min_G [\iint_{\substack{0 \le u \le M_a(n-1) \\ 0 \le v \le M_b(n-1)}} \{R(u,v) + V_n[M_a(n-1), M_b(n-1)]\} dF(v) dG(u)] = \min_G \max_F [...] \tag{2.23}$$

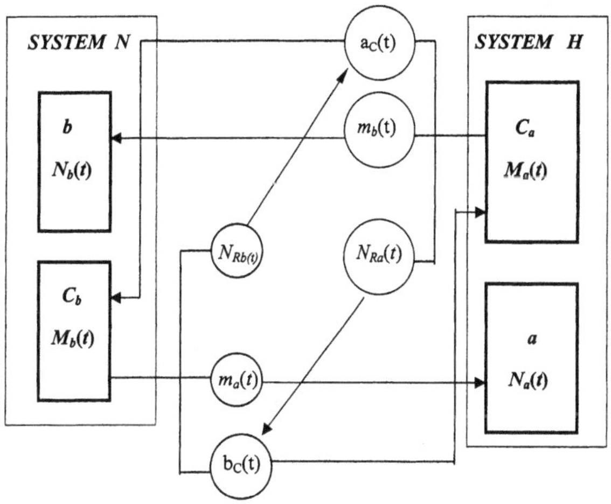

Fig. 2.10. Schematic diagram of the case of mutual interchangeability

where

$$V_1[M_a(0),M_b(0)] = \max_F \min_G [\iint_{\substack{0\le u\le M_a(0)\\ 0\le v\le M_b(0)}} R(u,v)dF(v)dG(u)] = \min_G \max_F [...] \qquad (2.24)$$

The solution of these functional equations is a difficult problem. Therefore, not without reason in this connection, the expression *curse of dimensionality* was introduced. In our study, we shall present the solution of some specific cases. This task is solved in more detail by Nitu et al. (2000a). For the solution of the proposed problem, we shall begin with the case where in both systems the C and R elements are indistinguishable. The diagram of interaction between systems H and N can then be represented in such a form as shown in Fig. 2.10. By taking into account the designations used in this diagram, we see that the number of C_a and C_b elements that have reached the C_b and C_a elements of the opposite system in a one-step operation will be respectively: $\max\{0, a_C - N_{Rb}\}$ and $\max\{0, b_C - N_{Ra}\}$. Therefore, the numbers of C_a and C_b elements that are put out of operation on the average are $\max\{0, b_C - N_{Ra}\}p_1[N_{Ra}]$ and $\max\{0, a_C - N_{Rb}\}p_2[N_{Rb}]$. Consequently, after one step of the game, there remain in the system the following numbers of C_a and C_b elements that are not put out of operation: $\max[0, M_a - \max(0, b_C - N_{Ra})p_1[N_{Ra}]]p_1[N_{Ra}]$, $\max[0, M_b - \max(0, a_C - N_{Rb})p_2[N_{Rb}]]p_2[N_{Rb}]$. Going over to the multistep situation, we introduce the win function (2.16). The aim of system N is to destroy system H, or more precisely, to put out of operation all of the latter's a elements. For this purpose system N cannot use all C_b elements, since it will then be left defenseless against the C_a elements. An analogous situation holds true for system H. Each system is compelled to release, after each successive step of the game, a certain number of working a and b elements. The magnitude of this payoff is proportional to the difference $M_b - b_C - N_{Rb} - (M_a - a_C - N_{Ra})$. Now the problems on both sides become clear. System N must strive to provide such a number of C_b elements in all directions so as to maintain the maximum value of this difference. That is, it must provide the largest possible number of elements m_a for the destruction of a elements and thus increase its winnings. However, one shall also consider the presence of an analogous distribution of C_a elements and provide sufficient protection for system N through the maximum distribution of C_a elements. An analogous situation exists for system H. For $(n-1)$ moves left before the end of the game, we have:

$$M_a(n-1) = \max\{0, M_a(n) - \max[0, b_C(n) - N_{Ra}(n)p_1[N_{Ra}(n)]\}p_1[N_{Ra}(n)],$$
$$M_b(n-1) = \max\{0, M_b(n) - \max[0, a_C(n) - N_{Rb}(n)]p_2[N_{Rb}(n)]\}p_2[N_{Rb}(n)] \quad (2.25)$$

The payoff for the entire game according to Eq. (2.20) will be:

$$Q_k = \sum_{n=1}^{k} \left\{ M_b(n) - b_C(n) - N_{Rb}(n) - \left[M_a(n) - a_C(n) - N_{Ra}(n) \right] \right\} \quad (2.26)$$

The functional Eq. (2.23) will acquire the following form:

$$V_{n+1} = \left\{ M_b(n+1) - b_C(n+1) - N_{Rb}(n+1) - \left[M_a(n+1) - a_C(n+1) - N_{Ra}(n+1) \right] \right.$$
$$\left. + Q_k\left[M_a(n), M_b(n) \right] \right\} = \min_{N_B} \max_{H_B} \{\ldots\} \quad (2.27)$$

Since at the end of the game $Q_0 = 0$, we obtain from Eq. (2.26) for $k = 1$:

$$Q_1 = M_b(1) - b_C(1) - N_{Rb}(1) - M_a(1) + a_C(1) + N_{Ra}(1) \quad (2.28)$$

From Eq. (2.28), one step before the end of the game, we obtain the following optimal strategies:

$$N_{Rb}(1) = N_{Ra}(1) = 0,\ a_C(1) = b_C(1) = 0 \quad (2.29)$$

and the prize of the game is $V_1 = M_b(1) - M_a(1)$. This means that in the last step of the game both systems direct all their C_a and C_b elements that have remained from the previous steps towards the destruction of b and a elements, respectively. Analogously, two steps before the end of the game, we have:

$$Q_2 = M_b(2) - M_a(2) + a_C(2) - b_C(2) + N_{Ra}(2) - N_{Rb}(2) + V_1[M_a(1), M_b(1)] \quad (2.30)$$

where, $V_1 = M_b(1) - M_a(1)$, and

$$M_a(1) = \max\{0, M_a(2) - \max[0, b_C(2) - N_{Ra}(2)]p_1[N_{Ra}(2)]\}p_1[N_{ra}(2)],$$
$$M_b(1) = \max\{0, M_b(2) - \max[0, a_C(2) - N_{Rb}(2)]p_2[N_{Rb}(2)]p_2[N_{Rb}(2)] \quad (2.31)$$

It is obvious that at this step of the game the participants have no pure strategies. Therefore, the solution of this game is impossible in an analytic form, and it can be obtained in a concrete case only by the numerical method. The modeling of the game provides some understanding of the nature of its solution. Indeed, with the aid of a computer it is possible either to construct a model of the game and to gather statistics or to solve the functional Eq. (2.23) numerically and with the aid of heuristic concepts to investigate the dependence of strategies on the initial conditions. Of course, such an approach for a short interval of modeling cannot give any significant information concerning the solution. Nevertheless, this is the only possible approach at the present time. The feeling of hopelessness in a specific situation should not deter us from seeking a solution by analytical methods. The importance of obtaining analytical solutions is obvious, since they have the advantage over the numerical solutions in that they make possible the detection of the general regularities of the optimal behavior of complex systems in antagonistic situations. The importance of the analytical solutions has been pointed out by Krapivin (1978), who has shown that a single numerical solution cannot replace an analytical solution in which the quantitative description of the phenomenon is most concentrated. In the case examined here, when the participants have no information concerning the action of the opponent in the process of the entire game, the solution of particular cases with the aid of a computer enables us to obtain the following quantitative description of optimal strategies.

Let $M_b(n)/M_a(n) = \delta_n$. Then if $\delta_n >> 1$, system N in the initial stage of the game has a pure strategy, but for system H it is more advantageous to adopt a mixed strategy, using the tactics of deception. During the first steps, system N destroys only C_a elements and only during the last steps does it destroy the a elements. By using the corresponding probability mechanism, system H must direct all its C_a elements, with probability p_1 towards the destruction of the C_b elements, with probability p_2 towards the defense of its own elements, and with probability $1 - p_1 - p_2$ towards the attack of the b elements. In the case when $\delta_n \cong 1$, then at this stage of the game $p_1 + p_2 = 1$, $m_b(n) = 0$, and the behavior of the system becomes symmetrical. During the last steps of the game systems H and N, independent of the magnitude of δ_n, change over to the strategies: $m_a(t) \neq 0$, $m_b(t) \neq 0$, and $b_C = a_C = N_{Ra} = N_{Rb} = 0$, that is, to the destruction of working elements.

Let us consider a particular case where the systems have no protective elements, that is, $N_{Ra} = N_{Rb} = 0$, and therefore, $p_1 = p_2 = 1$. Then we obtain:

$$M_a(n-1) = \max\{0, M_a(n) - b_C(n)\},$$
$$M_b(n-1) = \max\{0, M_b(n) - a_C(n)\} \tag{2.32}$$

One step before the end of the game, the payoff, according to Eq. (2.31), is $V_1 = M_b(1) - M_a(1)$ and the optimal strategies $a_C^*(1) = b_C^*(1) = 0$; that is, at the last step the systems release all C elements for the destruction of a and b elements, respectively.

Next, by assuming $n = 2$, according to Eqs. (2.30) and (2.31), we obtain:

$$Q_2[a_C(2), b_C(2)] = \begin{cases} Q_{21}, & \text{when } b_C(2) \geq M_a(2), a_C(2) \geq M_b(2); \\ Q_{22}, & \text{when } b_C(2) < M_a(2), a_C(2) \geq M_b(2); \\ Q_{23}, & \text{when } b_C(2) \geq M_a(2), a_C(2) < M_b(2); \\ Q_{24}, & \text{when } b_C(2) < M_a(2), a_C(2) < M_b(2); \end{cases} \tag{2.33}$$

where $Q_{21} = M_b(2) - b_C(2) - M_a(2) + a_C(2)$, $Q_{22} = M_b(2) - 2M_a(2) + a_C(2)$, $Q_{23} = 2M_b(2) - b_C(2) - M_a(2)$ and $Q_{24} = 2M_b(2) - 2M_a(2)$.

The solution of the game with the win function (2.33) has the following form:

$$V_2 = 2M_b(2) - 2M_a(2),\ b^*_c(2) = M_a(2);\ a^*_c(2) = M_b(2); \tag{2.34}$$

Actually, if an account is made of the real situation, the optimal strategies of both systems two steps before the end of the game will be as follows:

$$a_C^*(2) = \min\{M_a(2), M_b(2)\},\ b_C^*(2) = \min\{M_a(2), M_b(2)$$

Therefore, at the penultimate step when $\delta_2 > 1$, system N releases a portion of its force and system H releases all its forces against the c elements of the other system.

By reasoning in an analogous way, we obtain that n-steps before the end of the game the strategies of systems H and N will be:

$$a_C^*(n) = \min\left\{M_a(n), M_b(n)\right\};\ b_C^*(n) = \min\left\{M_a(n), M_b(n)\right\} \tag{2.36}$$

It is evident that if in the process of the game no replacement of C elements takes place in systems H and N, then it follows from Eqs. (2.35) and (2.36) that it makes sense to conduct the game in two steps; and the two following cases are distinguishable:

1. When $M_a(0) = M_b(0)$, both systems release all their C elements into battle against the C elements of the other system, but in so doing the systems themselves survive.
2. When $M_a(0) > M_b(0)$, system N releases all the elements into battle against C_a elements, and system H releases $M_b(0)$ of C_a elements into battle against C_b elements, and $M_a(0)–M_b(0)$ of C_a elements into battle against b elements. Consequently, system H survives in every case, while system N survives only when $M_a(0) - M_b(0) < \theta_b N_b(0)$.

Now, the case of mutual indifference of C_a and C_b elements is considered. In this case, by supposing $r = \rho = 0$ in Eqs. (2.17) and (2.18), we obtain:

$$N_{Ra}(n) = N_{Ra}(0) - \sum_{i=0}^{n-1} m_{Ra}(i)p_1[N_{Ra}(i)],\ N_{Rb}(n) = N_{Rb}(0) - \sum_{i=0}^{n-1} m_{Rb}(i)p_2[N_{Rb}(i)] \tag{2.37}$$

and

$$\sum_{n=0}^{k-1}\left[m_a(n) + m_{Ra}(n)\right] = M_b(0),\ \sum_{n=0}^{k-1}\left[m_b(n) + m_{Rb}(n)\right] = M_a(0) \tag{2.38}$$

The solution of the game of the two systems, H and N, with the win function as shown in Eq. (2.16) and under conditions as expressed in Eqs. (2.37) and (2.38), is reduced to the problem of maximizing two functions:

$$Q_1 = \max_{Ra} \sum_{n=0}^{k-1} m_a(n)p_1[N_{Ra}(n)] \tag{2.39}$$

$$Q_2 = \max_{Rb} \sum_{n=0}^{k-1} m_b(n)p_2[N_{Rb}(n)] \tag{2.40}$$

From Eqs. (2.39) and (2.40) it is evident that when $p_1[N_{Ra}(n)] = p_2[N_{Rb}(n)] = 1$, the optimal strategies of both sides consist in the destruction of a and b elements. If the C elements act independently of one another, and the probability of putting the C elements out of operation by a single R element is equal to a constant value D_a and D_b for H and N systems, respectively, then

$$p_1[N_{Ra}(n)] = \exp[-d_a N_{Ra}(n)], p_2[N_{Rb}(n)] = \exp[-d_b N_{Rb}(n)] \tag{2.41}$$

where, $d_a = -\theta_1 \ln(1 - D_a)$, $d_b = -\theta_2 \ln(1 - D_b)$ are the *effectiveness* coefficients of the R_a and R_b elements; and θ_1 and θ_2 are the *densities* of the location of elements in the H and N system, respectively.

When $k = 1$, the optimal strategies of the behavior of the systems will be $m_a^*(0) = M_b(0)$ and $m_b^*(0) = M_a(0)$. This is natural, since in the last step it makes no sense to destroy protective elements. When $k = 2$, from Eqs. (2.37)–(2.40) we obtain the optimal strategies in the form:

$$\begin{aligned} &m_a^*(0) = m_b^*(0) = m_{Ra}^*(1) = m_{Rb}^*(1) = 0,\ m_a^*(1) = d_a^{-1}\exp[d_a N_{Ra}(0)],\\ &m_b^*(1) = d_b^{-1}\exp[d_b N_{Rb}(0)],\ m_{Ra}(0) = M_b(0) - d_a^{-1}\exp[d_a N_{Ra}(0)],\\ &m_{Rb}(0) = M_a(0) - d_b^{-1}\exp[b_b N_{Rb}(0)] \end{aligned} \tag{2.42}$$

The cost of the game is:

$$\begin{aligned} V_2 = {} & d_a^{-1}\exp\left\{-1 + d_a M_b(0)\exp[-d_a N_{Ra}(0)]\right\} - \\ & d_b^{-1}\exp\left\{-1 + d_b M_a(0)\exp[-d_b N_{Rb}(0)]\right\} \end{aligned} \tag{2.43}$$

From Eq. (2.43) it is evident that for system N to destroy system H it is necessary that the following inequality holds:

$$d_a M_b(0)\exp[-d_a N_{Ra}(0)] \geq l_n[d_a e(1-\theta_a)N_a(0)] \tag{2.44}$$

from which we have:

$$M_b(0) \geq d_a^{-1}\exp[d_a N_{Ra}(0)]\left\{1 + l_n[d_a(1-\theta_a)N_a(0)]\right\} \tag{2.45}$$

Similarly, we obtain the condition for system H:

$$M_a(0) \geq d_b^{-1}\exp[d_b N_{Rb}(0)]\left\{1 + l_n[d_b(1-\theta_b)N_b(0)\right. \tag{2.46}$$

Thus for system N to destroy system H it is necessary that the inequality (2.45) holds. In an analogous way, for system H it is necessary that the inequality (2.46) holds. The number of a and b elements, as a rule, is a fixed number, as follows from considerations related to the work of the system. The number of R elements, which perform protective functions, can be best selected when certain physical parameters in the problem are fixed and when a priori information concerning the amount of C elements of the opposite system is available. For example, if the effectiveness of all R elements is constant independent of their number, then in order for system H to survive, the necessary number of protective elements must satisfy the inequality:

$$N_{ra}(0) \geq d_a M_b(0) l_n / \{d_a(1 + l_n[d_a(1 - \theta_a)N_a(0)])\} \tag{2.47}$$

It appears clear that when the number of C_b elements is constant and the survivability of system H decreases, the necessary number of protective elements increases rapidly. In this case, the greater their effectiveness, the smaller the number of protective elements is required for carrying out one and the same task. With an increase in survivability, the necessary number of protective elements can be decreased.

Now let systems H and N have a fixed amount E_a and E_b of a certain substratum (for instance, energy) and be able to distribute it evenly between their protective elements, so that for each fraction of R_a and R_b elements there is $E_{1a} = E_a/N_{ra}(0)$ and $E_{1b} = E_b/N_{rb}(0)$ respectively. Then the efficiency coefficients of the protective elements must increase with the increase in the E_{1a} and E_{1b} portions, since in this case the probabilities of D_a and D_b increase. Therefore, let $D_a = 1 - \exp\{-F_a F_{1a}^{\alpha}\}$, $D_b = 1 - \exp\{-F_b F_{1b}^{\beta}\}$ where, F_a, F_b, α and β are independent of the number of protective elements. This gives then the following expressions for the coefficients of effectiveness of the protective elements:

$$\begin{aligned} d_a &= -\theta_a \ln(1 - D_a) = -\theta_a F_a E_{1a}^{\alpha} = G_a N_{Ra}^{-\alpha}(0), \\ d_b &= -\theta_b \ln(1 - D_b) = -\theta_b F_b E_{1b}^{\beta} = G_b N_{Rb}^{-\beta}(0) \end{aligned} \tag{2.48}$$

where $G_a = \theta_a F_a E_a^{\alpha}$, $G_b = \theta_b F_b E_b^{\beta}$.

From Eqs. (2.47) and (2.48) we obtain the following transcendental equations for the amount of the necessary protective elements in systems H and N:

$$N_{Ra}(0) = N_{Ra}^{\alpha}(0) G_a^{-1}\{\ln[G_a M_b(0)/f_1] - \alpha \ln N_{Ra}(0)\} \tag{2.49}$$

$$N_{Rb}(0) = N_{Rb}^{\beta}(0) G_b^{-1}\{\ln[G_b M_a(0)/f_2] - \beta \ln N_{Rb}(0) \tag{2.50}$$

where

$$\begin{aligned} f_1 &= \ln[eG_a(1 - \theta_a) N_a(0) N_{Ra}^{-\alpha}], \\ f_2 &= \ln[eG_b(1 - \theta_b) N_b(0) N_{Rb}^{-\beta}(0)] \end{aligned}$$

An analysis of Eqs. (2.49) and (2.50) reveals that in this case the number $N_{ra}(0)$ is very sensitive with respect to changes in the quantity $\ln[(1 - \theta_a)N_a(0)]$. This is natural, since with the increase in the number of R_a elements their effectiveness sharply decreases. It is clear that a certain optimal level exists for the number of protective elements. This level is defined by the assigned survivability of the system. It is better to have a small number of R_a elements of high effectiveness, than a large number of R_a elements of low effectiveness. When $k = 3$, from Eqs. (2.37) - (2.39) we have:

$$Q_1 = \sum_{i=1}^{3} m_a(i-1)\exp[-\alpha N_{Ra}(i-1)] \tag{2.51}$$

where,

$$\begin{aligned} N_{Ra}(1) &= N_{Ra}(0) - m_{Ra}(0)\exp[-d_a N_{Ra}(0)], \\ N_{ra}(2) &= N_{ra}(0) - m_{Ra}(0)\exp[-d_a N_{Ra}(0)] - m_{Ra}(1)\exp[-d_a N_{Ra}(1)] \\ & m_a(1) + m_a(0) + m_a(2) + m_{Ra}(1) + m_{Ra}(0) + m_{Ra}(2) = M_b(0) \end{aligned} \tag{2.52}$$

From Eqs. (2.51) and (2.52) we observe that the function Q_1 reaches its maximum value when $m_a^*(0) = m_{Ra}^*(2) = m_{Ra}^*(1) = 0$, with this maximum value being independent of the distribution of C_a elements during the last two steps. In this case $Q_1^*(3) = Q_1^*(2)$. An analogous result is also obtained for system N.

Thus for an identical game, for instance, system N can destroy at the most $Q_1^*(2) = M_b(0)\exp[-d_a N_{Ra}(0)]$ of a elements; when the number of steps is greater than 2, system N can destroy not more than $Q_1^*(2) = (ed_a)^{-1}\exp[d_a Q_1^*(1)]$ of a elements. Therefore, by comparing $Q_1^*(1)$ with $Q_1^*(2)$ we obtain that in the analyzed antagonistic situation the best strategy for both systems is the performance of all allowable operations in two steps. And that is, during the first step the action of a portion of each system's C elements is to be set against the protective elements of the opposite system, and during the second step the action of the remaining force is to be set against the working elements of the opposite system. This conclusion completely agrees with the conclusions reached by Krapivin (1978), which were obtained by different methods.

In the above-examined models of the interaction of the two systems, it was assumed that the effectiveness of the protective elements does not change with respect to time. However, this assumption in many real situations must be withdrawn. Let us consider a case where both systems can vary the effectiveness of the protective elements from step to step, so that at each step the effectiveness is independent of the number of protective elements. Let the effectiveness of R_a and R_b elements be equal to d_{1a} and d_{1b}, respectively, during the first step in the two-step case. Similarly, during the second step the effectiveness acquires values d_{2a} and d_{2b} so that $d_{1a} + d_{2a} = 2d_a$, $d_{1b} + d_{2b} = 2d_b$; that is, the summed value of the effectiveness does not exceed a constant value. As a result, we obtain the following matrix game:

$$\begin{array}{c|ccc|}
 & d_{1a} = d_{2a} & d_{1a} < d_{2a} & d_{1a} > d_{2a} \\
d_{1b} = d_{2b} & Q_{11} & Q_{12} & Q_{13} \\
d_{1b} < d_{2b} & Q_{21} & Q_{22} & Q_{23} \\
d_{1b} > d_{2b} & Q_{31} & Q_{32} & Q_{33}
\end{array},$$

where

$$\begin{aligned}
Q_{11} &= (ed_a)^{-1}\exp\{M_b(0)d_a\exp[-d_a N_{Ra}(0)]\} \\
&\quad - (ed_b)^{-1}\exp\{M_a(0)d_b\exp[-d_b N_{Rb}(0)]\},\\
Q_{22} &= Q_{33} = Q_{23} = Q_{32} = (ed_{2a})^{-1}\exp\{(d_{1a}-d_{2a})N_{Ra}(0) \\
&\quad + d_{2a}M_b(0)\exp[-d_{1a}N_{Ra}(0)]\} - (ed_{2b})^{-1}\exp\{(d_{1b}-d_{2b})N_{Rb}(0) \\
&\quad + d_{2b}M_a(0)\exp[-d_{1b}N_{Rb}(0)]\},\\
Q_{12} &= Q_{13} = (ed_{2a})^{-1}\exp\{(d_{1a}-d_{2a})N_{Ra}(0) + d_{2a}M_b(0)\exp[-d_{1a}N_{Ra}(0)]\} \\
&\quad - (ed_b)^{-1}\exp\{M_b(0)d_b\exp[-d_b N_{Rb}(0)]\},\\
Q_{21} &= Q_{31} = (ed_a)^{-1}\exp\{M_a(0)d_a\exp[-d_a N_{Ra}(0)]\} \\
&\quad - (ed_{2b})^{-1}\exp\{(d_{1b}-d_{2b})N_{Rb}(0) \\
&\quad + d_{2b}M_b(0)\exp[-d_{1b}N_{Rb}(0)]\}
\end{aligned} \tag{2.53}$$

The matrix of this game has a saddle-shaped point in the case of following conditions: $M_a(0) \le N_{Rb}(0)\exp[d_b N_{Rb}(0)]$, $M_b(0) \le N_{Ra}(0)\exp[d_a N_{Ra}(0)]$ and that it takes place in the real system.

Thus, for both systems, it is advantageous during the first step to provide the protective elements with a small effectiveness and to increase their effectiveness during the second step. This is natural, since it is better to lose a greater number of protective elements during the first step, thus securing a reliable defense of the working elements during the second step of the game.

From Eq. (2.53) it follows that the loses of a and b elements by system H and N, respectively, will amount to:

$$Q_{1*} = \exp\{-1 + (d_{1a}-d_{2a})N_{Ra}(0) + d_{2a}M_b(0)\exp[-d_{1a}N_{Ra}(0)]\}/d_{2a},$$
$$Q_{2*} = \exp\{-1 + (d_{1b}-d_{2b})N_{Rb}(0) + d_{2b}M_a(0)\exp[-d_{1b}N_{Rb}]\}/d_{2b} \quad (2.54)$$

In order to determine the optimum value of the effectiveness of the R_a and R_b elements for each step of the game, we must find:

$$\min_{(d_{1a},d_{2a})} Q_{1*}(d_{1a},d_{2a}) \text{ and } \min_{(d_{1b},d_{2b})} Q_{2*}(d_{1b},d_{2b})$$

From Eq. (2.54) we obtain:

$$\partial Q_{1*}/\partial d_{1a} = 1 + (2d_a - d_{1a})\{2N_{Ra}(0) - M_b(0)[1 + N_{Ra}(2d_a-d_{1a})]\exp[-d_{1a}N_{Ra}(0)]\} = 0, \quad (2.55)$$

$$\partial Q_{2*}/\partial d_{1b} = 1 + (2d_b - d_{1b})\{2N_{Rb}(0) - M_a(0)[1 + N_{Rb}(0)(2d_b - d_{1b})] \exp[-d_{1b}N_{Rb}(0)]\}. \quad (2.56)$$

From Eq. (2.55) it is evident that if the equation:

$$\{(2d_a - d_{1a})/(d_a - d_{1a})\}\exp[-d_{1a}N_{Ra}(0)] = 0.5\, N_{Ra}(0)/M_b(0) \quad (2.57)$$

has a real root $0 \le d_{1a}^* \le 2d_a$, then system H, by utilizing this root for its own optimum strategy, can guarantee on the average losses of elements not exceeding $Q_1^* = [e(2d_a - d_{1a}^*)]^{-1}$. If Eq. (2.57) does not have a root in the interval $[0, 2d_a]$, the optimum strategy is then determined either by the root of Eq. (2.55) or by $d_{1a}^* = 0$. In particular, when $M_b(0) = N_{ra}(0)$, then $d_{1a}^* = 0$.

Similar calculations are realized for Eq. (2.56).

2.5.3 Methods for Determining Stable Strategies

The task of Eqs. (2.11)–(2.13) is solved by means of game theory algorithms. There are many models to simulate the above situations describing the system interaction. Below, the two-player antagonistic game is considered.

In game $\Gamma(M,[0,1])$ with gain function $M(x,y)$ $(0 \le x,y \le 1)$ player I can receive no less than

$$v_1 = \max_F \min_G \int_0^1 E_1(F)dG(y) = \max_F \min_y E_1[F(y)],$$

where F and G are strategies of the first and second players, respectively:

$$E_1[F] = \int_0^1 M(x,y)dF(x)$$

Similarly, player **II** can receive no more than

$$v_2 = \min_G \max_F \int_0^1 E_2(G)dF(x) = \min_G \max_x E_2[G(x)],$$

where

$$E_2[G] = \int_0^1 M(x, y)dG(y)$$

Obviously if there exist F^* and G^* such that

$$E_1[F^*] = E_2[G^*] = v_1 = v_2 = v \tag{2.58}$$

then the functions F^* and G^* are optimal strategies of the players.

Let us consider the game with the gain function:

$$\begin{aligned}
&M(x-y) = \alpha_{k+1} \text{ for } x - y \leq -\sum_{j=1}^{k} \varepsilon_j,\\
&M(x-y) = \alpha_i \text{ for } -\sum_{j=1}^{i} \varepsilon_j < x - y \leq -\sum_{j=1}^{i-1} \varepsilon_j, (i = \overline{1,k});\\
&M(x-y) = \beta_j \text{ for } \sum_{i=1}^{j-1} \delta_i < x - y \leq \sum_{i=1}^{j} \delta_i, (j = \overline{1,s});\\
&M(x-y) = \beta_{s+1} \text{ for } x - y \succ \sum_{i=1}^{s} \delta_i
\end{aligned} \tag{2.59}$$

where α_i and β_j are arbitrary real numbers, ε_j and δ_i are positive quantities representing the step lengths of function M, and k and s are natural positive numbers. The number of steps for function M is $k + s + 2$.

Let the players **I** and **II** choose arbitrary values x and y from $[0,d]$, respectively. The next theorem is valid.

Theorem 2.1. If equation

$$\sum_{i=1}^{k} (\alpha_{i+1} - \alpha_i)\lambda^{-\psi_i} + \sum_{j=1}^{s} (\beta_j - \beta_{j+1})\lambda^{\gamma_j} = \beta_1 - \alpha_1 \tag{2.60}$$

where

$$\psi_i = \sum_{j=1}^{i} \delta_j \qquad \gamma_j = \sum_{i=1}^{j} \delta_i$$

has at least one root $\lambda^* = \rho \exp(i\omega)$ such that $\omega \leq \pi/(l+r)$, then stable strategies exist in the game with gain function (2.59) and they are optimal. Under this the game solution has the form:

$$\begin{aligned}
&F^*(x) = (\rho_0)^x [C_1\cos(\omega_0 x) + C_2\sin(\omega_0 x)] + B, -r \leq x \leq l;\\
&G^*(y) = D[1 - (\rho_0)^y \cos(\omega_0 y)], 0 \leq y \leq n;\\
&v = [\beta_{s+1}(\rho_0)^n - \alpha_{k+1}\cos(\omega_0 n)]/[(\rho_0)^n - \cos(\omega_0 n)],
\end{aligned}$$

where the root of Eq. (2.60) $\lambda^* = \rho_0\exp(i\omega_0)$ has minimal argument and maximal module; r, l, and n are minimal natural numbers which are greater than ψ_k, $d + \gamma_s$ and d, respectively;

$D = (\beta_0)^n/[(\rho_0)^n - \cos(\omega_0 n)]$; $B = \cos(\omega_0 n)[\cos(\omega_0 n) - (\rho_0)^n]^{-1}$;
$C_1 = [(\rho_0)^{n-l}\sin(\omega_0 r) + (\rho_0)^{n-r}\cos(\omega_0 n)\sin(\omega_0 l)]/\{\sin[(r+l)\omega_0][(\rho_0)^n - \cos(\omega_0 n)]\}$;

and

$$C_2 = [(\rho_0)^{n-r}\cos(\omega_0 r) - (\rho_0)^l\cos(\omega_0 n)\cos(\omega_0 l)]/\{(\rho_0)^{l-r}\sin[(r+l)\omega_0][(\rho_0)^n - \cos(\omega_0 n)].$$

The proof of the theorem is based on the solution of Eq. (2.58) under function (2.59).

Now we consider the game with gain function $M(x-y)$. In analogy with (2.58), we have:

$$\int_0^1 M(x-y)p_1(x)dx = \int_0^1 M(x-y)p_2(y)dy = v \quad (2.61)$$

where $dF(x) = p_1(x)dx$, $dG(y) = p_2(y)dy$.

Equations (2.61) are solved by means of the Fourier transform:

$$R_2(-p^2)v = R_1(-p^2)p_1(y) \quad (2.62)$$

$$R_2(-p^2)v = R_1(-p^2)p_2(x) \quad (2.63)$$

where $R_1(\omega^2)$ and $R_2(\omega^2)$ are polynomials with real coefficients defined from Eqs. (2.61).

Theorem 2.2. If the solution $\overline{p_1}(x) \geq 0$ of (2.62) exists, the game with gain function $M(x-y)$ has optimal stable strategies $p_1^*(x)$ and $p_2^*(y)$ such that $p_1^*(x) = p_2^*(x) = K\overline{p_1}(x) + \sum_{i=1}^{\infty}[A_i\delta^{(i-1)}(x) + B_i\delta^{(i-1)}(x-1)]$, where $\delta(x)$ is the delta-function and the constants K, A_i, and B_i are determined from (2.61) and the conditions:

$$\int_0^1 p_1^* dx = \int_0^1 p_2^* dy = 1.$$

Example 2.1. Let us focus on the game $\Gamma(M,[0,1])$ with gain function $M(x-y) = a[\delta(x-y) + d\exp\{-b|x-y|\}]$, where a, b, and d are arbitrary constants.

It is easy to see that theorem 2.2 gives the following solution for this game:

$p_1(x) = p_2(x) = vb^2(1 + A\exp\{\gamma x\} + B\exp\{-\gamma x\})[a\gamma^2]^{-1}$,
$v = a\gamma^2 b^{-2}[1 + A(e^\gamma - 1)/\gamma - B(e^{-\gamma} - 1)/\gamma]^{-1}$,

where $\gamma^2 = b^2 + 2db$,

$A = (\gamma - b)b^{-1}[(\gamma + b)^2 - (\gamma^2 + b^2)e^{-\gamma}][(\gamma + b)^2 e^\gamma - (\gamma - b)^2 e^{-\gamma}]^{-1}$,
$B = -(\gamma + b)b^{-1}[(\gamma^2 + b^2)e^\gamma - (\gamma - b)^2)][(\gamma + b)^2 e^\gamma - (\gamma - b)^2 e^{-\gamma}]^{-}$

Example 2.2. It is easy to find the solution of game $\Gamma(M,[0,1])$ with gain function

$$M(x-y)=\sum_{i=1}^{n} a_i \exp(-b_i|x-y|).$$

From theorem 2.2 it follows:

$$p_1(x)=p_2(x)=v\{a[\delta(x)+\delta(x-1)]$$

$$+\sum_{j=1}^{n-1} c_j[\exp(-b_j x)+\exp\{-b_j(1-x)\}]+\theta_{2n}/\rho_{2n}\}$$

$$v=\{2a+\sum_{j=1}^{n-1} 2a_j[1-\exp(-b_j)]/b_j+\theta_{2n}/\rho_{2n}\}^{-1},$$

where a is an arbitrary root of $R_2(\omega^2)$ and c_j $(j = 1,...,n)$ are roots of $R_1(\omega^2)$.

Example 2.3. Let consider the game with gain function:

$$M(x,y)=\begin{cases} 1 & for \quad (x,y)\in G_1, \\ a & for \quad (x,y)\in G_a, \\ b & for \quad (x,y)\in G_b, \end{cases}$$

where $x \in X, y \in Y, G_1 \cup G_a \cup G_b = X * Y, G_1 = \{(x,y): \rho(x,y) < \varepsilon, x \in K_m\}, G_a = \{(x,y): \rho(x,y) \geq \varepsilon\}, G_b = \{(x,y): \rho(x,y)<\varepsilon, x \in X \setminus K_m\}$, ε is an arbitrary value, X and Y are m-dimensional bit cubes, K_m is an arbitrary set having volume 0.5, $\rho(x,y)$ is the distance between x and y, and $a = k\varepsilon^m$.

It is obvious that,

$$p_1^*(x)=p_2^*(x)=\begin{cases} 2/(1+\lambda) & for \quad x\in K_m, \\ 2\lambda/(1+\lambda) & for \quad x\in X\setminus K_m, \end{cases}$$

$$v = \varepsilon^m (2\lambda b\pi^{0.5m}\,\Gamma(m/2)(1+\lambda)^{-1} - kb) + O(\varepsilon^m),$$

where $\lambda = (1-a)/(b-a)$ and $\Gamma(m/2)$ is a gamma function.

CHAPTER 3

Mathematical Model for Global Ecological Investigations 3

3.1 Conceptual Aspects of Global Ecological Investigations

Science and technical progress and, most importantly, rapid growth in the use of energy, generates numerous problems that require fundamental scientific analysis. They arise primarily from the direct impact of human activity on the environment. The problem concerning not only the coexistence, but also the coevolution and harmonic union of man (system H) and the biosphere (system N), is becoming one of the most important social issues of concern nowadays. However, the investigation into the conditions of mutual evolution of man and the environment does not fit in with any of the traditional scientific trends. First, this problem is diversified. It links up the natural dynamic processes with those processes which are taking place in the society. Its solution in the final analysis is the elaboration of the conditions that would provide a further development of human civilization and the entire species of man (Kondratyev et al. 1997; Kondratyev 1998a, 1999b–d; Moiseyev 1998; Arbatov 2000; Bratimov et al. 2000; Kondratyev and Varostos 2000; Andreev 2001).

Thus, the emerging field of investigations requires joint efforts not only of specialists-naturalists working in various spheres, but also a fundamental synthesis of natural-scientific and humanitarian knowledge (Shah 1998; Lavrov 1999; Sarkisyan 2000; Barnett 2001; Carvalho 2001; Martin and Shumann 2001).

The elaboration and analysis of mathematical models, describing the dynamic processes in the biosphere, will play a substantial role here, while the computerized simulation of these processes under study is acquiring a special importance. It will carry diverse loads. Moreover, it must be made an instrument that will provide a means for linking and coordinating information of various physical content, which is indispensable in complex interdisciplinary investigations. Such investigations necessitate a unified language. Those participating in ecological studies – physicist, biologist, economist, mathematician, etc. – should understand each other, and only the language of mathematical modeling can serve as such a unified starting principle.

The models of global processes expounded here are integrated into a single system: the global model. In this model, the land and ocean are divided into n and m regions, respectively.

Certainly, in describing human activity, we cannot transgress the laws of physics. The laws of conservation should underlie any model. These laws should be formulated in terms which define the most substantial aspects of human activity. For instance, in studies of social macrosystems, the laws of conservation must be formulated in economic terms. These laws are well known and are referred to as balance correlations. However, such correlations are insufficient for constructing the model: many values (for instance, price level, investment structure, etc.) cannot be determined from these balance correlations. Similar difficulties are faced in describing the biotic level. However, in the latter case, it is sufficient to use either trophic functions, which determine feedback connections, or competition functions to build the model.

In describing the social processes, the situation becomes more complicated. Man predicts the results of his own activity and organizes the processes of treatment of information. This implies that alongside the description of the processes for transformation of material, energy and motion we must have an information processing model. Apart from this, it is necessary to define the goal of humanity and set up strategies for reaching this goal (Kondratyev et al. 1996b; Inozemtsev 1999; Girusov et al. 2000; Kondratyev 2000b, 2001b; Kondratyev and Demirchian 2001). This problem under modern conditions cannot be tackled unambiguously. One of its possible solutions has been suggested in the volume by Krapivin et al. (1982), where mankind is regarded as a single whole. Other solutions are also possible in the framework of the survivability model described in Chapter 2 (this Vol.).

At present, the industrial and agricultural activities of man are becoming the main forces that change (locally, regionally, and globally) both the quantitative characteristics of global material circulation and the structure of the entire biosphere as a whole.

The consequences of anthropogenic activity are manifested in:

1. depletion of energy resources (Hotuntsev 2001; Klyuev 2001),
2. pollution of the environment with industrial wastes (Luecken et al. 1991; Macdonald et al. 1997; Doos 2000; Gerard et al. 2001),
3. destruction and changes of the natural ecological systems of land, fresh-water basins and the ocean (Blumenstein et al. 1999; Trofimov et al. 1999; Nikanorov and Horuzhaya 2000; Balvanera et al. 2001; Kondratyev and Krapivin 2001b),
4. changes in the physical structure of the Earth's surface (Gorshkov 1998; Ernst 1999; Grigoriev and Kondratyev 2001),
5. changes in climate, etc. (Gupta 2001; Kondratyev et al. 2001a,b; Kondratyev 2002).

All these changes occur interdependently and are manifested on the local, regional, and global scale.

Though the consequences of resource depletion and environmental pollution are more or less obvious, it is difficult to say what the destruction and changes of ecosystems or modifications of climate will lead to. The answer is still ambiguous due to the fact that we know too little about these processes. For instance, anthropogenic activity affects the cycles of practically all chemical elements.

The problem regarding the consequences of the anthropogenic modification of climate is also very complicated. For example, CO_2 emissions into the atmosphere lead to an increase in the average global air surface temperature (*greenhouse effect*). This temperature also rises with the increase in energy production (direct heating). However, emissions of aerosols may cause a *backscattering effect*, and as a consequence, a decrease in temperature. Therefore, in our model, we have singled out a separate unit simulating these processes.

Disregarding the details, we will consider some general problems associated with man's impact on the environment of his inhabitancy in the biosphere, primarily relating to the stability of the current conditions of the biosphere. Mathematical investigation includes two important approaches. The first deals with the biosphere as a statistical ensemble of weakly interacting elementary structures, the biomes. This approach involves the use of the thermodynamic measure of stability, entropy. In the second approach, we consider the biosphere as a complex dynamic system and employ the ideas and methods of Lyapunov for the characteristic of its stability. These approaches are expounded by Krapivin et al. (1982), where an hypothesis of the possible mechanisms of destruction of the biosphere by man has been suggested. The development of such approaches lays in the synthesis of the survivability criteria.

Indeed, despite the fact that the energy power of the H system is insufficient for a considerable reduction in the total amount of living matter, it is already enough for a substantial decrease in species diversity (e.g., due to urbanization, pollution, the use of the one-crop system in agrocenoses, and the transformation of soil-plant formations). The decrease in diversity may proceed at more rapid rates than the reduction of the total biomass (and with the intensification of agriculture, the diminishing of diversity is accompanied even by an increase in overall biomass). All this may lead to the destruction of the current stationary condition of the biosphere even with rather slight (energy) impacts on the part of man (Kondratyev 1990; Kuznetsov et al. 2000; Kondratyev and Krapivin 2001a; Losev 2001).

The new feature of the relationship of *Nature* and *Humanity* is that the history of interaction between N and H is passing into such a period when the population of the Earth is becoming a determining factor in the existence and evolution of the biosphere. The most important point of this transition appears to be a scientific and technical revolution, the most characteristic product of which is the *economic crisis*. There are numerous publications which deal with the problem arising here (Dobson 1998; Ursul 1998; Kotlyakov 2000; Koptyug et al. 2000; Lisichkin and Shelepin 2001; Clark et al. 2001; Losev 2001; Stroev 2001; Vaughan et al. 2001). They consider at large the social, geographical, historical and other aspects of human ecology. As Odum (1971) reported, "at present, the representatives of almost all disciplines and professions in the field of both natural and social sciences are striving to find a common platform for tackling the problems of man's ecology". As far back as 1864, Marsh wrote on this subject the classical treatise "Man and Nature. Physical Geography and its Change under the Influence of Man". Further, *Homo sapiens;* as a particular species on Earth, has attracted the attention of an increasing number of researchers (Kondratyev et al. 1997; Fisher and Hajer 1999; Gorshkov et al. 2000; Uvarov 2000; Subetto 2000; Danilov-Danilian et al. 2001).

Of particular interest at present is the problem of human population growth, this being one of the most contradicting problems of today. In discussing the *problems of population* in the press, many authors often turn to the models of Forrester (1971) and Meadows et al. (1972). Actually, these are economic and socio-economic models. Typically, a small number of ecological variables (for instance, the level of population life) are in certain hypothetical dependence on the economic and social indices of human activity.

The other approach, which is realized in the works by Krapivin (1978) and Krapivin et al. (1982), considers the problem of interaction between the systems H and N as a problem of biospheric stability. In other words, the primary task is to construct qualitatively adequate models for the processes of physico-chemical and biological nature taking place in the biosphere of the Earth, and then comes the task of studying the character of anthropogenic influence on these processes in order to determine (within the scope of the available models) the limits of permissible influence of man on the environment. According to this concept, the model of global biosphere processes appears as the model of biogeocenosis distributed on the Earth's surface with several layers of productivity and with man as a consumer of certain parts of the production at each level.

The unit of human population dynamics, or the demographic unit, represents an important part of this model. In the present work, the demographic unit of the global model has been considered from the viewpoint of its computerized realization in the form of a dialogue between the researcher and the model.

The concept of the biosphere as a *field of life* and the Earth's external shell was suggested for the first time by Lemark early in the nineteenth century. In geology, the term and idea *biosphere* were introduced in 1875 by Suss, who regarded it as one of the Earth's shells. However, the founder of the current concept of the biosphere was Vernadski. He expounded these concepts in his two lectures published in 1926. The global view of the problem, bold and rational aggregation of descriptions, historicity, and a new scientific methodology, which he called "the method of empirical generalization" – all this points to a systematic approach to the studies of the biosphere in Vernadski's works (1926, 1940, 1944).

What Vernadski formulated for the biosphere is commonly called at present a conceptual model in the systemic analysis, while a further formalization of the model's separate units was carried out by his follower and disciple Kostitzyn (1937). Vernadski pointed to the great role of the global circulation of oxygen, carbon and nitrogen in the geological history of the planet, specifically in the evolution of the atmosphere and climate. The volume by Kostitzyn, published in 1937, presents a mathematical model for the first time which describes these global cycles. Based on the balance correlations, the model provided a means for estimating the periods of the global cycles and their connection with periodic changes of the planet's climate. Despite the fact that this work was published more than 40 years ago, it is quite up to-date with a modern methodology, being a typical systemic analysis (though the absence of computers at that time limited the possibilities of investigating the model, basically it is a qualitative analysis).

Vernadski's concept is characterized by a maximum degree of aggregation and global scale of description (actually, this is an outward view of the biosphere), while the concept of biocenosis formulated and developed by Sukachev (1945) discriminates the elementary biospheric unit, existing in reality in the surrounding nature. This concept could be arbitrarily called the *atomistic* concept of the biosphere. Proceeding with this analogy, it can be said that Vernadski's school is characterized by a macroapproach, whereas Sukachev's school, by a microapproach to the analysis of the biosphere.

According to Vernadski, the biosphere is an external shell of the Earth, the encompassing sphere of life. However, as Vernadski noted, the definition of the biosphere as a sphere of life is not complete. In Vernadski's opinion, it includes:

1. *Living matter.*
2. *Biogenic matter,* i.e., organo-mineral or organic products produced by living matter (coal, peat, bedding, humus, etc.).
3. *Dead organic matter* produced by living organisms together with inorganic nature (water, atmosphere, sedimentary rock).

Vernadski regarded the biosphere as a form of cosmic organization: "Creatures of the Earth are involved in the complex cosmic processes, making up an indispensable and regular part of the harmonious cosmic mechanism, in which as we know, there are no accidents". He noted that due to cosmic (solar) radiation, the substance of the biosphere becomes active: "Actually, the biosphere can be regarded as a sphere of the Earth's crust with transformers converting cosmic radiations into effective terrestrial energy – electrical, chemical, mechanical, thermal, etc." The GSM structure described in Chapter 2 is one of the ways to solve the problem of valuation of the biosphere survivability.

Vernadski attached an exclusive role to the living matter of the biosphere in the formation of the Earth's crust and atmosphere: "There is no chemical force on the terrestrial surface that could be more constantly acting, and hence more powerful in this effect than the living organisms taken as a whole."

Vernadski noted that the living organisms were integral systems: "The living organism of the biosphere should be studied empirically now as a special body that is not to be completely reduced to the known physico-chemical systems. The possibility of it to be completely reduced to them some time in the future is still an open question for science."

However, he regarded the biosphere as an integral system: "In studying the manifestations of life in its environment on the planetary scale, we should detach ourselves from the conventional aspect of the organism... Life makes up an indispensable part of biospheric organization." Thus, Vernadski recognized various levels of the integrity of the living, and hence recognized the hierarchic subordination of biological systems.

While other researchers considered a separate organism as a fundamental principle of biological organization, Vernadski regarded life from the viewpoint of the *cosmic mechanism* and maintained that "the solution of life cannot be made only by studies of the living organism." He recognized the biosphere as an element of life's organism.

The chief requirement of the system method is a study of the entire phenomenon as a whole. It is impossible to single out one or several important features of the phenomenon, disregarding its other aspects, and form one's own opinion of it based solely on studies of these separate, though important characteristics. The method of *empirical generalization* suggested by Vernadski is actually such a system method (Sukachev 1945).

In its structure, the biosphere is a complex hierarchic system. The processes of energy and substance transformation are relatively autonomous within each structural unit. At the same time, all living matter is united by a specific form of organization and by common mechanisms regulating the flows of energy and the circulation of substances. The direction and intensity of energy and substance flows depend upon the structure of the geochemical cycles and the circulation of the atmosphere and water. Damages and changes of these cycles, which result from anthropogenic activity with increasing frequency, may alter both the local and global characteristics of the biosphere.

Several constituents may be discriminated in the structure of the biosphere, such as the atmosphere and hydrosphere, autotrophic (plants) and heterotrophic organisms, soil, etc. In the biosphere, it is natural to differentiate continents and oceans, which in their turn are successively divided into increasingly smaller elements (natural zones, regional units, biogeocenoses, communities, etc.).

Accepting Vernadski's postulate that the total amount of living matter in the biosphere remained constant for almost the whole geologic period, the only alternative in this case is to assume that the biosphere could have evolved only by way of structural complication of the living matter. This postulate provides the idea for the synthesis of the survivability model considered in Chapter 2.

Over hundreds of millions of years the Earth accumulated solar energy in various fossil hydrocarbons. At present, these reserves are involved in energy circulation, changing eventually, the biogeochemical and thermal balance of the biosphere. This change depends to a large measure upon human activity both in direction and rate. Large amounts of waste resulting from industrial activity of man are not only involved in the biogeochemical cycles, but also frequently change the characteristics of landscapes. Agricultural biocenoses are beginning to play a global role. An increase in their productivity is accompanied by a drastic decrease in the diversity of biological species, and hence in the stability of cenosis. Large amounts of chemical substances – fertilizers and toxic chemicals – introduced into the soil also markedly change the structure of the biogeochemical cycles. In other words, the complex and hierarchic biogeocenotic system – biosphere – should include inputs of anthropogenic origin, the consideration of which should be one of the starting principles for making up global models and models of the entire biosphere.

The concepts expounded above make it possible to take up a discussion of the last stage in the given problem, that is, constructing the conceptual model of the biosphere. Figure 2.7 presents the basic units of this model.

The biogeocenotic process of the biosphere is based on energy exchange described by the equation of energy conservation. The energy coming into the biosphere is consumed for maintaining the biogeocenotic processes, which finally leads to the conservation of a part of the energy in organic residues and to the formation of sedimentary rocks; part of the energy returns to the cosmos due to reflection and thermal emission. We will consider only two sources of energy flux to the biosphere: solar radiation and combustion of fuel of biogenic origin.

In the total energy balance of the biosphere, a special place belongs to the consumption of solar energy. The balance of solar energy consumption can be presented by the following equation: $E = E_1 + E_2 + E_3$, where E is the total amount of solar energy flux to the Earth, E_1 is the energy reflected back to the cosmos, E_2 is the part of the energy that goes for heating the atmosphere, and E_3 is the solar energy coming in contact with photosynthesizing elements of the biosphere. The value of E_1 depends on the Earth's albedo which, in its turn, is determined by the area of sea and ice, clouds and aerosols in the atmosphere and the structure of

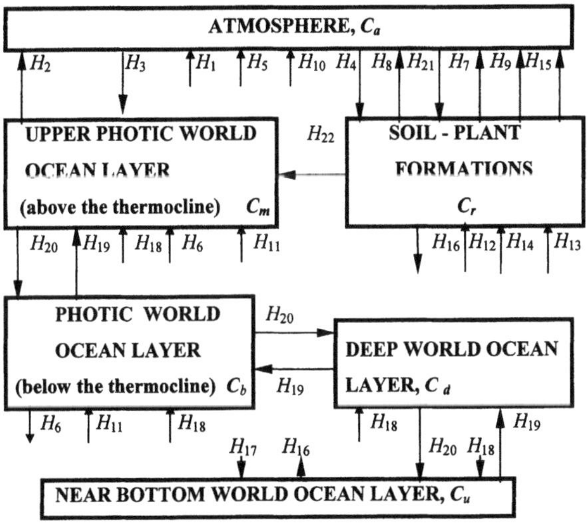

Fig. 3.1. Block diagram of the carbon biogeochemical cycle in the biosphere. The list of *symbols* is given in Table 3.1. The dynamic equations describing the carbon balance are given by Nitu et al. (2000a)

vegetation cover. The value of E_2 depends on the aerosol content of the atmosphere and the amount of cloud cover. A cloudless, clear atmosphere is rather transparent for short-wave solar radiation and can be heated mainly as a result of heating from the surface of land and ocean. However, due to dustiness and cloudiness of the atmosphere, a certain part of solar energy does not reach the terrestrial surface (Kondratyev 1999a; Kondratyev and Demirchian 2000; Asrar et al. 2001; Harris and Hudson 2001).

Investigation of the photosynthetic processes serves as a starting point for studies on the geochemical cycles of the biosphere. First of all, the oxygen cycle is worthy of notice. Estimates show that the amount of free oxygen over the periods of hundreds of years is practically stable. However, the oxygen cycle cannot be disregarded, insofar as it carries control functions in the model. This is not the case with the carbon cycle, the diagram of which is presented in Fig. 3.1

Table 3.1. Carbon flows included in Fig. 3.1

Flow	Origin of flow	Flow	Origin of flow
H_1	Fuel burning	H_{12}	People vital functions
H_2	Desorption	H_{13}	Animal vital functions
H_3	Sorption	H_{14}	Mortality of plants
H_4	Rock erosion	H_{15}	Secretion by roots
H_5	Volcanic emission	H_{16}	Deposition
H_6	Photosynthesis in the ocean	H_{17}	Ocean depositions dissolving
H_7	Respiration of plants	H_{18}	Detritus decomposition
H_8	Burning of plants	H_{19}	Water rising
H_9	Decomposition of humus	H_{20}	Water descending and sedimentation
H_{10}	People activities	H_{21}	Photosynthesis on land
H_{11}	Vital functions of biota in the ocean	H_{22}	River flow
		H_{23}	Decomposition of litter

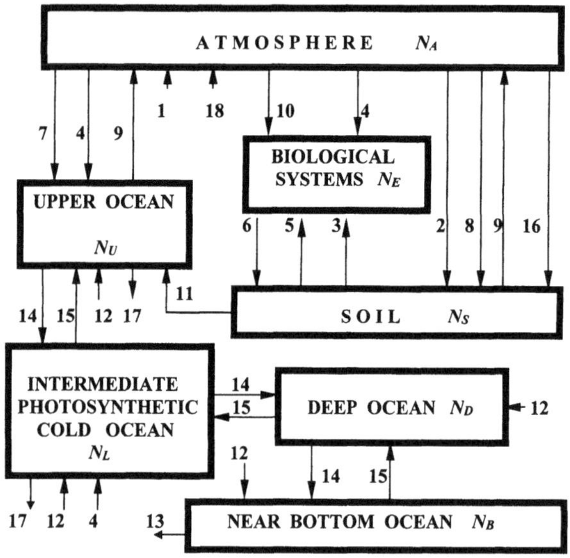

Fig. 3.2.
Block diagram of the nitrogen biogeochemical cycle in the biosphere. The list of *symbols* is given in Table 3.2

(Table 3.1). The amount of CO_2 in the atmosphere varies greatly. Thus, over the last 25 years, the amount of CO_2 in the atmosphere has increased on the average by 0.29% annually.

The next cycle to be considered in the GSM is the nitrogen circulation (Fig. 3.2, Table 3.2). The productivity of all natural and artificial cenoses is associated with this cycle. The nitrogen cycle model is one of the most complex models in the entire biogeocenotic model of the biosphere. To begin with, the nitrogen cycle is intimately connected with the processes occurring in the soil and with the chain of chemical transformations in which the dying organic matter is converted into humus and nutrients for plants. Since the processes in the soil differ with the type of soils, climatic conditions, the structure of moisture circulation and other factors, any equation describing the nitrogen circulation can be related only to a given region. Apart from this, it is very hard to separate nitrogen circulation from moisture circulation. Under certain conditions the circulation of oxygen and carbon can be regarded separately, but the nitrogen cycle always has to be considered together with the water cycle.

In a similar manner to the nitrogen cycle, the model takes into account the phosphorus (Fig. 3.3, Table 3.3) and sulfur (Fig. 3.4, Table 3.4) cycles.

The climate unit holds a special place in the GSM. Climate is responsible for the possibility of life on Earth, and hence for the existence of human civilization. Moreover, the range of climate parameters admitting the existence of civilization is extremely narrow.

Climatic processes are characterized by instability and imbalance. In both, fluctuation processes may be differentiated of external origin (for instance, with a duration from 10 to 100 years, associated with changes in the level of solar activity) and internal origin (for instance, glaciation of the Earth with a duration of about 10^4 years, of self-fluctuating character). The presence of a great number of feedback connections in climatic processes with biospheric processes, in-

Table 3.2. Nitrogen flows included in the GSM (Fig. 3.2)

Flow	Origin and symbol of flow	Flow	Origin and symbol of flow
1	Geospheric sources, H_1^N	10	Biological assimilation, H_{10}^N
2	Fuel burning, H_2^N	11	Water flows, $H_{8,i}^N$
3	Soil decomposition, H_3^N	12	Detritus decomposition, H_{12}^N
4	Fixation, H_4^N	13	Bottom sediments, H_{13}^N
5	Photosynthesis on land, H_5^N	14	Sedimentation, H_{14}^N
6	Plant functions, H_6^N	15	Ascent in the ocean, H_{15}^N
7	Precipitation on the ocean, H_7^N	16	Precipitation on land, H_{16}^N
8	Fertilizer production, H_8^N	17	Photosynthesis in the ocean, H_{17}^N
9	Denitrification, H_9^N	18	Anthropogenic sources, H_{18}^N

Table 3.3. Phosphorus flows included in the GSM (Fig. 3.3)

Flow	Origin of flow	Flow	Origin of flow
H_1^P	Volcanic	H_2^P	Fertilizers
H_3^P	Assimilation by plants	H_4^P	Dead biomass
H_5^P	Vital functions on land	H_7^P	Erosion
H_{20}^P	Vital functions in the ocean	H_6^P	Transformation to the unassimilated form
H_{13}^P	Sedimentation		
H_{16}^P	Wet sedimentation in the ocean	H_{17}^P	Photosynthesis
H_8^P	Wet deposition on land	H_{11}^P	Washing into the ocean
H_9^P	Removal by fishing	H_{10}^P	Removal by birds
H_{18}^P	Detritus decomposition in the photic layer	H_{15}^P	Ascent
		H_{12}^P	Detritus decomposition in deep waters
H_{14}^P	Descent		

cluding climatic connections with physico-chemical processes in the ocean, is responsible for a certain degree of climatic stability relative to the results of anthropogenic activity (Kondratyev 1999a).

For instance, due to the ability of the ocean and land ecosystems to absorb part of anthropogenic CO_2 emissions into the atmosphere, the amount of CO_2 in the atmosphere increases less rapidly. This leads to less heating of the atmosphere, thereby stabilizing the climate.

The scientific literature contains conclusions that the presence of a powerful positive connection between atmospheric temperatures and the albedo of the terrestrial surface is responsible for the instability of ice cover in the Polar regions. A slight increase or fall in atmospheric temperature may cause an irreversible process of ice melting and atmospheric heating, or complete glaciation of the Earth and climatic cooling. However, this result is purely an implication of the model. As the article by Krapivin (1999) shows, there are other sufficiently powerful feedbacks which prevent the indicated catastrophic phenomena. For instance, the formation of ice cover on the dry land depends not only upon temperature, but also upon the amount of precipitation. A decrease in precipitation associated with climatic cooling prevents the spread of glaciation of the con-

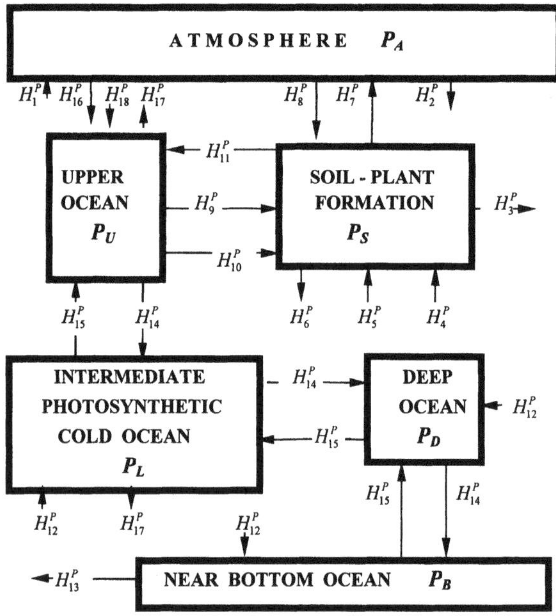

Fig. 3.3.
Block diagram of the phosphorus biogeochemical cycle in the biosphere. The list of *symbols* is given in Table 3.3

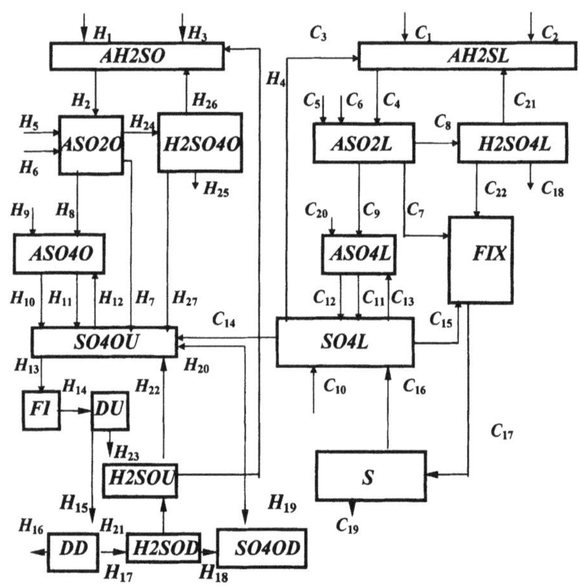

Fig. 3.4.
Block diagram of the sulfur biogeochemical cycle in the biosphere. The list of *symbols* is given in Table 3.4

Table 3.4. Sulfur flows included in the GSM (Fig. 3.4)

Origin of sulfur flow	Symbol of the GSM	
	Land	Ocean
Volcanic		
H_2S	C_1	H_3
SO_2	C_5	H_5
SO_4^{2-}	C_{20}	H_9
Anthropogenic contamination		
H_2S	C_2	H_1
SO_2	C_6	H_6
SO_4^{2-}	C_{10}	
Oxidation of H_2S to SO_2	C_4	H_2
Oxidation of SO_2 to SO_4^{2-}	C_9	H_8
Dry sedimentation of SO_4^{2-}	C_{12}	H_{11}
Wet sedimentation of SO_4^{2-}	C_{11}	H_{10}
Discharge of H_2S to the atmosphere at the expense of biological decomposition	C_3	H_4
Accumulation of SO_4^{2-} by biota	C_{15}	H_{13}
Formation of SO_4^{2-} at the expense of biological decomposition	C_{16}	H_{17}, H_{23}
Settling and sedimentation	C_{18}, C_{19}	H_{15}, H_{16}
Return to the atmosphere at expense of wind	C_{13}	H_{19}, H_{25}
Addition of sulfur to the environment at the expense of dead biomass	C_{17}	H_{12}
Adsorption of SO_2 from the atmosphere	C_7	H_{14}
Washing out of SO_2 from the atmosphere	C_8	H_7
SO_4^{2-} in river flow to the ocean	C_{14}	H_{24}
Transformation of H_2SO_4 gas phase to H_2S	C_{21}	H_{26}
Adsorption of atmospheric SO_2 by biota	C_{22}	H_{27}
Oxidation of H_2S to O_2		H_{18}, H_{22}
SO_2 advection		H_{20}
H_2S advection		H_{21}

tinents. An increase in cloudiness with warmer climate should prevent atmospheric heating and impede ice melting. However, despite the presence of a certain degree of climatic stability with respect to anthropogenic impacts, there is no doubt that sufficiently strong and persistent impacts may give rise to an irreversible transition of climate into a new functional state with a new regime of heat and moisture distribution on the planet.

Apparently, the climatic model should be related to the basic biosphere models, without which the atmospheric dynamics cannot be described. However, the current state of climatology is such that not enough data for the practical forecasting of the main climatic parameters yet exist. At present, the climatic models are mainly of solely theoretical importance.

There are two current trends in climatology. One of them is related to the construction of sufficiently simple energy balance models. However, these models are insufficient for estimating climatic variations in separate regions. The other trend is to develop models of general circulation in the atmosphere and ocean.

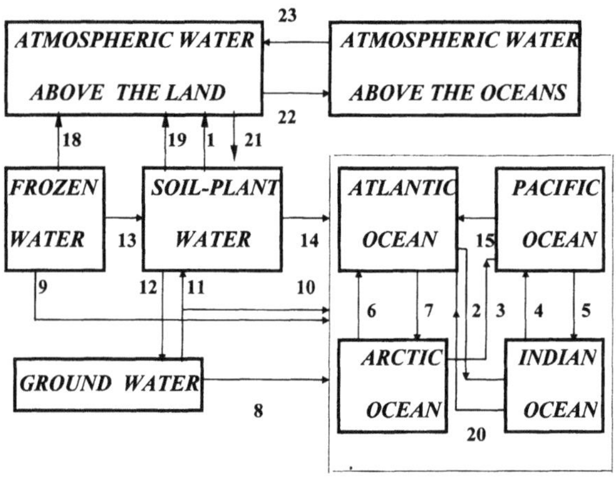

Fig. 3.5.
Block diagram of the water biogeochemical cycle in the biosphere. The list of *symbols* is given in Table 3.5

Table 3.5. List of water flows in Fig. 3.5

Flow	Origin of flow	Flow	Origin of flow
1	Soil evaporation	12	Infiltration
2	Antarctic current between the Atlantic and Indian Oceans	13	Freezing
		14	Land-surface flow
3	Flow in the Bering strait	15	Current across the Drake Passage
4	Antarctic current between Indian and Pacific Oceans	16	Precipitation over the oceans
		17	Evaporation from the oceans
5	East-Australian current	18	Evaporation from ice surface
6	Flow across straits from Arctic Ocean to Atlantic	19	Transpiration
		20	Agulhas headland current
7	Flow across straits from Atlantic Ocean to Arctic Ocean	21	Precipitation over land
		22	Atmospheric water flows from the oceans to land
8	Underground flow		
9	Melted waters	23	Atmospheric water flows from land to the oceans
10	Part of the domestic-use waters		
11	Irrigation waters		

These models describe at great length the basic physical processes in the atmosphere and ocean, but require much computer operating time for obtaining statistically and physically significant predictions. Really, using the evolutionary technology for the synthesis of a climatic model may help to solve the problem arising here. The model of biospheric processes presented in this volume employs a sufficiently simple climatic model including the global water cycle (Fig. 3.5, Table 3.5).

Finally, the construction of the global model faces the problem of modeling human activity. The problem of simulating the human activity in rational form in the biogeocenotic model is open to discussion. It is more or less obvious that it is necessary to use certain macroeconomic models. The unit "human activity" as incorporated in the model is schematically described by Krapivin et al. (1982)

and Krapivin (1993). Substantially, it deals with combining the models of demographic and production processes. The model of scientific and technical progress holds a special place here. Its inclusion is an indispensable condition for any promising assessment of global processes.

The current literature suggests several methods for the formalized description of scientific and technical progress. However, they require an economic analysis too delicate for our tasks. The main requirements for the unit of scientific and technical progress in the GSM are simplicity of description and availability of statistical data, where relatively rough approximations are permissible.

3.2 General Description of the Global Model

A conceptual diagram and unit contents of the GSM are given in Fig. 3.6 and Table 3.6. Unit-to-unit connections provide for information exchange through their inputs and outputs. Such a structure makes the model amenable to the addition of new units or their modifications. The switching over of interunit links (feedbacks) is performed automatically, entailing no changes in the model structure. When units or links are removed, stand-by connections to the global data sets are turned on. Thus, the interrelation between the global model parts is ensured through parametric compatibility of the units.

The spatial structure of the model is determined by the data sets available. The simplest version of a one-dimensional model is realized when the input information is entered in the form of values averaged over the planet surface. The basic type of spatial discretization of the Earth's surface is a uniform geographic grid with latitude and longitude steps of $\Delta\varphi$ and $\Delta\lambda$, respectively. Realization of an actual version of application of the existing database within the framework of the GSM leads to nonuniform regional structures for different units. Thus, in a version of the global model described by Krapivin (1993), the

Table 3.6. The GSM units

Unit	Description
CARBON	Model of the global biogeochemical carbon cycle
NITROG	Model of the global biogeochemical nitrogen cycle
OXYGEN	Model of the global biogeochemical oxygen cycle
SULFUR	Model of the global biogeochemical sulfur cycle
WATER	Model of the global water balance
VEGET	Model of the soil-plant formations
PELAG	Model of the World Ocean pelagic ecosystem
ARCTIC	Model of the World Ocean arctic ecosystem
SHELF	Model of the World Ocean shelf ecosystem
DATA SETS	Environmental database and the necessary software
SCENARIOS	Scenarios for the human activities
SERVICE	Computer programs for the service functions
CLIMATE	Climate model or scenario

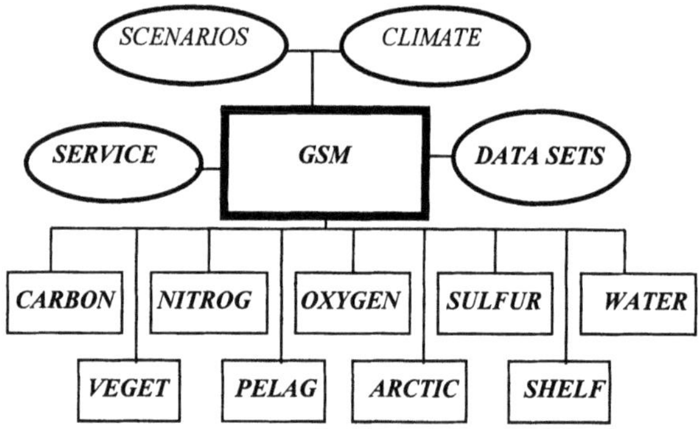

Fig. 3.6.
Structure of the global simulation model. Description of identifiers is given in Table 3.6

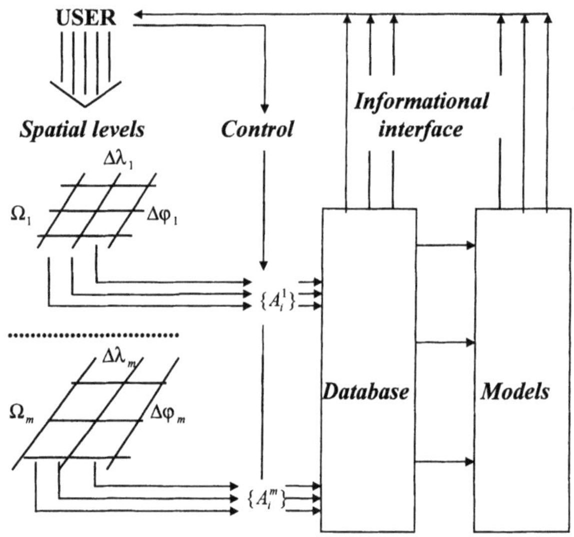

Fig. 3.7.
Block diagram for control of simulation experiment in the problem of environment investigations by means of the GSM

Table 3.7. Set of main identifiers in the GSM structure

Identifier	Description
A_1	Adaptation of the CBS model to the territory
A_2	Localization of the natural and anthropogenic elements on the territory
A_3	Formation of the distribution of typical elements of the hydrosphere
A_4	Definition of the soil-plant formation distribution
A_5	Formation of the pollution sources distribution
A_6	Determination of the pollutant type
A_7	Formation of the relief structure
A_8	Assignment of the socio-economical structure
A_9	Control of the simulation experiment structure

global socio-economic structure is represented by seven regions; the climate and biogeocenosis units have $\Delta\varphi = \Delta\lambda = 4°$; and in the unit of the global hydrological cycle, the grid for the Russian territory has $\Delta\varphi = \Delta\lambda = 2°$ while that for the rest of the earth's area has $\Delta\varphi = \Delta\lambda = 10°$.

Depending on specific features of the natural processes under consideration, the regional division structure may be linked to climatically identical zones, continents, latitude zones, natural zones, and socio-administrative formations. The interactive mode of the GSM provides the adaptation regime to the real natural system through the set of identifiers mentioned in Table 3.7. The regime of such adaptation is demonstrated by Fig. 3.7. An example of the identifier structure is given in Fig. 3.8 and Table 3.8. The database of the GSM has identifier sets for the spatial distribution of soil-plant formations. The user can form such identifiers according to the concrete discretization of space $(\Delta\varphi, \Delta\lambda)$ using his own data sets. To understand the structure of the identifiers level of the GSM database the

Table 3.8. Identifiers adopted in the GSM for designation of soil-plant formations: σ is the area of soil-plant formation (10^6 km^2), P is the production of soil-plant formation (kg C · m^{-2} year^{-1})

Identifier	Soil-plant formation	σ	P
A	Arctic deserts and tundras	2.55	0.17
B	Alpine deserts	1.15	0.47
C	Tundra	2.93	0.36
D	Mid-taiga forests	5.73	0.63
E	Pampas and grass savannas	1.66	1.11
F	North-taiga forests	5.45	0.54
G	South-taiga forests	6.60	0.65
H	Subtropical deserts	7.16	0.12
I	Subtropical and tropical thickets of the grass-tugai type	0.90	1.96
J	Tropical savannas	17.10	1.35
K	Solonchaks	0.38	0.18
L	Forest tundra	1.55	0.65
M	Mountain tundra	2.23	0.38
N	Tropical xerophytic open woodlands	9.18	1.42
P	Subtropical broad-leaved forests and coniferous forests	5.75	1.72
Q	Alpine and subalpine meadows	3.54	0.76
R	Broad-leaved coniferous forests	2.12	0.87
S	Subboreal and saltwort deserts	2.69	0.25
T	Tropical deserts	11.50	0.18
U	Xerophytic open woodland and shrubs	3.91	0.56
V	Dry steppes	2.66	0.38
W	Moderately arid and arid steppes	4.29	0.79
X	Forest steppe (meadow steppe)	3.72	0.74
Y	Variable-humid deciduous tropical forests	7.81	2.46
Z	Humid evergreen tropical forests	10.40	3.17
+	Broad-leaved forests	7.21	1.25
&	Subtropical semideserts	2.08	0.45
@	Subboreal wormwood deserts	2.77	0.61
*	Absence of vegetation	14.6	0.00

```
   W    150   120     90     60     30     0      30     60     90    120    150     E
N ......................****. **********................................................ N
  .......................AAA.. **********.....**.....................A...................
  ..............AAA. AAAA.... *********.................C.....AAAAA.....A........
  ....CCCC.....CCA. A. AAAC...*****.........LLL.....C. CCCCCMMMMCAAAAAAAA.....
  MC. FFFMFMFFFFCCCCCAA.. MM. ***...L......DDDFF. FFFFCLLLFFFMFFFFFMAAAMMMMMM
  ....LLFFFFCCCCCCCC...C.....L..........RG. DDDDDDDDDFFFFDDDDDDDDDMMFFMCCM.
  .....L...FFFFGGGFF...CFC...........+...R. RGGGGGGGDGGGGGGGGDDDDDDM..L....
  ...L......+RGGGGGDDF. FFFF..........++. ++RRRRXXXXGXXXXGGGGGGDDGDDD...L....
50...........RRXWXXGGDDDDD. D............+++++XXWWVVVVVVVGGGGGGGGGDDD........50
  ............RXWWX++. +++G............+++++XXWWV. S. @@QQSGVVVVGXRR..........
  ............BBBXXX++++..............++..P++.. WQ. KS&QQQSSSSSSXX++..+.......
  ............RBUWXP+PP.............PP....P. WWWW&K&&QQSSSSSQXX. +..P........
  .............@UUUPPPP..............&&&&.....&&WUUUUUQBBQQ+++...P.........
30.............@@@UP..P.............HHHHHHHHHHHHHHTHHHJPQQQQPPPP............30
  ..............@@@.................HHHHHHHHH. HH. TTTHNNYZZPPPP.............
  ...............U@...J............TTTTTTTTTTTTTTT...NNI. YYYI..............
  ................YYZ...J..........TTTTTTTTTTT. TT....JN.. YYY..............
  ..................YZ.............JJJJJJTJJJNTT.....NN...YY..Y...........
10...................Z. JJ..........JJJJJJJJINNTT.....N....Y..............10
  ....................ZJJZ.........YJZYJJJJJJT..........Z...............
  ....................ZZZZJ............ZYYYYJNN..........Z..Z............
  ...................QZZZZJZZ..........YYZYJJJ............Z. Z............
  ...................WZZZZJJJJJ.........JYYNNJ..................ZZ......
10...................ZZZZJJJJ..........JNNNN.......................Z.....10
  ....................WWZZZJJY..........JNNNN..Z...............NN........
  .....................WWNZYJY..........JJJNJ. J...............NNNNN......
  .....................TNNYYY...........TJJJJ. N.............JJTTJJJ.......
  .....................TNNYP............HHJE...............JJTTTTNN.....
30.....................UUEY.............HEE.................HHTTTJJ......30
  .....................UUEI..............E...................U...UUP.....
  .....................UEE.........................................P.
  ....................+VV....................................P....+..
  ....................+V...............................................
50..................+E...............................................50
  .....................+...............................................
S ..................................................................... S
   W    150   120     90     60     30     0      30     60     90    120    150     E
```

Fig. 3.8. Spatial distribution of soil-plant formations in the biosphere. The list of *symbols* is given in Table 3.8

example of identifier A_4 is given. For a complete explanation of the identifier structure many other processes have to be taken into account. Figures 3.9 and 3.10 show the idea of using the identifier sets in the GSM database. Each of the identifiers or their group plays the role of a driver which connects the database and the user with the real system. Really, anyone of the identifiers has a matrix structure. For instance, identifier $A_5 = \left\| a_{ij}^{5} \right\|$, where indexes i and j correspond to the spatial structure of the natural object studied in the framework of the

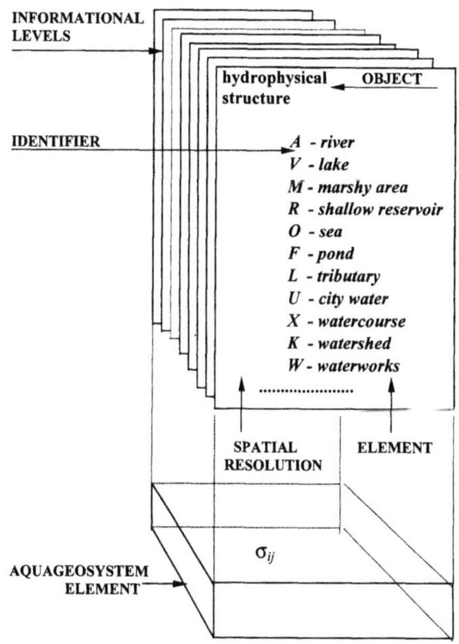

Fig. 3.9.
Cartographic identification and formation scheme of the GSM database

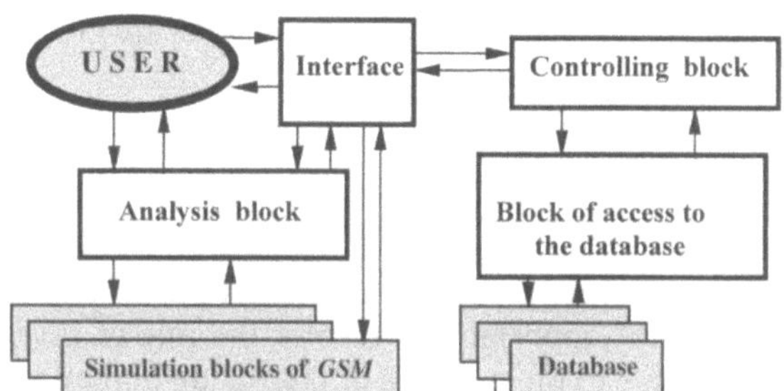

Surface	Speed of dry sedimentation (*cm/s*)	
	SO_2	NO_2
Water surface	0.2 - 1.5	0.01 - 0.04
Alkali surface	0.3 - 1.0	0.03 - 0.1
Acid dry soil	0.1 - 0.5	0.01 - 0.05
Acid wet soil	0.1 - 0.8	0.01 - 0.08
Grass (≤ 10 *cm*)	0.1 - 0.8	0.02 - 0.25
Vegetation (< 1 *m*)	0.2 - 1.5	0.4 - 0.7

Fig. 3.10.
Informational flows into the GSM

concrete project: $a_{ij}^5 = 0$ when pollution sources are absent, = 1 for chemical plant, = 2 for crop spraying and dusting, = 3 for crushing grinding and screening, = 4 for demolition, = 5 for field burning, = 6 for frost-damage control, = 7 for fuel burning, = 8 for fuel fabrication, = 9 for ink manufacturing, = A for metallurgical plant, = B for milling, = F for ore preparation.

As follows from Fig. 3.9, the assignment of $\|a_{ij}^5\|$ signifies the determination of the pollutant types.

3.3 Biogeochemical Cycles

The GSM comprises units which describe the biospheric cycles of carbon, nitrogen, sulfur, phosphorus, oxygen, and methane. These units are described in detail by Krapivin et al. (1982), Krapivin and Vilkova (1990), and Krapivin (1993). They are based on the balance relations between the flows of a given element α in the biosphere. Thus, if $\alpha_S(t)$ represents the α element content in media S at moment t, then the following relation is valid:

$$d\alpha_S(t) / dt = \sum_j H_{jS} - \sum_i H_{Si}$$

where H_{jS} and H_{Si} are the input and output flows of element α, respectively. Summation is performed over media i and j which interact with the S medium. Typical flow charts for various elements are given in Figs. 3.1–3.4.

3.3.1 Carbon Unit

Understanding of the basic regularities of carbon exchange in the biosphere, with due regard for spatial and temporal nonuniformity, is of great practical and theoretical importance. Recently, the role of investigations in this respect has grown due to the pressure of increasing human activity, since the basic characteristics of the carbon cycle in terrestrial ecosystems and the hydrometeorological conditions of the territory are most sensitive to the economic activity of humans, resulting in adverse effects.

The components of the small biological carbon cycle in the biosphere and the basic processors in it, such as photosynthesis, the formation of dead organic matter and its decomposition, have been well described in the literature (Williamson and Platt 1991; Vloedbeld and Leemans 1993; Zonneveld 1998; Bolin and Sukumar 2000) and may be taken into account in a global model.

For stationary systems, the amount of organic matter resulting from the annual dying of plant structures may be determined with a high degree of precision to be equal to the annual primary production and the soil-formation processes may be considered to be balanced. At the same time, growing ecosystems are characterized by the predominance of production over destruction. It is possible to take account of these differences and reflect all the natural diversity in the model only by introducing the territorial distribution of soil-plant formations (Apps et al. 1993; Turner et al. 1993; Watson et al. 2000; Kogan 2001). Every model is constructed in such a way as to specify the carbon cycle aspect under study more comprehensively.

The carbon exchange model in the *atmosphere-plant-soil* (APS) system takes into consideration spatial diversity by dividing the land surface into areas of $5° \times 4°$ of longitude and latitude, respectively. According to Bazilevich's classification (Table 3.8), each of these corresponds to one of the 30 ecosystems. Exchange processes are considered with regard to the effect of increased CO_2 content and the climatic change associated with it. This model makes it possible to define the parameters of the small biological-carbon cycle and to calculate their changes upon transition into new quasi-stationary states due to increasing or decreasing the average annual temperature and annual precipitation and doubling the CO_2 content in the atmosphere. The adequacy of this model has been studied within the framework of these scenarios.

The present paragraph considers the APS system together with the four-box ocean model. The four-box ocean model can be adequate provided so that the terrestrial-carbon-cycle model is given a more detailed elaboration. The exchange processes in the ocean also depend on the changes of CO_2 content in the atmosphere and the related change of temperature in the surface ocean. The calculations presented here have been carried out by using the carbon exchange model in the atmosphere-plant-soil-ocean (APSO) system. They enable one to follow the distribution dynamics of industrial CO_2 between the atmosphere, ocean and vegetation, as well as the response dynamics of biota in various latitudes and individual ecosystems to one of the strongest impacts on the carbon cycle – emission of industrial CO_2 and human interference in terrestrial vegetation.

Calculations of the effects of industrial CO_2 emission on the carbon cycle, beginning in 1860, have been calculated for various hypothetical amounts of its absorption by biota, disregarding forest cutting and burning. The paper of Kondratyev (2000b) indicates that until 1960 the carbon release resulting from the disturbance of vegetation and soil was greater than that from the combustion of fossil fuel. Conversely, Broecker et al. (1981) maintained that there was no special reason to believe that the terrestrial biomass had been decreasing at a rate comparable to that of the increase in fossil fuel combustion. Apparently, the processes of forest destruction and their regeneration are more or less balanced. This is supported by Stuiver's (1978) findings suggesting that from 1950 on the terrestrial biomass neither considerably increased nor decreased, as well as the conclusion of Kohlmaier et al. (1987) that the climax forest system does not markedly differ in biomass reserves as compared to the forest ensemble harvested at a constant age. Consideration of industrial CO_2 emission together with deforestation of land indicates that its future transformation would apparently change the carbon dynamics in reservoirs. Despite the fact that deforestation is recognized as the most dangerous factor of human activity this process increases at a growing rate (Turner et al. 1993; Hidayat 1994).

The proposed model of the global carbon cycle considers the atmosphere, ocean, and land as the main reservoirs (Fig. 3.11). The ocean is subdivided into two reservoirs: surface layer and deep water mass. Land surface in the model is divided into areas each with $\Delta\lambda = 5°$ in longitude and $\Delta\varphi = 4°$ in latitude. Each

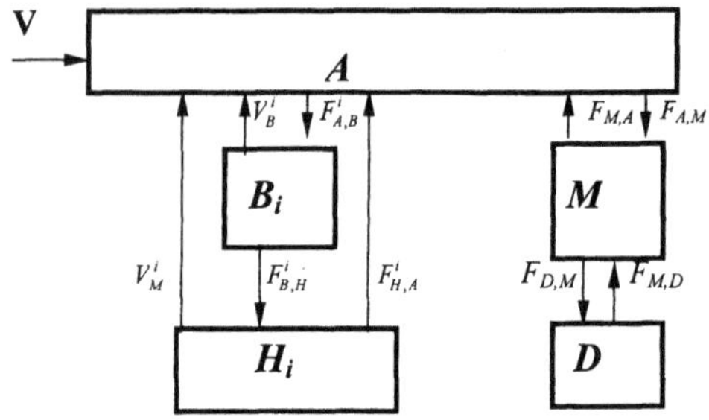

Fig. 3.11.
Diagram of the carbon cycle in the model of the "atmosphere-plant-soil-ocean" system

area belongs to one of the 30 types of ecosystems, according to the classification of Table 3.8, or has no vegetation (Antarctica, part of Greenland). They were referred to as a certain ecosystem on the basis of a diagrammatic map of terrestrial plant-zones. Land areas belonging to several types of ecosystems were referred to as the ecosystem with average annual temperature and total annual precipitation typical of a specific land area. The spatial distribution of the ecosystems is presented in Fig. 3.8.

The functioning of the land biota is thus described by a set of models for soil-plant formation. Since we regard the biota from the standpoint of CO_2 source and discharge, it is convenient to divide the biomass into living and dead. Only living biomass can act to discharge CO_2, whereas both living and dead biomass may serve as a source, but since primary production, rather than complete assimilation, appears to be related to sink intensity, the main CO_2 source is soil humus.

The model uses the following designations: A, carbon amount in the atmosphere; M and D, carbon amount in the surface and deep oceans, respectively; B_i and H_i, carbon content per unit of area in the biomass of living plants and dead organic matter of soil in the ith land area; T_i and W_i, average annual temperature and total precipitation; A^0, M^0, D^0, B^0, H^0, T^0, W^0, values of the respective parameters in the preindustrial epoch.

The carbon dynamics is described by the following system of equations:

$$
\begin{aligned}
\frac{dB_i}{dt} &= F_{A,B}^i - F_{B,H}^i - V_B^i, \qquad (i = 1,\ldots,N)\\
\frac{dH_i}{dt} &= F_{B,H}^i - F_{H,A}^i - V_H^i, \qquad (i = 1,\ldots,N)\\
\frac{dM}{dt} &= F_{A,M} - F_{M,A}\\
\frac{dD}{dt} &= F_{M,D} - F_{D,M}\\
\frac{dA}{dt} &= \sum_{i=1}^{N} (F_{H,A}^i - F_{A,B}^i + V_B^i + V_H^i)S_i - F_{A,M} + F_{M,A} + V
\end{aligned}
\tag{3.1}
$$

where V, $V^i{}_B$, $V^i{}_H$ are the carbon fluxes into the atmosphere resulting from industrial CO_2 emission and human impacts on vegetation and soil; S_i is the size of the ith land area; and N the number of land areas ($N = 752$).

The carbon fluxes in the ocean are described as follows:

$$F_{A,M} = K_{A,M}A,\ K_{A,M} = F^0_{A,M}/A^0,\ F_{D,M} = K_{D,M}D,\ K_{D,M} = 1/T_D,$$
$$F_{M,D} = K_{M,D}M,\ K_{M,D} = D^O/\left(M^O T_D\right), \tag{3.2}$$
$$F_{M,A} = K_{M,A}P\left(M,T\right),\ K_{M,A} = K_{M,A}A^O/P^{-1}\left(M^O, T^O\right)$$

where $P(M,T)$ is the carbon partial pressure in the surface ocean layer, which is a function of dissolved inorganic carbon content and temperature; T_D is the carbon lifetime in the deep ocean water mass.

To compute $P(M,T)$, we use a result obtained by the basis of data on the constants of carbonic acid dissociation for various temperatures: $P(M,T) = [CO_2]/\alpha_s$, where α_s is the coefficient of gas solubility in the seawater. The concentration of dissolved CO_2, $[CO_2]$ is obtained from the solution of equations:

$$C = (1 + kx + x^2)\,[CO_2],$$
$$a = (kx + 2x^2)[CO_2] + W(x),$$

where

$$C = [CO_2] + [HCO_3^-] + [CO_3^{2-}],\ x = (K_1K_2)^{1/2}/[H+],\ k = K_1/K_2,$$
$$a = [HCO_3^-] + 2[CO_3^{2-}] + [B(OH)_4^-] + [OH^-] - [H+],$$
$$W(x) = B_T/\{1 + (K_1K_2)^{1/2}x^{-1}/k_B\} + xk_W/(K_1K_2)^{1/2} - (K_1K_2)^{1/2}/x,$$
$$k_W = [H+][OH^-],\ B_T = [B(OH)_3] + [B(OH)_4^-],$$
$$k_B = [H+][B(OH)_4^-]/[B(OH)_3],$$
$$K_1 = [H+][HCO_3^-]/[CO_2],\ K_2 = [H+][CO_3^{2-}]/[HCO_3^-].$$

The regularity of the geographical distribution of annual production, reserves of phytomass, and dead organic matter points to their close relationship with climatic factors. Therefore, the terrestrial carbon fluxes under study are the functions of climatic factors, types of ecosystems and carbon concentration in the atmosphere:

$$F^i_{A,B} = F\left(T_i, W_i\right)\left(l + \beta\left(A/A^O - l\right)\right)$$
$$F^i_{B,H} = K_iB_i,$$
$$K_i = F\left(T_i^O, W_i^O\right)/B_i^O \tag{3.3}$$
$$F^i_{H,A} = E_i\left(T_i, W_i\right)H_i,$$
$$E_i\left(T_i^O, W_i^O\right) = F\left(T_i^O, W_i^O\right)/H^O\left(T_i^O, W_i^O\right)$$

where function $F(T_i, W_i)$ represents the annual production of plants per unit area computed in terms of carbon; $E(T_i, W_i)$ is the specific decomposition rate of dead organic matter in soil; and coefficient K_i is the specific die-off rate of plant biomass.

All the components of the carbon cycle in the APS system are determined for each land area (Fig. 3.8). The results of estimating the carbon cycle components for all the land areas for a given year are recorded in a file, from which the changing values are used as an entry for the subsequent year.

Data on average annual temperature and total annual precipitation, on the one hand, and primary and humus production for the basic terrestrial soil-plant formations, on the other hand, help to establish a quantitative dependence of the latter on the former. Walter's ratio and Van't Hoff's rule appear to be one of the first of such relationships obtained experimentally; however, they have a limited range of adequacy. Lieth's formula of primary production is affected in a limiting way by Liebig's principle incorporated in the indicated formula. A regression formula for the primary production as a polynomial to the third power in T and W was obtained from the data of Tables 3.9 and 3.10 by Vilkova et al. (1998) using the least-squares technique:

$$F(T,W) = a_1T^3 + a_2W^3 + a_3T^2W + a_4TW^2 + a_5T^2 + a_6W^2 + a_7TW + a_8T + a_9W + a_{10},$$

$$H^O(T,W) = b_1T^3 + b_2W^3 + b_3WT^2 + b_4TW^2 + b_5T^2 + b_{6W}^2 + b_7TW + b_8T + b_9W + b_{10},$$

$$B^O(T,W) = c_1T^3 + c_2W^3 + c_3WT^2 + c_4TW^2 + c_5T^2 + c_6W^2 + c_7TW + c_8T + c_9W + c_{10}$$

where coefficients $\{a_i, b_i, c_i\}$ are given in Table 3.11. This gives a poor prediction in those regions where either temperature or precipitation appear to be a limiting factor. All the aforementioned relationships formed the basis for the relationship $F(T,W)$ (Table 3.9).

In small regions, the annual production is considerably influenced by the local features of the ground. Therefore, when the annual production on a limited land area is to be determined, it is necessary to introduce its fertility factor into the formulae. By using the dependence $F(T,W)$, we can compute the average

Table 3.9. Dependence of annual production F (kg m^{-2} year^{-1}) on average annual temperature T and total annual rainfall W

W (mm year^{-1})	T (°C) -10	-6	-2	2	6	10	14	18	22	26
3125						3.4	3.5	3.7	3.8	3.9
2875						3.2	3.3	3.5	3.6	3.7
2625						3.0	3.2	3.3	3.4	3.5
2375						2.8	2.9	3.0	3.1	3.2
2125						2.5	2.6	2.7	2.9	2.9
1875					1.6	2.3	2.3	2.4	2.5	2.6
1625			0.4	0.6	1.3	2.0	2.1	2.1	2.2	2.3
1375	0.2	0.3	0.4	0.7	1.1	1.7	1.9	1.9	2.1	2.1
1125	0.3	0.3	0.4	0.6	1.0	1.6	1.8	1.9	1.8	1.8
875	0.3	0.4	0.6	0.8	0.9	1.5	1.4	1.3	1.3	1.2
625	0.3	0.5	0.6	0.9	0.9	0.9	0.8	0.7	0.7	0.7
375	0.4	0.5	0.7	0.6	0.6	0.6	0.5	0.5	0.5	0.4
125	0.3	0.3	0.2	0.2	0.2	0.2	0.2	0.2	0.1	0.1

Table 3.10. Dependence of humus amount H (kg · m^{-2}) in a 100-cm layer on average annual temperature T and total annual rainfall W: $H(T, W)$

W (mm year^{-1})	T (°C) -10	-6	-2	2	6	10	14	14	22	26
250	6.1	7.5	11.0		13.5	14.3	10.1	5.2	5.2	1.1
500	6.1	7.4	9.1		19.1	58.3	45.1	23.3	23.3	14.3
750	5.5	6.0	6.6		11.0	21.5	35.1	25.1	25.1	23.1
1000	5.0	5.0	6.5		6.1	16.9	25.8	24.1	24.1	22.5
1250	–	–	6.0		6.1	16.2	23.2	23.0	23.0	21.7
1500	–	–	–		–	15.6	22.8	22.8	22.8	21.6
1750	–	–	–		–	–	22.8	22.7	22.7	21.6
2000	–	–	–		–	–	22.7	22.7	22.7	21.5
2250	–		–		–	–	22.6	22.6	22.6	21.5
2500	–	–	–		–	–	22.5	22.5	22.5	21.3
2750	–	–	–		–	–	22.4	22.4	22.4	21.2
3000	–	–	–		–	–	21.9	21.8	21.8	21.1

Table 3.11. Coefficients of correlations $F(T,W)$, $H(T,W)$ and $B(T,W)$

i	a_i	b_i	c_i
1	4.25×10^{-4}	-5.16×10^{-2}	-9.02×10^{-3}
2	-8.76	-161.39	225.79
3	-1.99×10^{-2}	-9.41×10^{-2}	1.11
4	4.29	6.79	-29.39
5	2.29×10^{-2}	-9.47×10^{-1}	-5.87
6	19.05	199.51	-511.72
7	-8.79×10^{-2}	-4.37	41.29
8	4.56×10^{-1}	7.47×10^{-2}	-11.37
9	-14.16	-44. 17	356.97
10	4.18	4.93	-62.94

annual production for a sufficiently large land area. In this case the local features of the ground, which cause deviations from the mean production index, are usually equalized and the fertility factor practically becomes equal to one. Soil fertility, the amount of phytomass and other factors are implicitly incorporated into the relationship $F(T, W)$, since the data on annual production considered for the corresponding climatic variables depend on these factors.

Each ecosystem is characterized by a certain mean factor of biomass growth β_j which is hard to estimate due to scarce experimental data. For the short-term effect of higher CO_2 content in the atmosphere, Kohlmaier et al.(1987) provide experimental values for certain plant species which range from 0.19 to 1.03. Depending on model structure, various hypotheses on the photosynthesizing properties of terrestrial vegetation and with various carbon-cycle parameters, in a number of studies the value of β varies over a wide range from 0.005 to 0.3. In our model, $\beta = 0.173$ for all land areas. The factor of biomass increase β and the depth of the ocean surface layer have been selected in such a manner that the model can simultaneously predict Suess' effect and the atmospheric part. The preindustrial CO_2 level was taken to be 290 ppm.

There are many models of the CO_2 cycle in the biosphere. The GSM has two such subunits. The block diagram of second unit is represented in Figs. 3.1 and 3.12. A complete global cycle of carbon in the biosphere consists of two different cycles, terrestrial and oceanic. Of all the carbon-containing gases in the atmosphere, only CO_2 is present in a sufficiently high concentration; therefore we will consider, as before, that the carbon cycle is determined only by the CO_2 cycle. Since the atmosphere is rapidly mixed, it is regarded as an integral CO_2 reservoir. In compliance with the regional division of the land, accepted in the present work, we will consider in each region three internal uniform reservoirs characterized by forest, agricultural and grass vegetation. In the ocean, we will deal with biota and detritus. In conformity with Fig. 3.1 we will differentiate four layers in the ocean. Figure 3.1 exhibits a general diagram of CO_2 flows.

The model is described by the following equations:

$$\frac{dC_A}{dt} = H_5^C + \sum_{k=1}^{m}(H_{2,k}^C - H_{3,k}^C)\frac{\sigma_{o,k}}{\sigma} + \sum_{i=1}^{n}(H_{1,i}^C - H_{4,i}^C + H_{7,i}^C + H_{8,i}^C +$$
$$H_{9,i}^C + H_{10,i}^C + H_{11,i}^C - H_{21,i}^C)\sigma_i / \sigma$$
$$\frac{dC_{U,k}}{dt} = H_{19,k}^C - H_{20,k}^C - H_{6,k}^C + H_{22,k}^C + H_{18,k}^C$$
$$\frac{dC_{L,k}}{dt} = \Delta H_{20,k}^C + H_{18,k}^C + \Delta H_{19,k}^C - H_{16,k}^C + H_{23,k}^C$$
$$\frac{dC_{S,i}}{dt} = H_{12,i}^C + H_{13,i}^C + H_{14,i}^C + H_{15,i}^C + H_{24,i}^C + H_{4,i}^C - H_{8,i}^C - H_{7,i}^C - H_{9,i}^C - H_{22,i}^C$$
$$\frac{dC_{D,k}}{dt} = \Delta H_{20,k}^C + H_{18,k}^C + \Delta H_{19,k}^C$$
$$\frac{dC_{B,k}}{dt} = H_{20,k}^C - H_{19,k}^C - H_{16,k}^C + H_{18,k}^C + H_{17,k}^C$$

where σ is the area of the biosphere, σ_i is the area of the ith land region and $\sigma_{o,k}$ is the kth water area of the ocean.

Using this version of GSM, we will try to assess the greenhouse effect caused by emissions of industrial CO_2 in the atmosphere. The greenhouse effect problem involves several critical scientific, technical and political questions. There are many discussions of this problem (Goudriaan et al. 1990; Kondratyev 1992, 1998a, 1999a, 2000a; Adamenko and Kondratyev 1999; Demirchian and Kondratyev 1999; Bolin and Sukumar 2000; Kondratyev and Demirchian 2000; Watson et al. 2000). The main opinion of many experts is that an objective estimation of the consequences of human activity concerning CO_2 emissions does not exist. The special report of the International Panel on Climate Change (Watson et al. 2000) and the Kyoto Protocol are based on rather contradictory data. Many recent studies show that the greenhouse effect is mainly a natural phenomenon. The anthropogenic component of the greenhouse effect changes at the present time is still comparatively small. That is why understanding the basic regularities of carbon exchange in the NS system with due regard for spatial and temporal non-uniformity is of great practical and theoretical importance.

Recently, the role of investigations in this respect has grown due to the pressure of increasing human activity, since the basic characteristics of the carbon cycle in terrestrial ecosystems and hydrometeorological conditions of the territory are most sensitive to the economic activity of humans, resulting in adverse effects. It is important to select key questions relating to the exchange of carbon between the atmosphere, terrestrial biosphere and ocean ecosystems. It is important to examine how carbon flows between different pools and how carbon stocks change in response to ocean pollution, land-use activities and other anthropogenic processes. The principal task is the assessment of the role of various ecosystems in these flows. Many experts attach great importance to the Arctic ecosystems dynamics (Riedlinger and Preller 1991; Kelley and Gosink 1992; Kelley et al. 1992a, 1999; Legendre and Krapivin 1992; Krapivin 1995, 1999; Phillips et al. 1997; Legendre and Legendre 1998; Kondratyev and Varostos 2000; Krapivin and Phillips 2001a).

Studies of the climate-forming processes in the Arctic, whose principal feature is determined by the isolation of the ocean from the atmosphere by the ice cover, have long been of particular concern. The most important impacts of sea ice on the climate, as revealed through numerical modeling, are the following: (1) maximum climate warming with increasing CO_2 concentration in the wintertime Arctic due to increased heat input from the ocean through the thinner ice as a result of the warming; (2) the effect of albedo of the more extended sea ice 18,000 years ago on the atmospheric temperature, which is compatible with the impact of continental glaciers; (3) the possible reversal of the conventional relationship between the amplitude of the annual change of temperature and the depth of the oceanic mixed layer when the sea ice dynamics are taken into account. A thinner mixed layer favors the strengthening of the wintertime sea ice, which causes a delay of the springtime melting and produces a colder summer. The latest data of numerical modeling have confirmed, on the whole, these conclusions (Kondratyev and Johannessen 1993).

When discussing the problem of the role of the polar regions in the formation of global changes, one should have in mind that two aspects of the global ecology are of paramount importance: (1) the anthropogenically induced redistribution of the components of the heat balance of the Earth as a planet (with emphasis on the problem of the atmospheric greenhouse effect and its climatic impact); (2) anthropogenically induced breaking of the global biogeochemical cycles (primarily, this refers to carbon, nitrogen, and sulfur). That is why polar regions are very specific as a component of the global ecosystem. These aspects are not taken into account completely by the existing global models of the Nature/Society system. A new approach to the synthesis of geoinformation monitoring systems, proposed by Kondratyev et al. (2000), overcomes this shortfall. In the framework of this approach the interchange of CO_2 between high latitude vegetation, the North Ocean and the atmosphere is considered as part of a global biogeochemical cycle described within the framework of a GSM. The GSM development will provide effective technology for the assessment of climate change.

The GSM makes it possible to compute the dynamics of the industrial CO_2 distribution between the oceans, terrestrial biota and the atmosphere. The GSM

Fig. 3.12.
Integrated flow chart of the CO_2 flows in the land ecosystems

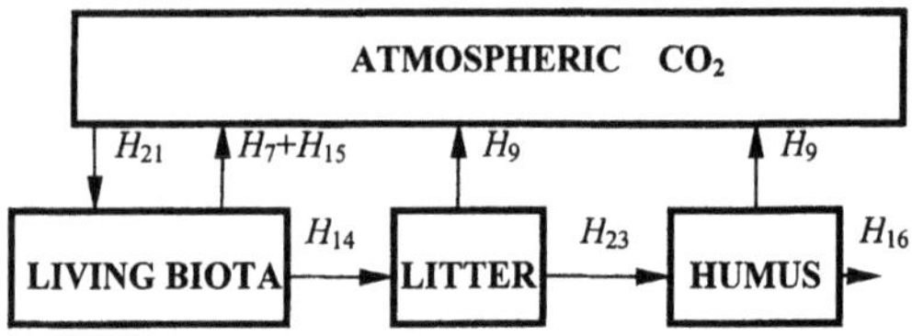

describes the World Ocean by a spatial four-layer model with due regard for water chemistry. Spatial inhomogeneity of the World Ocean is represented by the structural distribution of surface temperature applied to the upwelling and convergence zones. Sea ice in the polar regions is considered through use of CIESIN data. The biogeochemical cycle of CO_2 is described by balance equations in accordance with Figs. 3.1 and 3.12 The GSM CO_2 unit takes into consideration the dependence of flows H_2 and H_3 from water surface processes (wind-wave mixing, rough seas, foaming waves). Simulation experiments have shown that these flows have variations from 16 to 1250 mol m^{-2} $year^{-1}$. In the Arctic Ocean during the June – September period, the partial pressure p_a of CO_2 in the atmosphere exceeds the partial pressure of CO_2 in the seawaters by $20 \div 110$ ppm. These variations in the partial pressure of CO_2 have specific distributions for the arctic seas. For instance, the Norwegian Sea and Bering Sea have CO_2 deficits of $18 \div 54$ ppm and $33 \div 69$ ppm, respectively. The average CO_2 deficit reaches 450 g C m^{-2}. The flow H_3 changes between 1.5 and 4.1 g C m^{-2} day^{-1}. In addition, linear correlations between the partial pressure of CO_2 and the water temperature T are observed, with a proportionality coefficient equal to 9.8 ppm °C^{-1}.

The flow H_3 is calculated by the formula

$$H_3 = \psi(T)(p_a)^{-1/2}(1 + 0.5\,S),$$

where S (‰) is the water salinity, Weis's function ψ describes the influence of the water temperature T (K) on the CO_2 solubility in the seawater. The carbon flows H_{22} between land and ocean are defined by the functions of runoff Q_U and a model of atmospheric transport. $Q_U = 0$ is taken for the World Ocean pelagic areas. For shelf water areas Q_U is calculated on the assumption of uniform distribution of the runoff from the Kth land area into the Mth water area.

The terrestrial carbon fluxes under study are the functions of climatic factors, types of ecosystems and carbon concentration in the atmosphere. The annual primary production of plants per unit of area computed in terms of carbon is represented in the form of Table 3.9. In small regions the annual production is considerably influenced by the local features of the ground. Therefore, when the annual production on a limited land area is to be determined, it is necessary to introduce its fertility factor into the formulae. By using the dependence F we can compute the average annual production for a sufficiently large land area. In this case the local features of the ground, which cause deviations from the mean production index, are usually equalized and the fertility factor, practically, becomes equal to one. Soil fertility, the amount of phytomass and other factors are implicitly incorporated into relationship F, since the data considered on annual production for the corresponding climatic variables depend on these factors. Dependence of humus amount from rainfall and temperature is based on the data on humus reserves in various ecosystems and the data on soil types and ecological boundaries.

The oceanic part of the global carbon cycle is represented by three structures having a specific trophic pyramid: (1) pelagic tropical latitudes (long trophic chains), (2) tropical-latitude shelf zones and mid-latitude aquatic zones (medium-length trophic chains), and (3) arctic latitudes (short trophic chains).

The problem of the greenhouse effect is discussed by many authors within the framework of different anthropogenic scenarios (Kondratyev 1998a; Gorshkov et al. 2000). The main conclusion is that global climate change brought about through CO_2 dynamics will be insignificant during the next century if: (1) the world ocean pollution, especially by oil products, does not exceed the level of 1990 by 10%; (2) agricultural land areas do not expand at the expense of forests; (3) the rate of fossil fuel consumption remains at the level of 1990 with a dispersion of 15%; and (4) alternative energy sources (atomic, wind, etc.) are developed at a rate which does not hinder the food production.

The role of soil-plant formation for the absorption of excess atmospheric CO_2 under the above scenario is estimated with the data displayed in a geographic grid of 4° × 5° lat./long. (see Fig. 3.8). The role of the World Ocean is considered by taking into account the water temperature in the surface layer. It has been shown that the atmospheric CO_2 concentration can reach a mixing level of 556.7 ppm during the twenty-first century. The dynamics of industrial CO_2 distribution between the atmosphere, oceans and vegetation will fluctuate with an amplitude less than 25%. With an increase in industrial CO_2 emission over the period from 1990 to 2090, the atmospheric CO_2 portion will rise and the contribution of the oceans to its absorption will increase; whereas that of the biota will stabilize after going through a small maximum. At the end of the 21st century and beginning of the twenty-second century, during the highest level of human economic activity, the contribution of the oceans (especially the Arctic Basin) to absorption of industrial CO_2 will be considerably higher than that of vegetation. This is due to the fact that, with an increase in CO_2 concentration in the atmosphere, the ability of the upper ocean layers to absorb industrial CO_2 will be enhanced by the transformation of biogeochemical processes in the deep ocean layers. The restoration of plant cover and reduction of ocean pollution are the main near-time problems (Watson et al. 2000). For example, it was shown that if the natural/disturbed land relation changes from 2/3 in 1990 to 3/2 in 2050 then the atmospheric CO_2 concentration would reach no more 497.3 ppm during the twenty-first century. This illustrates that the role of the biospheric system in global change needs to be investigated more thoroughly (Krapivin and Vilkova 1990; Gorshkov and Makarieva 1999; Krapivin 1999).

Simulation results show that there is an overall exaggeration of the importance of the problem of global climate change caused by anthropogenic CO_2 emissions. Thus, the existing arguments for climate change are not reliable enough. It is necessary to develop the GSM by inserting new correlations between the elements of the NS system and taking into account the biotic regulation processes.

This study showed that the role of the Arctic Ocean in the global CO_2 balance was estimated with low precision. Therefore the Kyoto Protocol, solving the problems of climate warming via reduction of emissions of greenhouse gases,

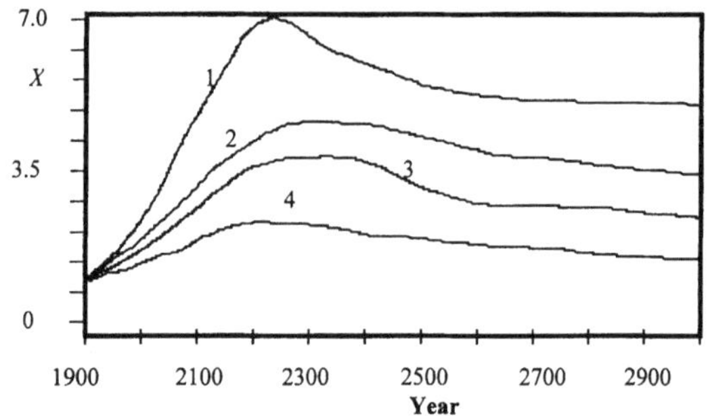

Fig. 3.13. Forecasts of the CO_2 content in the atmosphere obtained under different scenarios for anthropogenic impacts: *1* Keeling and Bacastow pessimistic scenario; *2* Bjorkstrom optimistic scenario; *3* IPCC scenario; *4* Kondratyev realistic scenario. The ordinate $X = C_a(t)/C_a(1900)$

suffers from a lack of objective reasons. In fact, existing climate models do not reliably simulate the correlation of global temperature with anthropogenic emissions of greenhouse gases. These models do not take into consideration the role of biospheric processes in the land and oceans (biotic regulation of the environment and the functioning of the ocean chemical system). The Millennium Ecosystems Assessment Program can solve the problem of global model synthesis in order to obtain forecasts of the dynamics of the NS system.

A summary of existing simulations of the global CO_2 cycle shows the existence of long-term studies concerning this problem. The main conclusion is that the exchange of carbon between the atmosphere and biosphere has spotty spatial characteristics. This spottiness, taken into account in the models of the CO_2 cycle, produces high errors in assessments of mean values. That is why the Kyoto Protocol conclusions were based on incomplete estimations. There are many carbon flows (natural and anthropogenic), mentioned in Fig. 3.1, which have not been satisfactorily described parametrically. The principal key questions relating to the exchange of carbon between the atmosphere and the terrestrial pool of the above-ground biomass, below-ground biomass, soils and hydrospheric systems are discussed by Martchuk and Kondratyev (1992), Kondratyev (1999c, 2000b), Gorshkov et al. (2000) and Watson et al. (2000). The Arctic basin is the most weakly studied part of the biosphere in order to understand its global role in the CO_2 absorption processes. Figure 3.13 demonstrates the high variability in the forecasting of the CO_2 dynamics under different anthropogenic scenarios: (1) the pessimistic scenario of Keeling and Bacastow (1977) describes a situation where the ocean's role in the carbon exchange with the atmosphere is restricted only to physical processes; (2) the optimistic scenario of Bjorkstrom (1979) takes into account the ocean-carbonate system under the parameterization of the H_2 and H_3 flows in Fig. 3.1; (3) the scenario of the IPCC, Intergovernmental Panel on Climate Change (Watson et al. 2000), is based on specific activity requirements concerning land-use strategy (e.g., planting versus regeneration through silvicultural activities); (4) the realistic scenario of Kondratyev (1999d) foresees the existing tendencies in the world's energetic, demographic and urbanization processes and is realized by the GSM.

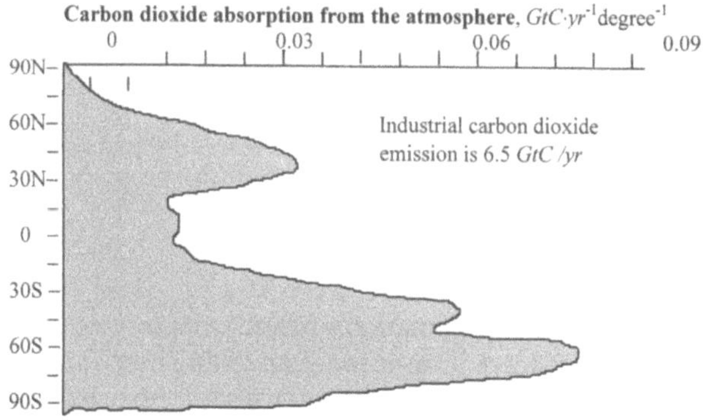

Fig. 3.14. The longitude averaged rate of industrial CO_2 absorption calculated by means of the GSM. The list of the soil-plant formations taken into consideration under the calculations is given in Table 3.8. The spatial distribution of soil-plant formations in the biosphere is given in Fig. 3.8. Table 3.12 demonstrates the vegetation role on Russian territory

Figure 3.14 gives the spatial distribution of the assimilated CO_2 from the atmosphere under the Kondratyev realistic scenario when the annual industrial carbon emission is 6.5 Gt C year^{-1} (Table 3.12). The distribution between the atmosphere, oceans and land vegetation is characterized by 0.6, 3.2 and 2.7 Gt C, respectively. An increase in the industrial CO_2 emission causes quasi-linear growth in the absorbing ability of the oceanic and land ecosystems. This growth is a complex function of $X = C_a(t)/C_a(1900)$, temperature and soil quality. The land vegetation photosynthesis is stimulated by an increase of ambient CO_2 only as much as $X \approx 3$. When $X > 3$ the saturation of photosynthesis is reached and the absorbing ability of the land ecosystems is stabilized.

It is obvious that in the framework of the fourth scenario a greenhouse effect via CO_2 is problematic. To estimate the dynamics of the atmospheric temperature dependence on CO_2 it is necessary to use interrelationships describing the greenhouse effect. There are many various empirical functional representations of the warming effects (Krapivin and Vilkova 1990; Krapivin 1999). The following simple correlation approximates the existing empirical dependence of the

Table 3.12. A model estimation of the surplus CO_2 absorption by vegetation on Russian territory. The anthropogenic emission of carbon is assumed to be 1.9 Gt C year^{-1} (global carbon emission is 6.5 Gt C year^{-1})

Soil-plant formation	Carbon flow absorbed by vegetation (10^6 t C year^{-1})
Arctic deserts and tundra, subarctic grasslands and marshes	2.2
Tundra, mountain- and forest-tundra	9.7
North-taiga forests	10.8
Mid-taiga forests	31.2
South-taiga forests	22.9

atmospheric temperature deviation ΔT_{CO_2} upon the variations in the atmospheric CO_2 parameter X:

$$\Delta T_{CO_2} = \begin{cases} L_1 & when \quad X \geq 1 \\ L_2 & when \quad X < 1 \end{cases} \tag{3.4}$$

where

$$L_1 = -0.847 + 4.528 \ln X - 1.25 \exp\{-0.82(X-1)\};$$
$$L_2 = -2.63\, X^2 + 6.27\, X + 1.509 \ln X - 3.988.$$

Finally, according to the GSM calculations, future emissions of CO_2 will commit Earth to a warm-up by 0.2–1.4 °C over the year 1900. These variations in ΔT_{CO_2} depend upon the assumptions about the intensity of the urbanization process and about land-use strategy. The calculation results show that the dynamics of industrial CO_2 distribution between the atmosphere and oceans changes with the increasing preponderance of flows H_3 in the northern aquatories. At the end of the twenty-first century and beginning of the twenty-second century, during the highest pressure of human economic activity, the contribution of the oceans to industrial CO_2 absorption will be considerably lower than that of vegetation. This is due to the fact that, with an increase in CO_2 concentration in the atmosphere, the partial pressure in the oceans rises and its ability to absorb industrial CO_2 decreases, whereas the productivity of vegetation does not fall. By the end of the twenty-second century, with a projected decrease in human impact, the contribution of the oceans increases due to the growing role of its deep layers. The atmospheric CO_2 will decrease at the same rate as the contribution of the oceans increases.

The entire area of the Arctic ocean is only 3.8% of the world's ocean surface, but its role in the CO_2 absorption varies from 23 to 38% (16.7 ÷ 28.9 Gt C year^{-1}). This role is greatly influenced by seasonal variations in the ice cover and in the ecosystem productivity. That is why the results of this study have a preliminary character. It is necessary to continue better definition of the parametrical descriptions of flows H_i from Fig. 3.1. It is clear that policy-oriented computer tools aimed at supporting decision making processes related to global change require the design of a new methodology in global modeling based on simulation models with detailed spatial descriptions of the biosphere systems and with operative informational service defining the parameters.

A final conclusion from this study is that existing statements about the anthropogenic cause of the CO_2 growth in the atmosphere are to be further considered. The Earth's climatic system has a complex character described by a set of feedbacks between its components. The level of investigation of these feedbacks is insufficient to make objective conclusions about atmospheric warming caused by emissions of greenhouse gases. The primitive interpretation of "global warming" as the general atmospheric temperature increase does not reflect the real processes in the Earth's climatic system (Kondratyev 1990, 1992; Adamenko and Kondratyev 1999; Kondratyev and Fedchenko 2001).

The study of long-term evolution of the dynamics of the Earth's system is realized by means of different models of the atmospheric circulation. The main drawback of these models lies in the inadequate consideration of biological processes. It is obvious that the spatial and temporal dynamics of the land and ocean biosphere have an important role in the interactions between all major components of the Earth's system. There is a set of Earth-System models taking into consideration the dynamic processes between its subsystems, the geosphere and the biosphere. The progress achieved in modeling the separate elements of the geosphere and the biosphere indicates the significance of these models. There are models focusing on atmospheric and ocean circulation, and on land-vegetation and ice-sheet dynamics. This approach has a principal restriction caused by unstable development of the circulation processes. It is evident that progress in global change modeling is expected by means of integrated models using combinations of parametrical descriptions and scenarios.

The GSM, Potsdam Earth System Modelling (POEM), Climate and Biosphere Model (CLIMBER) and the Simple Biosphere model (SiB2) are perspective models (Sellers et al. 1996; Boysen 2000) as regards the objective assessment of the global perspective.

A realistic approach was proposed by Kondratyev and Krapivin (2001b) whereby the global modeling procedure is connected with the global geoinformation monitoring system (GIMS). This approach makes possible the combination of theoretical descriptions of the natural and anthropogenic processes with observation data and the use of complex criteria to assess the state of the NS system. In the framework of the GIMS, an adaptive control of the environmental state is realized by coordinating the processes of the modeling and measurements. Progress lies in this strategy. A "global warming" pre-conception cannot be a reference point for future studies of the global carbon cycle (Soros 2000; Tol 2000).

3.3.2 Nitrogen Unit

According to the diagram of fluxes in Fig. 3.2 this unit takes into account the following processes: atmospheric fixation, denitrification, biological fixation on the land and in the oceans, nitrogen flux from geospheric sources to the atmosphere, turbulent mixing in the oceans, and anthropogenic impacts on the nitrogen cycle. Considering the designations in Fig. 3.2 and in Table 3.2, the synthesis of the integral model for the nitrogen cycle in the biosphere is reduced to a system of equations which has the same form of balance correlations as in the carbon unit. Let us describe in more detail the components of the nitrogen cycle with this unit serving as an example. A similar approach was used in selection of functional correlations in the subsequent two units.

3.3.2.1 Atmospheric Fixation

As a result of various physico-chemical processes in the atmosphere, free nitrogen can pass from the atmosphere into the soil and water reservoirs. The fixation of atmospheric nitrogen due to electrical charges and photochemical processes does not amount to more than 0.035 t km^{-2}. According to other estimates these values can be as low as 0.027 t km^{-2} for land and 0.01 t km^{-2} for the ocean. Since nitrogen flows, due to atmospheric fixation, depend to a large extent on meteorological conditions, it is natural to consider them independently for each region in the form of temperature functions and precipitation. The flow H^N_2 of nitrogen fixed in the atmosphere is described in the ocean by the following equation:

$$H^N_{4,k} = (n_1\theta_1^{\Delta T_{ok}} + n_2 R_W) N_A$$

where ΔT_{ok} is the deviation of the water temperature in the kth ocean water area, θ_1 is the index of temperature dependence for the rate of atmospheric nitrogen fixation, R_W is atmospheric precipitation, and n_1 and n_2 are experimentally determined coefficients. The value of R_W will be calculated from the formula:

$$R_W = W_{OAO} + (\sigma_i / \sigma_O) \sum_{j=1}^{n} W_{jAO} ,$$

where W_{OAO} is precipitation of local origin, and W_{jAO} is precipitation due to moisture brought from the jth land region.

Similarly, let us write down the equation for the process of atmospheric fixation over the ith land region:

$$H^N_{4,i} = (n_3\theta_2^{\Delta T_i} + n_4 R_A) N_A$$

where W_{jAi} is the precipitation due to moisture brought from jth region, W_{OAi} is the total precipitation due to oceanic moisture, and

$$R_A = (\sigma_{Ok} W_{OAi} - \sum_{j=1}^{n} \sigma_j W_{jAi}) / \sigma_i$$

3.3.2.2
Denitrification

Denitrification processes on the land and in the ocean H^N_9 appear to be important channels of nitrogen flux to the atmosphere. The intensity of these processes is dependent upon temperature, humidity and pH. The quantitative and functional characteristics of such dependences have been studied quite well (Sofiev and Galperin 1998; Wyers and Erisman 1998; Nielsen 1999; Lin et al. 2000; Pauer and Auer 2000). Within the scope of the global model, it is possible to consider only the factors of temperature and humidity. Therefore, let us assume the following approximations for denitrification processes:

$$H^N_{9,k} = n_5\theta_3^{\Delta T_k} N_{Uk} \qquad H^N_{9,i} = n_6\theta_4^{\Delta T_i} \frac{N_{Si}}{n_7 + N_{Si}} W_{Si}$$

where W_{si} is the soil moisture, and n_5, n_6, n_7, θ_3, and θ_4 are coefficients.

3.3.2.3
Biological Fixation

The processes of nitrogen fixation by microorganisms and plants play an important part in the biogeochemical cycle. Biological nitrogen fixation amounts to at least 148 million tons per year on the land and 10 million t/year in the ocean.

The fixation rate, depending on the environment, may vary over a wide range, reaching 3 billion tons per year on the land in highly productive regions and 20.7 million tons per year in the photic zone of the ocean.

The nitrogen flux $H^N_{5,i}$ depends upon the distribution of vegetation cover on the land and may be described by the following equation:

$$H^N_{5,i} = \sum_{i=1}^{n} \left(H_{xi} + H_{yi} + H_{li}\right)$$

where $H_{Xi} = n_8 R_{Xi}$, $H_{Yi} = n_9 R_{Yi}$, $H_{li} = n_{10} R_{Li}$; and the coefficients n_8, n_9 and n_{10} are determined from the experimental data.

Nitrogen fixation by plants directly from the soil through root systems (flux $H^N_{3,i}$ also plays an important part in the nitrogen cycle, especially on the areas under agricultural use. This process may be included in the model in the following way:

$$H^N_{3,i} = \sum_{i=1}^{n} \left(d_1 R_{Xi} + d_2 R_{Li} + d_3 R_{Yi}\right)$$

where d_i (i = 1,2,3) are the coefficients.

The process of nitrogen fixation in water ecosystems has been studied insufficiently and the available quantitative data on this process are rather approximate. Thus, in the photic zone of small lakes the rate of nitrogen fixation amounts to $(36 \div 1800) \cdot 10^4$ t km^{-3} year^{-1}. For the ocean, the rate averages 0.0277 t km^{-2} year^{-1}.

3.3.2.4
Nitrogen Loss due to Drainage from the Soil

On the global scale, the ways of nitrogen migration into the biosphere include transfer of nitrous compounds between land regions, as well as between the land and ocean due to river runoff. Annual nitrogen flux from the land into the ocean is estimated at 38.6 million tons. Let us assume that the rate of the total runoff to the ocean from the *i*th land region is described by the function W_{si}. Then the nitrogen flux $H^N_{11,i}$ may be approximated by the following equation:

$$H^N_{11,i} = \rho N_{si}\left[1 - \exp(-k W_{Si})\right]$$

This dependence provides for zero nitrogen flow from the land to the ocean if there is no runoff and stabilization at the level ρ when the volume of water drainage to the ocean drastically increases.

3.3.2.5
Nitrogen Flux to the Atmosphere from Geospheric Sources

The nitrogen flow H^N_1 is dependent upon the geothermal activity of the Earth. Let us take into account nitrogen flux to the atmosphere with volcanic gases. It is known, for instance, that the gases emitted by Mount Vesuvius volcano contained 96% of nitrogen by weight, while the lava gases of the Hawaiian volcanoes contain only 5.7% of nitrogen. Overall, the flux of volcanic nitrogen on the worldwide scale is estimated at 0.4 million t/year.

Let us consider the intensity of flow H_1^N in the model to be a function of time in the general case. The dependence $H_1^N(t)$ may be obtained from the data bases of the World Resources Institute data.

3.3.2.6
Nitrogen Flows in the Ocean

Alongside the processes considered above there are also nitrogen transformation processes taking place in the ocean, such as replenishment due to decomposition of detritus and vital activity of living organisms, and nitrogen exchange between the upper photic layer of the ocean and the deep ocean. Though these processes have not been studied sufficiently well, it is possible to consider them within the scope of the given model and to suggest the following functions for the description of these flows.

Nitrogen reserves in the ocean are supplemented through bacterial decomposition of organic residues and soluble organic substances. The model deals with the component *detritus D;* therefore, let us assume $H_{12,k}^N = \lambda k_U G_k$ where D_k are the reserves of detritus in the ocean, t km^{-2}; k_U is the rate of detritus decomposition in the ocean, year^{-1}; and λ is the coefficient characterizing the nitrogen content in the biomass of detritus.

The reserves of free nitrogen in the ocean are replenished also in the process of vital activity of the various organisms. These processes will be considered only in the photic zone of the ocean. Since the global model considers nekton r and phytoplankton Φ, then $H_{10,k}^N = a_1 T_{rk} + a_2 T_{\Phi k}$ where T_{rk} and $T_{\Phi k}$ are the characteristics of the metabolic processes in nekton and phytoplankton, respectively, and a_1 and a_2 are the coefficients of proportionality. Nitrogen exchange between the upper and lower layers in the ocean occurs at the rates $H_{14,k}^N$ and $H_{15,k}^N$. Nitrogen returns to the mixed layer due to rising of deep waters as a result of turbulence, turbulent diffusion and vertical convection on the average at the rate of 10^{-2} – 10^{-3} cm s^{-1}. On the contrary, the deep ocean is replenished with nitrogen due to the processes of convergence and gravitation settling. The processes of nitrogen transfer by migrating animals may be considered counterbalanced, hence they can be disregarded. Let us assume $H_{14,k}^N = b_1 N_{Uk}$, $H_{15,k}^N = b_2 N_{Lk}$.

3.3.2.7
Anthropogenic Processes of Significance for the Nitrogen Cycle

At present, the contribution of human activity to the total biospheric nitrogen cycle has become tangible, according to the estimates of many authors, though the reserve of nitrogen cycle stability is still high and in the next few centuries no nitrogen starvation threatens humanity. Nevertheless, the outlined tendencies of human intervention into biospheric processes may lead, by and large, to unpredictable consequences. For instance, the technological accumulation of nitrogen from the biosphere under conditions of fertilizer production, on the one hand, plays a positive role in increasing the productivity of terrestrial and water ecosystems and, on the other hand, causes undesirable eutrophication of reservoirs. The extraction of nitrogen from the atmosphere for industrial and agricultural purposes is compensated by technogenic nitrogen flux to the atmosphere upon combustion of solid and liquid fuel, amounting up to tens of millions of tons per year.

Let us assume that the intensity of anthropogenic processes in the nitrogen cycle primarily depends upon the population density G_i ($i = 1,\dots,n$), men · km^{-2}. The production of fertilizers $H^N_{8,i}$, technogenic accumulation of nitrogen upon fuel combustion $H^N_{2,i}$ and anthropogenic nitrogen flux of the land to the atmosphere $H^N_{18,i}$ will be described by the following equations:

$$H^N_{8,i} = \min\{u_i G_i, N_A \sigma_i / \sigma\},\ H^N_{2,i} = f_1 R_{Mi},\ H^N_{18,i} = f_2 Z_{Gi},$$

where R_{mi} and Z_{Gi} are the consumption rates of mineral resources and generation of wastes, respectively; f_1 and f_2 are the coefficients of proportionality.

3.3.3 Sulfur Unit

Taking into account the designations in Fig. 3.4 and Table 3.4, the equations of the sulfur unit of the GSM are written in the form of balance correlations (Krapivin and Nazaryan 1997):

$$
\begin{aligned}
&d\,AH2SL/dt = C_1 + C_2 + C_3 + C_{21} - C_4\\
&d\,ASO2L/dt = C_4 + C_5 + C_6 - C_7 - C_8 - C_9\\
&d\,ASO4L/dt = C_9 + C_{13} + C_{20} - C_{11} - C_{12}\\
&d\,S/dt = C_{17} - C_{16} - C_{19}\\
&d\,SO4L/dt = C_{10} + C_{11} + C_{12} + C_{16} - C_3 - C_{13} - C_{14}\\
&d\,FIX/dt = C_7 + C_{15} + C_{22} - C_{17}\\
&d\,H2SO4L/dt = C_8 - C_{18} - C_{21} - C_{22}\\
&d\,AH2SO/dt = H_1 + H_3 + H_4 + H_{26} - H_2\\
&d\,ASO2O/dt = H_2 + H_5 + H_6 - H_7 - H_8 - H_{24}\\
&d\,ASO4O/dt = H_8 + H_9 + H_{12} - H_{10} - H_{11}\\
&\partial\,SO4OU/\partial t + v_z\,\partial\,SO4OU/\partial z + k_z\,\partial^2\,SO4OU/\partial z^2\\
&\quad = H_7 + H_{10} + H_{11} + H_{20} + H_{22} + H_{27} + C_{14} - H_{12} - H_{13}\\
&\partial\,H2SOU/\partial t + v_z\,\partial\,H2SOU/\partial z + k_z\,\partial^2\,H2SOU/\partial z^2 = H_{21} + H_{23} - H_4 - H_{22}\\
&\partial\,H2SOD/\partial t + v_z\,\partial\,H2SOD/\partial z + k_z\,\partial^2\,H2SOD/\partial z^2 = H_{17} - H_{18} - H_{21}\\
&\partial\,SO4OD/\partial t + v_z\,\partial\,SO4OD/\partial z + k_z\,\partial^2\,SO4OD/\partial z^2 = H_{18} - H_{19} - H_{20}\\
&\partial\,DU/\partial t + v_z\,\partial\,DU/\partial z + k_z\,\partial^2\,DU/\partial z^2 = H_{14} - H_{15} - H_{23}\\
&\partial\,DD/\partial t + v_z\,\partial\,DD/\partial z + k_z\,\partial^2\,DD/\partial z^2 = H_{15} - H_{16} - H_{17}\\
&\partial\,FI/\partial t + v_z\,\partial\,FI/\partial z + k_z\,\partial^2\,FI/\partial z^2 = H_{13} - H_{14}\\
&d\,BOT/dt = H_{16} + H_{19}
\end{aligned}
$$

where v_z is advection velocity (m day^{-1}) and k_z is coefficient of turbulent mixing (m^2 day^{-1}). The reservoir designations are given in Table 3.13. Functional repre-

Table 3.13. Initial data taken into account under simulation experiments

Reservoir		Identifier of the GSM	Preliminary estimation of reservoir (mg m^{-2})
Atmosphere above the ocean			
	H_2S	*AH2SO*	10
	SO_2	*ASO2O*	5.3
	SO_4^{2-}	*ASO4O*	2
Atmosphere above land			
	H_2S	*AH2SL*	36.9
	SO_2	*ASO2L*	17.9
	SO_4^{2-}	*ASO4L*	12.9
Land			
	SO_4^{2-}	*SO4L*	11.2
	Biomass	*FIX*	600
	Soil	*S*	5000
Ocean photic layer			
	H_2S	*H2SOU*	1.9
	SO_4^{2-}	*SO4OU*	19×10^7
	phytomass	*FI*	66.5
	DOM	*DU*	730
Deep ocean layers			
	H_2S	*H2SOD*	2×10^6
	SO_4^{2-}	*SO4OD*	3.4×10^9
	DOM	*DD*	13120

sentations of the sulfur flows are given in the paper by Krapivin and Nazaryan (1997).

The discharge speed of H_2S to the atmosphere due to humus decomposition is described by a linear function $C_3 = \mu_1(pH) \cdot SO4L \cdot T_L$, where μ_1 is the proportionality coefficient depending on the soil acidity pH (day^{-1} K^{-1}) and T_L is soil temperature (K).

It is supposed that flow H_4 is a function of the rates of alignment for H_2S oxidation in the photic layer with the vertical velocity of water rising. Therefore, for the description of flow H_4 the parameter t_{H2SU}, which reflects the lifetime of H_2S in the water is used: $H_4 = H2SU/t_{H2SU}$, where t_{H2SU} is a function of the velocity of vertical advection u_z and of the oxygen concentration $O2$ in the upper layer having the thickness Z_{H2S}:

$$t_{H2SU} = H2SOU \cdot O2^{-1} u_z(\theta_2 + O2)(\theta_1 + u_z)^{-1}.$$

The constants θ_1 and θ_2 are defined empirically while the value of $O2$ is estimated by the oxygen unit of the GSM. Flows H_2 and C_4 reflect the correlation between the sulfur and oxygen cycles:

$$C_4 = AH2SL/t_{H2SA},\ H_2 = AH2SO/t_{H2SA},$$

where t_{H2SA} is the lifetime of H_2S in the atmosphere.

The mechanism of SO_2 removal from the atmosphere is described by flows H_7, H_8, H_{27}, C_7 and C_9. These flows are characterized by typical parameters t_{SO2L} and t_{SO2A1} which are the lifetimes of SO_2 above the land and water surface, respectively. SO_2 is absorbed from the atmosphere by minerals, vegetation and soil. Dry absorption of SO_2 by vegetation from the atmosphere is described by the model $C_7 = q_2RX$, where $q_2 = q_2' \cdot ASO2L/(r_{tl} + r_s)$, r_{tl} is the atmospheric resistance to SO_2 transport over the vegetation of lth type (day m^{-1}), r_s is surface resistance to SO_2 transport over the surface of sth type (day m^{-1}), RX is the production of X-type vegetation (mg m^{-2} day^{-1}) and q_2' is the proportionality coefficient. The production RX is calculated by the biogeocenotic unit of the GSM.

The process of SO_2 washing from the atmosphere is described by the model: $C_8 = q_{1l}W\ ASO2L$, where q_{1l} is the characteristic parameter for the surface of lth type and $W(t,\varphi,\lambda)$ is the precipitation intensity. The interaction of acid rain with the land surface was reflected in Fig. 3.4 by means of flows C_{18}, C_{21}, C_{22}, H_{25}, H_{26} and H_{27}. These flows are parameterized by models: $C_{18} = h_1\ H2SO4L$, $C_{22} = h_2 \cdot RX \cdot H2SO4L$, $C_{21} = h_3T_a \cdot H2SO4L$, $H_{25} = h_6 \cdot H2SO4O$, $H_{26} = h_4T_a \cdot H2SO4O$, $H_{27} = h_5 \cdot RFI \cdot H2SO4O$, where $T_a(t,\varphi,\lambda)$ is the atmosphere temperature, $h_1 + h_2 \cdot RX + h_3T_a = 1$, $h_4T_a + h_5 \cdot RFI + h_6 = 1$, and RFI is the production of phytoplankton.

Similarly, the flows H_8, C_9, H_7 and H_{24} are simulated by models:

$$H_8 = ASO2O/t_{SO2A1},\ C_9 = ASO2L/t_{SO2L},\ H_7 = ASO2O/t_{SO2A2},\ H_{24} = q_{1l}W \cdot ASO2O.$$

Physical mechanisms of sulfate transportation in the environment are described by models of Luecken et al. (1991), Krapivin (1993), Bodenbender et al. (1999), Park et al. (1999): $H_{10} = \mu W \cdot ASO4O$, $H_{11} = \rho v_0 \cdot ASO4O$, $C_{11} = b_3W \cdot ASO4L$, $C_{12} = d_1v_a \cdot ASO4L$, where v_0 and v_a are the rates of dry sedimentation of aerosols over the water surface and land, respectively.

For the flows C_{13}, H_{12}, C_{14}, C_{16} we consider the following models:

$$C_{13} = d_2 \cdot RATE \cdot SO4L,\ H_{12} = \theta \cdot RATE \cdot SO4L,$$
$$C_{16} = b_2ST_L,\ C_{14} = d_3W \cdot SO4L + (C_{11} + C_{12})\,\sigma,$$

where $RATE(t,\varphi,\lambda)$ is the wind velocity over the surface (m s^{-1}) and coefficient b_2 reflects the sulfur content in the dead plants.

The terrestrial part of the sulfur cycle correlates with the water part through flows in the system *atmosphere-hydrosphere-land*. We have:

$$H_{13} = \gamma \cdot RFI,\ H_{14} = b \cdot MFI,\ H_{15} = f \cdot DU,\ H_{16} = p \cdot DD,\ H_{17} = q \cdot DD,$$
$$H_{18} = H2SOD/t_{H2SOD},\ H_{19} = u \cdot SO4D,\ H_{20} = a_1v_D \cdot SO4D,$$
$$H_{21} = b_1v_D \cdot H2SOD,\ H_{22} = H2SOU/t_{H2SOU},\ H_{23} = g \cdot DU,$$

where MFI is the mass of dead phytoplankton, t_{H2SOU} and t_{H2SOD} are the characteristic times for H_2S total oxidation in the photic layer and deep waters, respectively.

3.4 Units of Biogeocenotic, Hydrologic and Climatic Processes

As shown in Figs. 3.8 and 3.9, the GSM comprises 30 models (or less) for soil-plant formations. In synthesizing these models, use was made of results obtained by Krapivin and Vilkova (1990), Papakyriakon and McCaughey (1991), Friend (1998), Yokozawa (1998), Holmberg et al. (2000), Peng (2000) and Wirtz (2000). All of these are based on the equation for the balance of the biomass $X(t,\varphi,\lambda)$: $\partial X/\partial t = \xi - \omega_X - \tau - \Sigma$, where ξ is the actual plant productivity, ω_X and τ are the quantities of mortality and the outlays for energy exchange with the environment and Σ are the biomass losses due to anthropogenic reasons. These functions are described in detail by many authors (the above mentioned, for instance). In the GSM the value of ξ is approximated as follows:

$$\xi = \delta_c\delta_o\,(1 + \alpha_T \cdot \Delta T/100)\exp(-\beta_1/X)\min\{\delta_e, \delta_Z, \delta_W, \delta_N, \delta_S, \delta_P\} \qquad (3.5)$$

where α_T and β_1 are indices of dependence of production on temperature and biomass, correspondingly; δ_e is the index of production limitation by the θ factor (e = illumination, Z = pollution, W = soil moisture, N, S and P are the nitrogen, sulfur and phosphorus in soil, respectively).

Formula (3.5) was chosen after performing numerous computational experiments taking into account various options for the limiting-factor dependence of the plant productivity. The δ_θ functions actually used were calculated based on data published in literature. Thus, the role played by C_A in photosynthesis is described by the relation $\delta_c = bC_A/(C_A + C_{0.5})$, where $C_{0.5}$ is the CO_2 concentration for which $\delta_c = b/2$. The influence of the solar radiation intensity $e(t,\varphi,\lambda)$ on photosynthesis is parameterized by the relation $\delta_e = \delta^* exp(1-\delta^*)$, where $\delta^* = e/e^*$, e^* is the optimal illuminance. In the soil-plant formations unit for which the maximum photosynthesis value d_1 and the initial slope of the photosynthesis curve m_1 are known, use was made of the relation $\delta_e = d_1e/(d_1/m_1 + e)$. The limiting of photosynthesis by pollution is defined by the exponential dependence $\delta_Z = \exp(-fZ)$, where f is a constant. The effect of soil moisture on photosynthesis is expressed by the function $\delta_W = 1 - \exp(-gW)$, where g is a constant. The biogenic-element dependence of plant production is represented in the form of $\delta_\theta = \theta/(\theta + \theta_A)$, where θ_A is the θ element concentration in soil for which $\delta_\theta = 0.5$.

Water is responsible for channels of interrelations in the biosphere between natural systems. The water cycle in the biosphere includes the exchange of water in its various phase states between the hydrosphere, atmosphere and living organisms. The reserves of water in its various forms are described in the literature at great length; therefore, there is every possibility of constructing a mathematical model of the global water cycle. Such a model was suggested in the study by Krapivin et al. (1982) in conformity with the diagram in Fig. 3.5. In this version of the model, the atmospheric water circulations are simulated by a simplified diagram of stable transports. Actually, the process of atmospheric circulation is far more complex in space and time. It is roughly characterized by alternation of zonal and meridian motions.

Satellite systems for measuring environmental parameters allow rapid acquisition of data pertaining to water content in various biospheric reservoirs, and in particular, the atmospheric moisture content. This information may be obtained

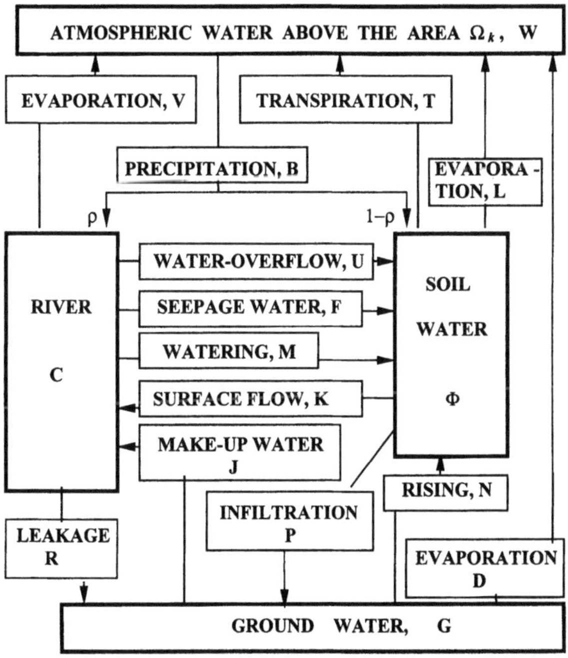

Fig. 3.15. Block diagram of the AYRS water regime in the area Ω_k (k = 1,...,N). The functions W, B, C, G, and Φ have a linear dimension (m). All the other functions are measured in m^3 day^{-1}

simultaneously with synoptic data on the temperature, velocity and direction of wind, atmospheric pressure, content of pollutants in the atmosphere and the ground surface heat-flux balance. Such measurements and published data on biospheric water distribution made it possible to construct a flow chart of the global water balance (Figs. 3.5 and 3.15). These flow charts are based on balance equations. The form in which the latter are written is illustrated by the following example:

$$dW_{iH} / dt = W_{SiH} - W_{iHG} - \sum_k W_{HiO}^k, \qquad (i = 1,...,n)$$

where W_{iH} is the level of underground waters, W_{SiH} is infiltration, W_{iHG} is irrigation waters, W_{HiO} is the runoff to the oceans, and n is the number of land regions.

The precipitation formation regime is represented by a threshold algorithm, namely, rain-fall for $T > T^*$, snow-fall for $T < T^*$, and snow thawing for $T > T_W$, where $T^* = 0$°C, and $T_W = 5.5$°C.

The most important climate factor that is responsible for the character of human activity in the various regions appears to be atmospheric temperature. A change in the atmosphere temperature leads to changes in the intensity of biological processes on land and causes disturbances in the biogeochemical cycles (Chen et al. 2000; Power 2000).

Atmospheric temperature is formed as a function of the carbon dioxide C_A and water vapor W_A content in the layer near the Earth's surface: $T = T_D + \Delta T(C_A, W_A)$, where T_D is the temperature in the parenthropogenic period. Estimations of ΔT are calculated by $\Delta T = \Delta T_C + \Delta T_W$, where ΔT_C and ΔT_W reflect changes in T caused by fluctuations of the CO_2 and W_A concentrations in the atmosphere, respectively. The value of ΔT_C has two components: $\Delta T_C = 0.5(\Delta T_{C1} + \Delta T_{C2})$. The values of ΔT_{C1} and ΔT_W are calculated using the climate model. The spatial distributions of ΔT_{C1} and ΔT_W are calculated as functions of C_A and W_A, respectively. The value of ΔT_{C2} is calculated using Eq. (3.4).

3.5
Other Units of the *Nature/Society* System Model

Over recent years a new understanding of complex natural systems and their dynamics and evolution has emerged, and these have been shown to provide a new basis for models of the changing patterns of natural and anthropogenic biosphere elements that co-evolve and shape the multiple landscapes (Shutko et al. 1994). The development of the GSM is realized by means of the insertion of new units expanding its functions. This section considers two additional units.

3.5.1
World Ocean Bioproductivity Unit

The world ocean ecosystems are represented by three trophic structures characteristic of (1) pelagial tropical (long trophic chains), (2) tropical-latitude shelf zones and mid-latitude aquatic zones (medium-length trophic chains), and (3) arctic latitudes (short trophic chains). In each of these structures, the water column is considered as a single biogeocenosis. The major factor ensuring this unity is the flow of organic matter which is produced in the surface layers and subsequently reaches maximum depths.

The functioning of the trophic pyramid is characterized by consumption intensity for the sth food variety at the ith level:

$$C_{is} = k_{is}\overline{B_s} / \sum_{j\in S_i} k_{ij}\overline{B_j}$$

where $\overline{B_j}$ is the effective biomass of the sth level, S_i is the food spectrum of the ith level, and k_{is} is the Ivlev coefficient used in the formula for the ith component ration:

$$R_i = k_i[1-\exp(-\sum_{j\in S_i} k_{ij}\overline{B_j})].$$

The equations used in describing the bioproduction process in the water column have the following forms:

$$\partial p / \partial t = P_p - \tau p - M_p - \beta \partial p / \partial z + \partial(A \partial p / \partial z) / \partial z - \sum_{i\in\Gamma_p} k_{ip}\overline{p}R_Z^i / (k_{ip}\overline{p} + k_{id}\overline{d} + \sum_{s\in S_i} k_{is}\overline{Z_s})$$

$$\partial Z_i / \partial t = (1-h_Z^i)R_Z^i - \tau_Z^i Z_i - M_Z^i - \beta_Z \partial Z_i / \partial z - \sum_{j\in\Gamma_i} k_{ji}\overline{Z_i}R_Z^i / (k_{jp}\overline{p} + k_{jd}\overline{d} + \sum_{s\in S_j} k_{js}\overline{Z_s})$$

$$\partial d / \partial t + \beta \partial d / \partial z + A \partial^2 d / \partial z^2 = M_p + \sum_{i=1}^{m} M_Z^i - \mu_d + \sum_{i=1}^{m} h_Z^i R_Z^i$$

$$\frac{\partial n}{\partial t} + \beta \frac{\partial n}{\partial z} + A \frac{\partial^2 n}{\partial z^2} = \lambda_0 d - \delta P_p + \rho \sum_{i=1}^{m} \tau_Z^i Z_i$$

where $M_\omega^i = \mu_\omega \max\{0, \omega_i - \omega_{i,\min}\}^{r_Z^i}$ is the mortality velocity of element ω (p, $Z_1, \ldots, Z_m$), p is the phytoplankton biomass, Z_i $(i = 1, \ldots, m)$ is the biomass of the ith component of the zooplankton, d and n are the concentrations of detritus and nutrients, respectively; τ_ω is the index of energy inputs of component ω; A is the turbulent diffusion coefficient; β_Z^i is the mobility index of the ith component of the zooplankton in vertical migrations; and β is the upwelling speed of the water.

3.5.2
Demographic Unit

The effect of numerous environmental and social factors on the population dynamics in the ith region G_i is reflected in the birth rate R_{Gi} and death rate M_{Gi}:

$$dG/dt = R_{Gi} - M_{Gi} \ (i = 1, \ldots, K).$$

In each of the K regions the birth and death rates depend on the food supply and quality, the environmental contamination, the gas composition of the atmosphere, the living standard, the power-resource sufficiency, and the population density as follows:

$$R_{Gi} = (1 - h_{Gi}) K_{Gi} G_i \min \{H_{GVi}, H_{GGi}, H_{GMi}, H_{AVi}\}$$

$$M_{Gi} = \mu_{Gi} G_i \max\{H_{\mu Ai}, H_{\mu Gi}, H_{\mu Vi}\}$$

where h_{Gi} is the quality coefficient representing the lack of nutrition in the food consumed by the population ("food inassimilability"), $H_{Gi} = 0$ for ideal conditions; K_{Gi} and μ_{Gi} are the constant components of birth rate and death rate, respectively; and the functions $H_{GVi}(H_{\mu Vi}), H_{GGi}(H_{\mu Gi}), H_{GMi}, H_{GAi}(H_{\mu Ai})$ indicate the effect on birth rate (death rate) of various factors respectively, such as food supply, population density, living standard, and the environment quality. Functional descriptions of these factors are related to the effects in human ecology. Thus, the function H_{GVi} is represented in the form of $H_{GVi} = 1 - \exp(-V_{Gi})$, where V_{Gi} is the effective food amount determined as a weighted sum of components in the *Homo sapiens* food spectrum:

$$V_{Gi} = K_{Gpi} p + K_{GFi}(F_i + \sum_{j \neq i} a_{Fji} F_j) + K_{Gri} I_i (1 - \theta_{Fri} - \theta_{Uri}) +$$
$$+ K_{GXi}[(1 - \theta_{FXi}) X_i + (1 - \nu_{FXi}) \sum_{j \neq i} a_{Xji} X_j]$$

Here coefficients $K_{Gpi}, K_{GFi}, K_{Gri}$, and K_{GXi} are defined following the method of Krapivin (1996), a_{Fji} and a_{Xij} are the portions of food of animal and plant origin, respectively, for the ith region; θ_{FXi} and ν_{FXi} are the portions of food of plant-origin used for cattle feeding produced in and imported to the ith region, respectively; and θ_{Fri} and θ_{Uri} are the portions of the fishery I_i allotted for cattle feeding and fertilizer production, respectively.

It is assumed that with the food supply increasing the population death rate drops down at a rate ρ_{Gi} to a certain level determined by the constant $\rho_{\mu i}$ so that

$$H_{\mu Hi} = \rho_{\mu i} + \rho_{Gi}/F_{RGi},$$

where $F_{RGi} = V_{Gi}/_{Gi}$.

Similarly, it is assumed that the birth rate dependence on the living standard M_{SGi} is described by a saturating function, so that the birth rate is maximum for low M_{SGi} values and falls as a_{Gi} grows, down to a certain level a_{GMi}. The rate at which the transition between the maximum and minimum birth rate levels occurs is denoted by α_G and defined by the equation

$$H_{GMi} = a_{Gi} + a_{GMi} \exp(-\alpha_G M_{SGi}).$$

The dependence of the birth rate on the population density is approximated by relation

$$H_{GGi} = g' + g^* \exp(-g'' G_i).$$

In general, the demographic unit has a ramified structure permitting the use of different parameters describing the population dynamics within the framework of a concrete computational experiment. The option used in the present mathematical model is described in detail in the publications by Krapivin (1978) and Krapivin et al. (1982). The demographic unit includes a matrix model comprising three population age groups (0–14, 15–64, and 64 and older) and a population of disabled persons. The unit structure also permits the use of different scenarios in describing both the population dynamics as a whole and its parts.

3.6 Biocomplexity Index

Biocomplexity refers to phenomena that result from dynamic interactions between the physical, biological, and social components of the *Biosphere/Society System* (BSS). The investigations of the processes of interaction between the *society* and *biosphere* are, as a rule, targeted at understanding and estimating the consequences of such interactions. The reliability and precision of these estimations depend on criteria founded on conclusions, expertise, and recommendations. At present, there is no unified methodology for selection between the set of criteria due to the absence of a common science-based approach to the ecological standardization of anthropogenic impacts on the natural environment. After all, the precision of the ecological expertise for the functioning and planning of anthropogenic systems, as well as the representativeness of the global geoinformation monitoring data, depend on these criteria.

3.6.1 Biocomplexity Indicator

The processes that have their origin in the environment can be presented as a combination of interactions between its subsystems. The human subsystem is a part of the environment and it is impossible to divide the environment into separate subsystems such as biosphere and society. The problem is to search for methodologies to describe existing feedbacks between nature and society (NS) and to simulate reliably the dynamic tendencies in the NS system. Unfortunately, the part of the NS system that is responsible for the quality of modeling the climatic processes introduces instability in the modeling results. That is why it is supposed below that the NS climatic component is replaced by a scenario describing stable climatic trends during the time interval of investigation. What is actually studied is the BSS (Kondratyev and Krapivin 2001a).

We introduce the scale symbol Ξ of biocomplexity ranging from the state where all interactions between the environmental subsystems are broken to the state where they correspond to natural evolution. In this case, we have an integrated indicator of the environmental state including bioavailability, biodiversity and survivability. It reflects the level of all types of interactions among the environmental subsystems. In reality, specific conditions exist where these interactions are changed and transformed. For example, under the biological interaction of the type *consumer/producer* or *competition-for-energy resources*, there exists some minimal level of food concentration where contacts between interacting components cease. In the common case, physical, chemical, and other types of interactions in the environment depend upon specific critical parameters. Environmental dynamics is regulated by these parameters and the main task is its parametrical description. Biocomplexity reflects these dynamics.

All of this corroborates the fact that biocomplexity is related to categories which are difficult to measure empirically and to express by quantitative values. However, we will try to transfer the truly verbal tautological reasoning to formalized quantitative definitions. For the transition to gradations of the scale Ξ with quantitative positions, it is necessary to postulate that relationships between two values of Ξ are of the type $\Xi_1 < \Xi_2, \Xi_1 > \Xi_2$, or $\Xi_1 \equiv \Xi_2$. In other words, a value of the scale ρ always exists that defines a biocomplexity level $\Xi \rightarrow \rho = f(\Xi)$, where f is a certain transformation of the biocomplexity concept to a number. Let us attempt to search for a satisfactory model to simulate the verbal biocomplexity image onto the field of conceptions and signs, subordinating to the formal description and transformation. With this purpose m subsystems of the BSS are selected. The correlations between these subsystems are defined by the binary matrix function: $X = \| x_{ij} \|$, where $x_{ij} = 0$, if subsystems B_i and B_j do not interact, and $x_{ij} = 1$, if subsystems B_i and B_j are interacting. Then any one point $\xi \in \Xi$ is defined as the sum $\xi = \sum_{i=1}^{m} \sum_{j>i}^{m} x_{ij}$. Certainly, there arises the need to over come uncertainty for which it is necessary to complicate the scale Ξ; for example, to introduce weight coefficients for all BSS subsystems. The origin of these coefficients depends on the type of subsystem. That is why three basic subsystem types are selected: living and nonliving subsystems and vegetation. Living subsystems are characterized by their density, determined by numbers of elements or by biomass value per unit area or volume. Vegetation is characterized by the type and portion of occupied territory. Nonliving subsystems are measured by their concentration per unit area or volume of the environment. In the common case, certain characteristics $\{k_i\}$, corresponding to the significance of the subsystems $\{B_i\}$, are assigned to every subsystem B_i ($i = 1,\ldots, m$). As a result, we obtain more adequate definition of the formula to move from the biocomplexity concept to the scale Ξ of its indicator:

$$\xi = \sum_{i=1}^{m} \sum_{j>i}^{m} k_j x_{ij}$$

It is clear that $\xi = \xi(\varphi,\lambda,t)$, where φ and λ are geographical latitude and longitude, respectively, and t is the current time. For the territory Ω the biocomplexity indicator is defined as the mean value:

$$\xi_\Omega(t) = (1/\sigma) \int_{(\varphi,\lambda)\in\Omega} \xi(\varphi,\lambda,t)d\varphi d\lambda$$

where σ is the area of Ω.

Thus, the indicator $\xi_\Omega(t)$ is the integrated BSS complexity characterization reflecting the individuality of its structure and the behavior at each time t in the space Ω. According to the natural evolution laws, a decrease (increase) in $\xi_\Omega(t)$ will correspond to an increase (decrease) in biocomplexity and the survivability of the nature-anthropogenic systems. Since a decrease in biocomplexity disturbs the biogeochemical cycles and leads to a decrease in stress on the nonrenewal of resources, then the binary structure of the matrix X is changed in the direction to intensify the resource-impoverishment technologies. The vector of energy exchange between the BSS subsystems is moved to the position where the survivability level of the BSS is reduced.

The global simulation model is constructed with the spatial discretization of the earth's surface with $\Delta\varphi$ in latitude and $\Delta\lambda$ in longitude. In other words, the BSS space Ω is divided into a set of cells Ω_{ij} ($\Omega = \cup\Omega_{ij}$; $\Omega_{ij} = \{((\varphi,\lambda);\ \varphi_i \le \varphi < \varphi_i + 1;\ \lambda_j \le \lambda < \lambda_j + 1;\ i = 1, ..., N; j = 1, ..., M; N = [180/\Delta\varphi]; M = [360/\Delta\lambda]\}$). Each cell Ω_{ij} has its biocomplexity indicator value:

$$\xi_\Omega(i,j,t) = (1/\sigma_{ij}) \int_{(\varphi,\lambda)\in\Omega_{ij}} \xi(\varphi,\lambda,t)d\varphi d\lambda \qquad (3.6)$$

The value $\xi\Omega(i,j,t)$ calculated by the formula (3.6) reflects the topological structure of the matrix $X(i,j,t)$. Consequently, $n = N\text{x}M$ matrixes and biocomplexity indicators exist to characterize the BSS biocomplexity. In the framework of the computer experiment a set of numerical characteristics of the BSS biocomplexity arises, distributed in space and time. Integrated BSS biocomplexity indicators can be calculated for any arbitrary area $\omega\in\Omega$:

$$\xi_\omega(t) = (1/\sigma_\omega) \sum_{(\varphi_i,\lambda_j)\in\omega} \xi_\Omega(i,j,t) \qquad (3.7)$$

This can be the average BSS biocomplexity by zone of longitude or latitude, by ocean or sea aquatory, by country or state territory, etc.

3.6.2
The BSS Biocomplexity Model

The BSS consists of subsystems B_i ($i = 1,..., m$) the interactions of which are formed during time as functions of many factors. The BSS biocomplexity indicates the structural and dynamic complexity of its components. In other words, the BSS biocomplexity is formed under the interaction of its subsystems $\{B_i\}$. In due course the subsystems B_i can change their state and, consequently, change the topology of the relations between them. The evolutionary mechanism of adaptation of the subsystem B_i to the environment allows the hypothesis that each subsystem B_i, independently from its type, has the structure $B_{i,S}$, behavior $B_{i,B}$ and goal $B_{i,G}$, so that $B_i = \{B_{i,S}, B_{i,B}, B_{i,G}\}$. The strivings of subsystem B_i to achieve certain preferable conditions are represented by its goal $B_{i,G}$. The expedience of the structure $B_{i,S}$ and the purposefulness of the behavior $B_{i,B}$ for subsystem B_i are estimated by the effectiveness with which the goal $B_{i,G}$ is achieved.

As an example, we consider the process of fish migration. The investigations of many authors have revealed that this process is accompanied by an external appearance of purposeful behavior. From these investigations it follows that fish migrations are subordinated to the principle of complex maximization of effective nutritive ration, given preservation of favorable environmental conditions (temperature, salinity, dissolved oxygen, pollution level, depth). In other words, the travel of migrating species takes place at characteristic velocities in the direction of the maximum gradient of effective food, given adherence to ecological restrictions. That is why we can formulate that the goal $B_{i,G}$ of the fish subsystem is toward the increase of their ration, the behavior $B_{i,B}$ consists in the definition of the moving trajectory securing the attainability of the goal $B_{i,G}$.

Since the interactions of the subsystems B_i ($i = 1,\dots, m$) are connected with chemical and energy cycles, it is natural to suppose that each subsystem B_i accomplishes the geochemical and geophysical transformation of matter and energy to remain in a stable state. The formalism of approach to this process consists in the supposition that the interactions between the BSS subsystems are represented as a process whereby the systems exchange a certain quantity V of resources spent in exchange for a certain quantity W of resources consumed. We shall call this process by the name (V, W) exchange.

The goal of the subsystem is the most advantageous (V, W) exchange, i.e., it tries to obtain maximum W in exchange for minimum V. The quantity W is a complex function of the structure and behavior of interacting subsystems, $W = W(V, B_i, \{B_k, k \in K\})$, where K is the space of subsystem numbers interacting with the subsystem B_i.

Designate $B_K = \{B_k, k \in K\}$. Then the following (V, W) exchange is the result of interactions between the subsystem B_i and its environment B_K:

$$
\begin{aligned}
W_{i,0} &= \max_{B_i} \min_{B_K} W_i(V_i, B_i, B_K) = W_i(V_i, B_{i,opt}, B_{K,opt}) \\
W_{K,0} &= \max_{B_K} \min_{B_I} W_K(V_K, B_i, B_K) = W_K(V_K, B_{i,opt}, B_{K,opt})
\end{aligned}
\tag{3.8}
$$

Hence, it follows that some range of the goal of the subsystem B_i exists which defines the levels of V_i and V_K. Since the limiting factors are defined by nature, then in this case it is natural to suppose that some level $V_{i,min}$ exists when the subsystem B_i ceases to spend its energy resource for obtaining the external resource, i.e., if $V_i \le V_{i,min}$, the subsystem B_i transfers to the regeneration of its internal resource. In other words, when $V_i \le V_{i,min}$, the decrease in the biocomplexity indicator $\le \Omega(t)$ takes place at the expense of breaking off interactions of the subsystem B_i with other subsystems. Commonly, the structure of $V_{i,min}$ is a checkered function, i.e., the changeover of x_{ij} from state $x_{ij} = 1$ to state $x_{ij} = 0$ is not realized for all j at the same time. Actually, in any trophic pyramid of living subsystems, the relationships of "producer/consumer" type cease under the decrease in the consumer biomass concentration below some critical level. In other cases, the interactions of the subsystems $\{B_i\}$ can be stopped at the expense of various combinations of its parameters. The parametrical description of possible situations of interactions of subsystems $\{B_i\}$ can be realized in the framework of the BSS simulation model.

3.7 Algorithms for the Data Processing

One of the GIMS-technology units provides the other units with remote monitoring data. There are many algorithms for processing and interpretation of these data (Engman and Chauhan 1995; Petty 1995; Schimel 1995; Strelkov 1995a,b; Sellers et al. 1996; Cherny and Raiser 1998; Ferm and Hultberg 1999; Klyuev 2000; Yakovlev 2001). However, many difficulties exist in this area arising from the solving of the inverse problems and from the fact that the measurement data usually are nonstationary. The functioning of the remote monitoring system in real time requires the time economy driver to be capable of the assessment of dynamic processes. That is why this section considers some algorithms which can help to overcome some of these difficulties.

3.7.1 Data Reconstruction Using the Harmonic Functions

The process of spreading of heat in the plane homogeneous media G with constant thermophysical characteristics (density ρ, specific heat C and conductivity coefficient K; ρ, $C, K = const > 0$) is described by the equation:

$$\partial T/\partial t = a^2(\partial^2 T/\partial \varphi^2 + \partial^2 T/\partial \lambda^2) \tag{3.9}$$

where $T = T(\varphi,\lambda,t)$ is the temperature of the media at the point with spatial coordinates $(\varphi,\lambda) \in G$ at time t; $a^2 = K/C\rho$ is the temperature-conductivity coefficient of G, φ and λ are the latitude and longitude, respectively.

If the thermal transport process is stationary then Eq. (3.9) transforms to the Laplace equation:

$$div \cdot grad\, T = \partial^2 T/\partial \varphi^2 + \partial^2 T/\partial \lambda^2 \tag{3.10}$$

The solution of Eq. (3.10) is the harmonic function of the spatial coordinates φ and λ. In accordance with the Rayleigh-Jeans law (DeWitt and Nutter 1988), the brightness temperature as a result of remote microwave measurements is $T_a(\varphi,\lambda) = T_a(\varphi,\lambda,\mu,\theta)$, where μ is the wavelength, θ is the observation angle and $(\mu,\theta = const)$. It is proposed that for any small area V_M at an arbitrary point $M \in G$ the brightness temperature is a linear function of the temperature of the media:

$$T_a(\varphi,\lambda) = A_M + B_M\, T(\varphi,\lambda);\ (\varphi,\lambda) \in V_M, \tag{3.11}$$

where A_M and B_M are constants.

Formula (3.11) follows from the theoretical and experimental estimations of T_a (Krapivin and Potapov 2001a). For media which are homogeneous in depth, the formula $T_a = \kappa T$ is valid. Here, $\kappa = \kappa(\mu,\theta,\varepsilon)$ is the media radiation coefficient where ε is the dielectric permeability. According to the experimental estimations, the radiance of fresh water measured by means of radiometers with wavelengths of 5–8 cm is a linear function of T. The slope of this dependence is 0.35–0.50 K/°C. An increase in the water salinity S from 0 to 16‰ is accompanied by a decrease in the sensitivity of the irradiation field to the temperature variations. This effect is observed for wavelengths from 10 to 50 cm. The sensitivity of the irradiation field to variations in T is a minimum when the following conditions are realized: $\mu S \cong 700$; $0 \le T \le 30\,°C$; $0 \le S \le 180‰$; $0 \le \theta \le 25\,°C$.

From Eq. (3.11) it follows that T_a at every point $M \in G$ satisfies the following condition:

$$T_a(\varphi,\lambda) = \frac{1}{2\pi}\int_0^{2\pi} T_a(\varphi + r\cdot\cos a, \lambda + r\cdot\sin a)da,$$

where the integral is over a circle of radius r centered at (φ,λ).

This condition is valid for any r $(0 < r < r_M)$ where r_M is the radius of the area V_M. Therefore, T_a is a harmonic function within G. A typical task here is the search of the harmonic function $T_a(\varphi,\lambda)$ within G when $T_a(\varphi,\lambda) = \tilde{T}_a(u)$ on the boundary Γ of G ($u \in \Gamma, u = \varphi + i\lambda$). Such a function is the real part of some analytic function $W(z)$ given by:

$$W(z) = \frac{1}{2\pi}\int_\Gamma \mu(\zeta)/(\zeta - z)d\zeta, \tag{3.12}$$

where $\mu(\zeta)$ is the real density, $\zeta \in \Gamma$ and $z = \varphi + i\lambda$ is an arbitrary internal point of G. We have Re[W(u)] and Im $[d\zeta/(\zeta - u)] = -\cos(r,n)\, d\sigma/r$ where r is the distance between ζ and u, $d\sigma$ is an element of Γ and n is the external normal to Γ. As $z \to u \in \Gamma, \mu(u)$ is approximated by the solution of the integral equation:

$$\mu(u) - \frac{1}{\pi}\int_\Gamma \mu(\zeta)r^{-1}\cos(r,n)d\sigma$$

When G is the circle $|z - z_o| < R$ the solution of this task is the Poisson integral:

$$T_a(r,\psi) = \frac{1}{2\pi}\int_0^{2\pi} \tilde{T}_a(a)\frac{R^2 - r^2}{R^2 + r^2 - 2Rr\cos(\psi - a)}da\ ,$$

where $\varphi + i\lambda = z_o + r\,e^{i\psi}$ $(r < R, 0 \le \psi \le 2\pi)$; and

$$\tilde{T}_a(a) = \tilde{T}_a(z_0 + R\cdot e^{ia}), 0 \le \alpha \le 2\pi.$$

The combination of this procedure with other algorithms of spatial-temporal interpolation gives a full representation of the environmental objects on the observation area by means of the parametrical estimations. For example, the TIM database was formed on the basis of the measurements traced by the in-flight laboratory using the set of analogous harmonic algorithms. Table 3.14 illustrates the precision of this algorithm as compared with the measured values and the differential approximation method.

Table 3.14. An example of the hydrophysical field reconstruction in the lagoon Nuoc Ngot. Field measurements were realized during March–April 2001. Signs correspond to the water inflow (+) and outflow (-). This lagoon is situated in central Vietnam, South China Sea: *U* is the measured value of water flow ($m^3\ s^{-1}$) on the boundary lagoon-sea, *M* and *R* are the calculated values of this flow with the differential approximation and harmonic functions methods, respectively

UM	*M*	Error (%)	*R*	Error (%)
10.3	11.3	10.1	11.6	12.9
-4.0	-3.6	9.5	-4.4	9.8
-8.7	-15.3	12.4	-9.9	13.3
-14.2	-12.3	13.1	-16.0	12.6
-24.1	-22.7	5.7	-22.2	7.7
-22.4	-24.4	8.9	-20.4	9.0
-16.3	-17.9	9.6	-18.2	11.5
2.3	2.6	11.5	2.0	11.4
15.0	13.2	12.3	12.6	15.8
35.1	30.9	12.2	30.1	14.2
39.2	35.1	10.4	43.0	9.7
38.2	42.6	11.6	33.2	13.1
34.8	39.6	13.8	30.0	13.7
26.5	25.2	4.8	28.5	7.6
19.2	21.1	9.9	21.2	10.3
Average error (%)		10.4		11.5
Maximal error (%)		13.8		15.8
Minimal error (%)		4.8		7.6

3.7.2
Method for Parametrical Identification of the Environmental Objects

Radiometers determine the brightness temperatures Z_{ij} ($i = 1,..., M; j = 1,..., n$) given by $Z_{ij} = T_j + \xi_{ij}$, where *M* is the number of measurements, *n* is number of radiometers, T_j is the real value of the brightness temperature for wavelength μ_j and ξ_{ij} is the noise with zero mean and dispersion σ_j. The problem is to determine the correlation function $T_j = f_j(X)$, where $X = \{x_1, ..., x_m\}$ are geophysical, ecological, biogeochemical or other parameters. There are many algorithms for the definition of the function *f*. As a general rule, the mean-square criterion is used for this purpose. However, such an approach has one defect consisting of the impossibility of taking into consideration the dispersion properties of the noise $E = \{\xi_{ij}\}$.

Let the function *f* be linear. Then we have the following system of equations for parameters A_{ij}:

$$||A_{ij}||\ X = T + E \tag{3.13}$$

It is necessary to solve Eq. (3.13) such that its solution has minimum dispersion. Such a solution is called the σ-solution.

The *i*th equation of system (3.13) is multiplied by the set of parameters $c_{1i}, ..., c_{mi}$. An additional condition is given:

$$\sum_{i=1}^{n} c_{ji} A_{il} = \delta_{jl} \tag{3.14}$$

where

$$\delta_{jl} = \begin{cases} 1 & for \quad j = l \\ 0 & for \quad j \neq l \end{cases} \quad (l, j = 1, ..., m) \tag{3.15}$$

Under the conditions (3.14) and (3.15) we have

$$x_1^o = \sum_{i=1}^{n} c_{1i} T_i \tag{3.16}$$

From (3.13) and (3.16) we obtain:

$$\tilde{x}_1 = \sum_{i=1}^{n} c_{1i} T_i + \sum_{i=1}^{n} c_{1i} \xi_i \tag{3.17}$$

The dispersion of solution (3.17) is

$$D[\tilde{x}_1] = \sum_{i=1}^{n} c_{1i}^2 \sigma_i^2 \tag{3.18}$$

Dispersions of $\tilde{x}_i \left(i = 2, ..., m\right)$ are calculated by analogy with (3.18). To calculate the min $D\left[\tilde{x}_1\right]$, the following additional equation is used

$$\psi(c_{11}, ..., c_{1n}) = \sum_{i=1}^{n} c_{1i}^2 \sigma_i^2 + \tau_1 \left(\sum_{i=1}^{n} c_{1i} A_{i1} - 1 \right) + \sum_{j=2}^{m} \tau_j \sum_{i=1}^{n} c_{1i} A_{ij}$$

The first derivatives of ψ are equal to zero, giving the following set of equations:

$$2c_{1k}\sigma_k^2 + \sum_{j=1}^{m} \tau_j A_{kj} = 0, \left(k = 1, \; ..., \; n\right) \tag{3.19}$$

The conditions of (3.14), (3.15) and (3.19) consist of a system of $(m + n)$ equations to be solved.

We have $D[x_j] = \tau_j/2$, where the set of τ_j are defined as solution of the following equations:

$$\sum_{j=1}^{m} \mu_j \sum_{i=1}^{n} \frac{A_{ij} A_{i1}}{\sigma_i^2} = -2; \; \sum_{j=1}^{m} \mu_j \sum_{i=1}^{n} \frac{A_{ij} A_{il}}{\sigma_i^2} = 0, \left(l = 2, \; ..., \; m\right)$$

These algorithms are used as subunits of the Aral-Caspian Expert System. The forecast of the ACS state is obtained from the TIM (see Chap. 9).

3.7.3 Method of Differential Approximation

Databases of the environment monitoring systems do not always correspond to the parametrical fullness in the framework of the GIMS technology standard. Therefore, an algorithm that allows us to adapt the database to this standard is considered. Let us suppose that N characteristics, x_i ($i = 1,...,N$), of environment are measured at the times t_s ($s = 1,...,M$). Formal dependence between x_i is represented by the system of differential equations with unknown coefficients $\{a_{ijk}, b_{ij}\}$:

$$\frac{d\xi_i}{dt} = \sum_{k,j=1}^{N} \left[a_{ijk}\xi_j(t)\xi_k(t) + b_{ij}\xi_j(t)\right] \tag{3.20}$$

Putting the initial conditions as

$$\xi_i(0)\ (i = 1,...,N) \tag{3.21}$$

the reconstruction task of $x_i(t)$ for the arbitrary time $t \in [0,T]$ is reduced to the simple task of the determination of unknown coefficients based on the criterion (Krapivin and Potapov 2001b):

$$E = \sum_{s=1}^{M} \left\{ \sum_{i=1}^{N} \left[\xi_i(t_s) - x_i(t_s)\right]^2 \right\} = \min_{\{a_{ijk}, b_{ij}, \xi_i(0)\}}$$

There are many methods to solve this task. One of them is based on Bellman's dynamic programming method (Krapivin and Kondratyev 2002).

3.7.4 Quasi-Linearization Method

A number of problems arising in ecoinformatics lead to the necessity of integrating generally nonlinear integro-differential equations; and in a majority of cases, these equations are not integrable by elementary and special functions. To solve them, as a rule, it is necessary to make use of the latest achievements of calculating methods and techniques. In many problems, the use of well-known numerical methods of solving initial value problems, even by modern high-speed electronic computers, does not lead to desirable results. The existing approximate methods of solving integro-differential equations, as a rule, are based on replacing the derivatives by the finite differences and represent a complicated multistep process, which in practical problems cannot be solved on computers in a reasonable time. Therefore, in solving practical problems, we have to search for other means of approximate solutions to integro-differential equations, without using the finite-difference methods.

In the method considered here, the integro-differential equation is substituted in each subinterval of the independent variable by an easily integrable ordinary differential equation with constant coefficients; this method is not a new theoretical idea for it was known to Euler. However, here, the error estimations are obtained for the first time, and methods applicable to various problems are developed in detail.

3.7.4.1 Method of Solution and Estimation of Error

Let us consider the equation

$$L[y] - \lambda W[y] = f(x,y), \tag{3.22}$$

$L[y]$ being the differential operator

$$L[y]=\sum_{i=0}^{n} P_i(x,y,y',\dots,y^{(m_i)})y^{(n-i)}, \qquad (m_i<n) \tag{3.23}$$

and $W[y]$ the generalized Volterra operator

$$W[y]=\int_a^x \sum_{j=0}^{r} K_j(x,\xi)y^{(j)}(\xi)d\xi, \qquad (r<n) \tag{3.24}$$

λ – a real number, $P_i(x,y,y',\dots,y^{(m_i)})$ and $f(x,y)$ – continuous functions with respect to their arguments in the finite interval $[a,b]$, $P_0 \neq 0$ and kernels $K_j(x,\xi)$, $j=0,1,\dots,r$ are continuous functions in the region $G\{a \le \xi \le x \le b\}$.

The initial conditions are

$$y^{(s)}(a)=y_0^{(s)}, \qquad s=0,1,\dots,n-1 \tag{3.25}$$

Assuming that Eq. (3.22) with the initial conditions (3.25) has a unique continuous solution $y(x)$, let us construct an approximate solution $\tilde{y}(x)$ in $[a,b]$. Let us divide the interval $[a,b]$ by a sequence of points $x_0=a,\ x_1,\ \dots,\ x_m=b,\ h_k=x_{k+1}-x_k$. On each subinterval $[x_k,x_{k+1}]$, $k=0,1,\dots,m-1$ let us replace Eq. (3.22) by the following linear differential equation of the nth order with constant coefficients

$$\tilde{L}_k[y]=\lambda\tilde{W}_k[y]+f(x_k,\tilde{y}_k) \tag{3.26}$$

with the initial conditions:

$$y^{(s)}(x_k)=\tilde{y}_k^{(s)}, \qquad s=0,1,\dots,n-1, \tag{3.27}$$

where

$$\tilde{L}_k[y]=\sum_{i=0}^{n} P_i(x_k,\tilde{y}_k,\tilde{y}_k',\dots,\tilde{y}_k^{(m_i)})\tilde{y}^{(n-i)}, \tag{3.28}$$

$$\tilde{W}_k[y]=\sum_{j=0}^{r}(K_{j,k,0}\tilde{y}_0^{(j)}h_0+K_{j,k,1}\tilde{y}_1^{(j)}h_1+\cdots+K_{j,k,k}\tilde{y}_k^{(j)}h_k) \tag{3.29}$$

The general solution of Eq. (3.26) is known:

$$\tilde{y}=\tilde{y}(x,c_1^{(k)},c_2^{(k)},\dots,c_n^{(k)}), \tag{3.30}$$

where the constants are determined from the initial conditions at the beginning of each interval $[x_k,x_{k+1}]$. The calculations are carried out successively beginning with the interval $(k=0)$.

Let us estimate the error in solution of Eq. (3.22). Let $y(x)$ and $\tilde{y}(x)$ by the exact and the approximate solutions, respectively. Let us denote

$$\tilde{P}_{ik} = P_i(x_k, \tilde{y}_k, \tilde{y}_k', ..., \tilde{y}_k^{(m_i)}),\ \tilde{f}_k = f(x_k, \tilde{y}_k),\ \varepsilon_k = y(x_k) - \tilde{y}(x_k) \tag{3.31}$$

Let us integrate times Eqs. (3.22) and (3.26) from x_k to x, and consider the final results for $x = x_k +_1$. For convenience and brevity, let us denote

$$\underbrace{\int_{x_k}^{x_{k+1}} \int_{x_k}^{x} \cdots \int_{x_k}^{x}}_{n} \varphi(x)\,dx \underbrace{\cdots}_{n} dx = \underbrace{\int_{x_k}^{x_{k+1}}}_{n} \varphi(x)dx,$$

we have

$$y_{k+1} = y_k + y_k' h_k + h_k^2 \sum_{s=2}^{n-1} y_k^{(s)} \frac{h_k^{s-2}}{s!} - \sum_{i=1}^{n} \underbrace{\int_{x_k}^{x_{k+1}}}_{n} P_i y^{(n-i)} dx + \underbrace{\int_{x_k}^{x_{k+1}}}_{n} f(x,y)dx + \lambda \underbrace{\int_{x_k}^{x_{k+1}}}_{n} W[y]dx \tag{3.32}$$

$$\tilde{y}_{k+1} = \tilde{y}_k + \tilde{y}_k' h_k + h_k^2 \sum_{s=2}^{n-1} \tilde{y}_k^{(s)} \frac{h_k^{s-2}}{s!} - \sum_{i=1}^{n} \underbrace{\int_{x_k}^{x_{k+1}}}_{n} \tilde{P}_{ik} \tilde{y}^{(n-i)} dx + \underbrace{\int_{x_k}^{x_{k+1}}}_{n} \tilde{f}_k dx + \lambda \underbrace{\int_{x_k}^{x_{k+1}}}_{n} \tilde{W}_k[y]dx \tag{3.33}$$

From (3.32) and (3.33) we get:

$$\varepsilon_{k+1} = \varepsilon_k + \varepsilon_k' h_k + h_k^2 \sum_{s=2}^{n-1} \varepsilon_k^{(s)} \frac{h_k^{s-2}}{s!} - \sum_{i=1}^{n} \underbrace{\int_{x_k}^{x_{k+1}}}_{n} [P_i y^{(n-i)} - \tilde{P}_i \tilde{y}^{(n-i)}]dx + \underbrace{\int_{x_k}^{x_{k+1}}}_{n} [f - \tilde{f}_k]dx + \lambda \underbrace{\int_{x_k}^{x_{k+1}}}_{n} \{W[y] - \tilde{W}_k[y]\}dx$$

We know that

$$\underbrace{\int_{x_k}^{x_{k+1}}}_{n} \tilde{f}_k dx = \frac{\tilde{f}_k h_k^n}{n!},\ \underbrace{\int_{x_k}^{x_{k+1}}}_{n} \tilde{W}_k[y]dx = \frac{\tilde{W}_k[y]}{n!}$$

Let us denote

$$E_k = \max_j \left|\varepsilon_k^{(j)}\right|,\ h_{\max} = \max_k h_k,\ p_i = \max_{[a,b]}\left|P_i\right|,\ M_{n-i} = \max_{[a,b]}\left|y^{(n-i)}\right|,$$
$$L_i = \max_{[a,b]}\left|\tilde{P}_{ik}\right|,\ N_{n-i} = \max_{[a,b]}\left|\tilde{y}^{(n-i)}\right|,\ F = \max_{[a,b]}\left|f\right|,\ G_0 = \max_{[a,b]}\left|\tilde{f}_k\right|,$$
$$T = \max_G\left|\tilde{W}_k[y]\right|,\ s = \left|b-a\right|\sum_{j=0}^{r}\max_G\left|K_j(x,\xi)\right|M_j \geq \max_G\left|W[y]\right|,$$
$$l = \frac{1}{n!}\left[\sum_{i=1}^{n}\left(p_i M_{n-i} + L_i N_{n-i}\right) + F + G_0 + \left|\lambda\right|(T+s)\right],$$
$$g = 1 + \sum_{s=2}^{n-1}\frac{h_{ax}^{s-1}}{s!}$$

and $M_{n-i} \approx N_{n-i}$. Then we get the following recurrent error estimation

$$E_{k+1} \leq (1 + g h_{\max})E_k + l h_{\max}^n \tag{3.34}$$

Hence, we get

$$E_k \leq (1 + g h_{\max})^k \varepsilon_0 + l h_{ax}^{n-1} g^{-1}[(1 + g h_{\max})^k - 1] \tag{3.35}$$

where ε_0 is the maximum error in the initial data. Obviously, if $\varepsilon_0 = 0$, then from (3.35) it follows that if $h_{\max} \to 0$ then $E_k \to 0$, i.e., $\tilde{y}(x_k) \to y(x_k)$.

In the case that the Eq. (3.22) has the form:

$$L[y] = f(x,y) + \int_a^x F(x, y, y', \ldots, y^{(m_l)})dx, \qquad x \in [a,b]\ ,$$

then the Eq. (3.35) will read

$$E_k \leq (1 + h p^{(0)})^k \varepsilon_0 + \frac{p^{(1)} h^n}{p^{(0)}}[(1 + h p^{(0)})^k - 1],$$

where

$$p^{(0)} = \sum_{s=1}^{n-1}\frac{h^{s-1}}{s!} + \frac{h^{n-1}}{n!}[\tilde{b} + \sum_{i=1}^{n}(p_i + (m_l+1)\gamma_i\beta_{n-i}) + (b-a)c(m_l+1)]\ ,$$

$$p^{(1)} = \frac{1}{(n+1)!}(\tilde{a} + \tilde{b}\beta_1 + 2B + \sum_{i=0}^{n} p_{0i}) + \frac{(b-a)q}{n!}\ ,$$

$$\tilde{a} = \max_{[a,b]}\left|\frac{\partial f}{\partial x}\right|,\ \tilde{b} = \max_{[a,b]}\left|\frac{\partial f}{\partial y}\right|,\ \gamma_i = \max_{s,[a,b]}\left|\frac{\partial P_i}{\partial y^{(s)}}\right|,\ \beta_i = \max_{[a,b]}\left\{\left|y^{(i)}\right|,\left|\tilde{y}^{(i)}\right|\right\},$$

$$c = \max_{s,[a,b]}\left|\frac{\partial F}{\partial y^{(s)}}\right|,\ q = 0.5(A + C\sum_{s=0}^{m_l}\beta_{s+1}),\ p_{oi} = \beta_{n-i}(l_i + \gamma_i\sum_{s=0}^{m_l}\beta_{s+1}) + 2\alpha_i\beta_{n+1-i},$$

$$A = \max_{[a,b]}\left|\frac{\partial F}{\partial x}\right|,\ B = \max_{[a,b]}\left\{\left|F\right|,\left|\tilde{F}\right|\right\}$$

3.7.4.2 Solution of Equation $y^{(n)} = f(x, y, y¢,\ldots,y^{(n-1)})$

We shall apply the approximate method of solution presented in Sect. 3.7.4.1 to integro-differential equations to solve the initial value problem:

$$y^{(n)} = f(x,y,y',\ldots,y^{(n-1)}),\ (x,y)\in G, \tag{3.36}$$

$$y(x_0) = y_0, y^{(j)}(x_0) = y^{(j)}{}_0, j = 1,\ldots,n-1,\ (x_0,y_0)\in G, \tag{3.37}$$

where the function f satisfies the Lipschitz condition

$$\left|f(x, y+\delta_0,\ldots, y^{(n-1)}+\delta_{n-1}) - f(x, y, y',\ldots, y^{(n-1)})\right| \le K\sum_{i=0}^{n-1}\left|\delta_i\right| \tag{3.38}$$

Let us divide the interval $[a,b]$ by a sequence of points $x_0 = a, x_1, \ldots, x_s = b$ into elementary intervals. Let $E = \{x_0, \ldots, x_s\}$. On each interval $[x_\nu, x_\nu + {}_1]$, let us solve the initial value problem:

$$y^{(n)} = f(x, \hat{y}_\nu, \hat{y}'_\nu,\ldots,\hat{y}_\nu^{(n-1)}), \quad (x,\hat{y}_\nu)\in G, \quad \nu = 0,1,\ldots,s-1;$$

$$y(x_\nu) = \hat{y}_\nu, \quad y^{(j)}(x_\nu) = \hat{y}_\nu^{(j)}, \quad j = 1,\ldots,n-1, \quad (x_\nu,\hat{y}_\nu)\in G$$

Then, if the function f satisfies condition (3.38) and $\max_i[x_{i+1} - x_i] = h$, the solution of this problem

$$\hat{y} = \{y_0, \hat{y}_1,\ldots,\hat{y}_s\}, \quad \hat{y}^{(j)} = \{y_0^{(j)}, \hat{y}_1^{(j)},\ldots,\hat{y}_s^{(j)}\}, \quad j = 1,\ldots,n-1$$

when $h \to 0$ tends to the solution of Eqs. (3.36), (3.37) and the estimation for the rate of convergence is as follows:

$$\max_l\left|y_r^{(n-l)} - \hat{y}_r^{(n-l)}\right| \le \varepsilon_0\left(1+h\alpha_0\right)^r + \frac{h\alpha_1}{2\alpha_0}\left[\left(1+h\alpha_0\right)^r - 1\right], r = 1,2,\ldots,s$$

where

$$\alpha_0 = \sum_{i=1}^{n-1}\frac{h^{i-1}}{i!} + K\left(n + \frac{h}{2}\sum_{j=0}^{n-1}\sum_{s=0}^{n-j-2}\frac{h^s}{s!}\right),$$

$$\alpha_1 = K\sum_{j=0}^{n-1}\left[M\frac{h^{n-j-1}}{(n-j-1)!} + \sum_{s=0}^{n-j-2}N_{s+j+1}\frac{h^s}{s!}\right],$$

$$M = \max_{[a,b]}\left|f(x, y, y',\ldots, y^{(n-1)}\right|, N_{s+j+1} = \max_r\left|\hat{y}_r^{(s+j+1)}\right|$$

If the initial conditions are exactly given, the error estimation has the form:

$$\max_{1\le l\le n-1}\left|y_r^{(n-l)} - \hat{y}_r^{(n-l)}\right| \le Dh^2, \quad r = 1,\ldots,s$$

where

$$N = \max_{1 \le r \le s} |N_r|, \; D = K(M + nD) \frac{\left[1 + he^h\left(1 + 0.5Knh\right) + Kn\right]^r - 1}{4\left[1 + Kh\left(0.5h + e^{-h}\right)\right]}$$

3.7.4.3
Solution of a System of Ordinary Differential Equations

For simplicity, let us confine ourselves to the important case of equations, having the canonical form

$$y_i^{(m_i)}(t) = f_i(t, y_1, y_1', \ldots, y_1^{(m_1 - 1)}, \ldots, y_n^{(m_n - 1)}), \quad i = 1, \ldots, n \tag{3.39}$$

The system (3.39) can be replaced by an equivalent system of $m = m_1 + \cdots + m_n$ equations of the first order, relative to the derivatives for all m unknown functions. Then one of the standard software can be used to solve the last system.

Let the functions f_i, $i = 1,\ldots,n$ be continuous and differentiable with respect to all arguments. Let us suppose that the solution of system (3.39) with the initial conditions

$$y_i(t_0) = (y_i)_0, \quad y_i'(t_0) = (y_i')_0, \quad \ldots \quad , \quad y_i^{(m_i - 1)}(t_0) = (y_i^{(m_i - 1)})_0$$

exists and is unique in $t_0 \le t \le T$.

Divide the interval $[t_0,T]$ into elementary intervals $\Delta k = [t_k, t_{k+1}]$ by a sequence of points $t_0 < t_1 < \ldots < t_l = T$. On each such interval, let us search the solution of system (3.39) in the form of a series:

$$\tilde{y}_i(t) = \tilde{y}_i(t_k) + \sum_{j=1}^{m_i - 1} \frac{(t - t_k)^j}{j!} \tilde{y}^{(j)}(t_k) + \frac{(t - t_k)^{m_i}}{(m_i)!} (\tilde{f}_i)_k \,,$$

$$\tilde{y}_i^{(j)}(t_k) = \tilde{y}_i^{(j)}(t_{k-1}) + \sum_{s=1}^{m_i - j - 1} \frac{(t_k - t_{k-1})^s}{s!} \tilde{y}_i^{(s+j)}(t_{k-1}) + \frac{(t_k - t_{k-1})^{m_i - j}}{(m_i - j)!} (\tilde{f}_i)_k \,, \quad (j = 1, \ldots, m_i - 1)$$

The error of such a solution can be easily estimated, considering the exact expansion of the functions $y_i(t)$ and $y_i^{(j)}(t)$ in a Taylor series:

$$\left|\varepsilon_i(t_{k+1})\right| \le \left|\varepsilon_i(t_k)\right| + \sum_{j=1}^{m_i - 1} \frac{h_k^j}{j!} \left|\varepsilon_i^{(j)}(t_k)\right| + M_i \frac{h_k^{m_i + 1}}{(m_i + 1)!} + M_i \frac{h_k^{m_i}}{(m_i)!} \sum_{j=1}^{n} \sum_{s=0}^{m_j - 1} \left|\varepsilon_j^{(s)}(t_k)\right| \tag{3.40}$$

$$\left|\varepsilon_i^{(j)}(t_k)\right| \le \left|\varepsilon_i^{(j)}(t_{k-1})\right| + \sum_{s=1}^{m_i-j-1} \frac{h_{k-1}^s}{s!}\left|\varepsilon_i^{(s+j)}(t_{k-1})\right| + M_i\left(\frac{h_k^{m_i-j+1}}{(m_i-j+1)!} + \frac{h_k^{m_i-j}}{(m_i-j)!}\sum_{j=1}^{n}\sum_{s=0}^{m_j-1}\left|\varepsilon_j^{(s)}(t_{k-1})\right|\right) \tag{3.41}$$

where

$$M_i = \max_{[t_0,T]}\left\{\left|\frac{\partial f_i}{\partial t}\right|,\left|\frac{\partial f_i}{\partial y_1}\right|,\ldots,\left|\frac{\partial f_i}{\partial y_n^{(m_n-1)}}\right|\right\}$$

Formulas (3.40) and (3.41) give a recurrent estimation for error. From them it is possible to obtain an error estimation applicable to the entire interval $[t_0,T]$:

$$E_k \le \varepsilon_0(1+hp_0)^k + \frac{hP_1}{p_0}\left[\left(1+hp_0\right)^k - 1\right],$$

where the following notations are introduced

$$h = \max_k h_k,\ E_k = \max_{i,j}\left|\varepsilon_i^{(j)}(t_k)\right|,\ M = \max_i M_i,\ \nu = \min_s m_s,$$

$$P_1 = M\frac{h^{\nu-1}}{(\nu+1)!},\ p_0 = mM\frac{h^{\nu-1}}{\nu!} + \sum_{s=1}^{\mu-1}\frac{h^{s-1}}{s!},\ \mu = \max_s m_s$$

3.7.4.4 Solutions of Equations with Known Moments on the Right Side

Let us consider a particular case of Eq. (3.22):

$$L[y] = y^{(n)} + \sum_{i=1}^{n} p_i y^{(n-i)} = f(x), \qquad x \ge 0 \tag{3.42}$$

where p_i is the constant coefficients, $f(x)$ is a single-valued and differentiable function, $f(x) \to 0$ as $x \to \infty$ and its moments are known:

$$M_\nu f(x) = \int_0^\infty x^\nu f(x)dx < \infty, \qquad \nu = 0,1,\ldots,m \tag{3.43}$$

It is necessary to solve the Eq. (3.42) with the following initial conditions:

$$y^{(s)}(x_0) = y_0^{(s)} \qquad (s = 0,1,\ldots,n-1)$$

We shall approximate $f(x)$ in the following manner:

$$f(x) \approx \exp(-kx)\sum_{i=0}^{m} a_i x^i = P_m(x)\exp(-kx), \tag{3.44}$$

where $m > 0$ is an integer, $k > 0$ and a_i are constants to be determined. Then, from (3.43) and (3.44) we have

$$\tilde{M}_\nu f(x) = \int_0^\infty x^\nu \left[\exp(-kx) \sum_{i=0}^m a_i x^i \right] dx = \sum_{i=0}^m a_i \frac{(\nu+i)!}{k^{\nu+i+1}}$$

k is fixed from the conditions of best approximation by (3.44). Then the Eq. (3.42) is replaced by the approximate equation:

$$L[\tilde{y}] = \exp(-kx) \sum_{i=0}^m a_i x^i, \tag{3.45}$$

which can be easily solved. For the error $\varepsilon(x) = y(x) - \tilde{y}(x)P_m(x)$, we obtain an equation from (3.42) and (3.45):

$$L[\varepsilon(x)] = f(x) - \exp(-kx)P_m(x) \equiv R_m(x) \tag{3.46}$$

Solving Eq. (3.46), we have

$$|\varepsilon(x)| \leq \frac{\eta}{n!} \sum_{k=0}^\infty \frac{(b-a)^{n+k}}{k!} M^k,$$

where

$$M = \max_{a \leq s \leq x \leq b} |K(x,s)|, \ \eta = \max_{[a,b]} |R_m(x)|, \ K(x,s) = \sum_{i=1}^n p_i \frac{(x-s)^{i-1}}{(i-1)!}$$

3.7.4.5
Refinements of Approximate Solutions of Volterra Integral Equations

Let us consider the Volterra integral equation of the first and second kind arising in remote monitoring problems:

$$\lambda \int_a^x G(x,y)\varphi(y)dy = g(x), \tag{3.47}$$

$$\varphi(x) - \lambda \int_a^x K(x,y)\varphi(y)dy = f(x), \tag{3.48}$$

where $x \in [a,b]$, the kernel $K(x,y)$ and its derivatives $K'_x(x,y)$ are continuous in the region $R\{a \leq y \leq x \leq b\}$, $f(x)$ is a continuously differentiable function in (a,b), the kernel $G(x,y)$ and $g(x)$ are twice continuously differentiable functions of x, $G(x,x) \neq 0$. Then, as it is known, Eqs. (3.47) and (3.48) have unique solutions $\varphi_1(x)$ and $\varphi_2(x)$, respectively, which are continuous and differentiable in $[a,b]$ for any value of λ. The case when $G(x,x) = 0$ for some point in the interval $[a,b]$ or for the entire interval needs special consideration. In our case, Eq. (3.47) is equivalent to the equation of the second kind:

$$\varphi(x) + \int_a^x \frac{G'_x(x,y)}{G(x,x)} \varphi(y)dy = \frac{g'(x)}{\lambda G(x,x)}$$

Therefore, the argument used to find an approximate solution of (3.48) is valid for Eq. (3.47) as well.

In the traditional way, the Eq. (3.48) is solved by replacing the integral of the equation by a finite sum of some quadratic formula. Applying this approach, let us divide the interval $[a,b]$ by a sequence of points $x_0 = a < x_1 < x_2 < ... < x_m = b$ into elementary intervals $\Delta j = [x_j, x_j + {}_1]$, and instead of (3.48) let us write the equation:

$$\varphi(x_j) - \lambda \sum_{i=0}^{j-1} \int_{x_i}^{x_{i+1}} K(x_j, y)\varphi(y)dy = f(x_j), \qquad (j = 0,1,...,m) \tag{3.49}$$

Further, because of the assumptions made on $\varphi(x)$ and $K(x,y)$, we can write

$$\varphi(x) = \varphi(x_i) + (x - x_i)\varphi'(x_i) + \frac{(x - x_i)^2}{2!}\varphi''(\xi_i), \tag{3.50}$$
$$(x_i \le \xi_i \le x \le x_{i+1})$$

and assuming the existence and differentiability of $K'_y(x, y)$ we have

$$K(x_j, y) = K(x_j, x_i) + (y - x_i)K'_y(x_j, x_i) + \frac{(y - x_i)^2}{2}K''_{yy}(x_j, \eta_i), \tag{3.51}$$
$$(x_i \le \eta_i \le y \le x_{i+1})$$

substituting (3.50) and (3.51) in (3.49) we get:

$$\varphi(x_j) - \lambda \sum_{i=0}^{j-1} \{K(x_j, x_i)\varphi(x_i) + \frac{h_i}{2}[K(x_j, x_i)\varphi'(x_i) + K'_y(x_j, x_i)\varphi(x_i)] +$$
$$\frac{h_i^2}{3}K'_y(x_j, x_i)\varphi'(x_i)\}h_i + R_j = f(x_j)$$

where

$$R_j = -\lambda \sum_{i=0}^{j-1} \{\frac{h_i^3}{2}\varphi''(\xi_i)[\frac{1}{3}K'_y(x_j, x_i) + \frac{h_i}{4}K'_y(x_j, x_i)] +$$
$$\frac{h_i^3}{2}K''_{yy}(x_j, \eta_i)[\frac{1}{3}\varphi(x_i) + \frac{h_i}{4}\varphi'(x_i)] + \frac{h_i^5}{20}\varphi''(\xi_i)K''_{yy}(x_j, \eta_i)\}$$

Neglecting the small quantity R_j in this expression, we get a recurrent formula for determining $\varphi(x_j)$ from the values of the function $\varphi(x)$ at $x = x_0, x_1, ..., x_{j-1}$. By differentiating (3.48), we get a formula to calculate the values of the derivative $\varphi'(x)$:

$$\varphi'(x) = f'(x) + \lambda K(x, x)\varphi(x) + \lambda \int_a^x K'_x(x, y)dy$$

From this equation, we have at $x = x_i$

$$\varphi'(x_i) = f'(x_i) + \lambda K(x_i, x_i)\varphi(x_i) + \lambda \sum_{s=0}^{i-1} \int_{x_s}^{x_{s+1}} K'_x(x_i, y)\varphi(y)dy \tag{3.52}$$

where

$$\varphi(x_0) = f(a), \ \varphi'(x_0) = f'(a) + \lambda K(a,a)\varphi(a)$$

Neglecting the quantity

$$r_i = \lambda \sum_{s=0}^{i-1} \{K''_{xy}(x_i, Q_s)\frac{h_s^2}{2}\left[\varphi(x_s) + \frac{2}{3}\varphi'(x_s)h_s + \frac{h_s^2}{4}\varphi''(\xi_s)\right] + K'_x(x_i, x_s)\frac{h_s^2}{6}\varphi''(\xi_s)\}, \quad (x_s \le \xi_s, Q_s \le y \le x_{s+1})$$

Eq.(3.52) can be written as:

$$\tilde{\varphi}'(x_i) = f'(x_i) + \lambda K(x_i, x_i)\tilde{\varphi}(x_i) + \lambda \sum_{s=0}^{i-1} h_s K'_x(x_i, x_s)\left[\tilde{\varphi}(x_s) + \frac{h_s}{2}\tilde{\varphi}'(x_s)\right]$$

We get finally the equation for determining $\tilde{\varphi}(x_j)$:

$$\tilde{\varphi}(x_j) - f(x_j) - \lambda \sum_{i=0}^{j-1} \left[\tilde{\varphi}(x_i)K(x_j, x_i)l_{ji} + h_i\gamma_{ji}\delta_i\right]h_i = 0, \tag{3.53}$$

where, for brevity, the following notations are introduced:

$$l_{ji} = 1 + \frac{h_i}{2}[\lambda K(x_i, x_i) + \frac{K'_y(x_j, x_i)}{K(x_j, x_i)}[1 + \frac{2}{3}\lambda h_i K(x_i, x_i)]]$$

$$\gamma_{ji} = \frac{1}{2}K(x_j, x_i)\left[1 + \frac{2}{3}h_i \frac{K'_y(x_j, x_i)}{K(x_j, x_i)}\right]$$

$$\delta_i = f'(x_i) + \lambda \sum_{s=0}^{i-1} h_s K'_x(x_i, x_s)\left[\tilde{\varphi}(x_s) + \frac{h_s}{2}\tilde{\varphi}'(x_s)\right]$$

Thus, starting the calculation by formula (3.53) from $j = 1$. We get the values of solution $\tilde{\varphi}_1, \tilde{\varphi}_2, \ldots, \tilde{\varphi}_m$ with the error $\varepsilon(x_i)$. Let us estimate the value of modulus of the error $\varepsilon(x)$ from above. Let us introduce the following notation:

$$M = \max_{[a,b]}|\varphi(x)|,\ N = \max_{[a,b]}|\varphi'(x)|,\ L = \max_{[a,b]}|\varphi''(x)|,$$

$$Q = \max_{R}\{|K'_y(x,y)|, |K'_x(x,y)|\},\ G = \max_{R}|K(x,y)|,$$

$$B = \max_{R}\{|K''_{yy}(x,y)|, |K''_{xy}(x,y)|\},\ h = \max_{s}|h_s|,$$

$$l = \frac{m}{6}\left[B\left(M + \frac{3h}{4}N\right) + L\left(G + \frac{3}{4}Qh + \frac{3}{10}Bh^2\right)\right],$$

$$n = \frac{|\lambda|Bm}{2}\left(M + \frac{2h}{3}N + \frac{h^2}{4}L\right) + \frac{h}{6}BL$$

Then we have

$$|\varepsilon_j| \le |\lambda| h\left(G + \frac{2}{3}Qh\right)\sum_{i=0}^{j-1}\left(|\varepsilon_i| + \frac{h}{2}|\varepsilon_i'|\right) + |\lambda| l h^3,$$

$$|\varepsilon_i'| \le |\lambda| G|\varepsilon_i| + |\lambda| Qh\sum_{s=0}^{i-1}\left(|\varepsilon_s| + \frac{h}{2}|\varepsilon_s'|\right) + nh^2,$$

where $\varepsilon_j = \varphi(x_j) - \tilde{\varphi}(x_j)$, $\varepsilon_i' = \varphi'(x_i) - \tilde{\varphi}'(x_i)$.

From these formulas, we get roughly

$$|\varepsilon_i'| \le |\lambda|\left(G|\varepsilon_i| + hQ\sum_{s=0}^{i-1}|\varepsilon_s|\right) + \left(n + |\lambda| QNm\right)h^2,$$

$$|\varepsilon_j| \le |\lambda| h\left(G + \frac{2}{3}Qh\right)\sum_{i=0}^{j-1}\left[\left(1 + \frac{|\lambda| Gh}{2}\right)|\varepsilon_i| + \frac{|\lambda| Gh}{2}\sum_{s=0}^{i-1}|\varepsilon_s|\right] + |\lambda| th^3,$$

where

$$t = l + \frac{h}{2}m\left(G + \frac{2}{3}Qh\right)\left(n + |\lambda QNm|\right)$$

Let us denote

$T = |\lambda|(G + 2Qh/3)$, $b = 1 + |\lambda|Gh/2$, $r = |\lambda|(t + QMTm^2)$

Then we get

$$|\varepsilon_j| \le hTB\sum_{i=0}^{j-1}|\varepsilon_i| + rh^3$$

or, finally, we have

$$|\varphi_j - \tilde{\varphi}_j| \le E_j = \varepsilon_0 Z_1^j + rh^3(1 - Thb_j)^{-1} \le \varepsilon_0 + rh^3(1 - Thb_j)^{-1},$$

where $h \le (Tb)^{-1}$, Z_1 is the real root of the equation

$Z^{j+1} - TbhZ^{j-1}/(Z-1) = 0$

between $Z = 1$ and $Z = hTb$.

3.8 Experiments Using the Global Simulation Model

Numerical experiments using the GSM require a set of scenarios including the parameterization of future development of society, science, engineering, and political structures based on the observable trends (Kondratyev et al. 1997).

The GSM variations were used in the evaluation of consequences of nuclear war, the forecast of estimates of the distribution of excess CO_2 in biospheric reservoirs and the examination of other scenarios of anthropogenic activities (Krapivin et al. 1990, 1996a,c, 1997a–c, 1998a; Krapivin and Phillips 2001a,b). By way of developing such studies we will consider a few examples of applying the GSM for solving some global – and regional – scale problems. The time intervals are one century with a time discretization step of 1 year. Some other scenarios are also discussed in various chapters of this volume (Turner et al. 1993).

The numerical experiments were performed using published data on natural processes. The data sets thus obtained reflect inconsistently the spatial structure of the components shown in Fig. 3.8. The data on soil-plant formations are displayed in a geographic grid of 4° × 5° lat./long. They include information concerning resources of the phytomass and dead organic matter as well as plant productivity.

The data sets on the ocean ecosystems are given in accordance with the four-part division of the World Ocean (corresponding to the Atlantic, Pacific, Indian, and Arctic Oceans). The data sets on the socio-economic and demographic structures are synthesized for seven regions: North America, the Russian Federation and other countries of the Former Soviet Union, Western Europe, Latin America, South and Southeast Asia, the Middle East and North Africa, and China and other Asian countries with centrally planed economies.

3.8.1 Plant Cover Restoration

One of the principal aspects of anthropogenic impact on the environment is the evaluation of the consequences of CO_2 emissions into the atmosphere. The published results estimating the greenhouse effect and excess CO_2 distribution in the biosphere, which bear on this problem, are widespread and sometimes contradictory or else too flatly stated. This is a natural consequence of all kinds of simplifications adopted in modeling the global CO_2 cycle. The GSM makes it possible to avoid excessive simplifications in modeling the system of biospheric relations. To demonstrate this possibility we shall consider a theoretical situation involving modification of the structure shown in Fig. 3.8. Specifically, we are going to see what global changes will result from changes in the structure of the soil-plant formations. In this experiment the symbols given in Fig. 3.8 are modified.

Table 3.15 shows the dynamic of the ratio between the rates of carbon absorption by the living phytomass and dead organic matter within the framework of the scenario suggested by Keeling and Bacastow (1977) for the structure of the soil-plant formations illustrated in Fig. 3.8. It can be seen that the role played by some soil-plant formations substantially changes with time. An increase in the atmospheric concentration of CO_2 entails its absorption by the forests of the northern hemisphere. At the same time, the CO_2 absorption rate of tropical xerophytic open woodland and tropical savannas decreases. It is also seen that the soil-plant formations of the northern latitudes are more stable than those of other latitudes.

Table 3.16 gives results of hypothetical changes in the plant-cover structure illustrating the role played by the soil-plant formation in the dynamics of carbon-dioxide. It can be seen that a structure of plant cover exists for which the dy-

Table 3.15. The role of soil-plant formations in the absorption of excess atmospheric carbon dioxide under Keeling-Bacastow conditions. [Σ is the carbon dioxide absorption by phytomass divided by carbon dioxide absorbed by dead organic matter (DOM) in given years]

Soil-plant formation (see Fig. 3.8)	Σ 1990	2020	2100	Soil-plant formation (see Fig. 3.8)	Σ 1990	2020	2100
A	0.44	0.41	0.43	*W*	0.19	0.14	0.08
C	0.52	0.46	0.37	*E*	0.36	0.33	0.22
M	0.51	0.45	0.37	*H*	0.29	0.27	0.25
L	0.56	0.51	0.42	*Q*	0.38	0.33	0.21
F	1.83	1.62	1.32	*Z*	2.94	2.78	2.45
D	3.04	2.89	2.43	*Y*	3.07	2.91	2.57
G	2.61	2.53	2.14	*N*	0.22	0.25	0.44
R	1.94	1.86	1.51	*J*	0.57	0.59	0.62
P	2.36	2.22	1.86	*T*	0.79	0.71	0.38
U	0.41	0.36	0.28	*I*	1.89	2.17	1.95

Table 3.16. Consequences of changes in the structure of the soil-plant formations for the dynamics of CO_2 distribution due to hypothetical substitutions in the soil-plant formation structure of Fig. 3.8

The symbols in Fig. 3.8		C_1/C_2 (C_1 and C_2 are the CO_2 absorptions for the new and the old structures of Fig. 3.8, respectively)		
Old	New	Years		
		1990	2020	2100
A	*L*	2.81	2.14	1.21
C	*L*	0.97	0.94	0.95
M	*L*	1.68	1.15	1.01
F	*D*	1.69	1.57	1.11
G	*D*	2.12	1.67	1.07
R	*D*	4.03	3.70	1.18
P	*Z*	3.19	2.68	1.24
U	*Z*	22.53	20.73	9.18
W	*Z*	23.14	19.44	5.63
E	*Z*	100.18	77.75	2.84
H	*Z*	194.61	155.50	1.86
Q	*Z*	801.43	777.50	3.47
Y	*Z*	1.44	1.39	1.27
N	*Z*	70.01	62.20	4.32
J	*Z*	-5.99	5.09	1.17
T	*Z*	26.56	25.92	1.04
I	*Z*	18.16	16.37	0.98
#	*Z*	0.94	1.12	0.97

namics of carbon dioxide concentration in the atmosphere does not produce long-term depressive effects. Even in the presence of an increasing rate of CO_2 emissions, contained within the scenario of Keeling and Bacastow (1977), mechanisms exist in the biosphere which are capable of damping this increase. With a certain delay (30 years for the dead organic matter), the soil-plant formations play a substantial role in the stabilization of the carbon balance in the biosphere.

Whereas the role of the ocean in excess CO_2 absorption so far has been exaggerated, that played by the land plants has remained underestimated. For instance, from Table 3.16, it can be seen that the maximum possible CO_2 concentration in the atmosphere may be reduced by 80% solely through changing the plant cover structure. In other words, if within the land ecosystem structure existing today the maximum CO_2 concentration in the atmosphere grows fivefold in comparison to the preindustrial level (Krapivin and Vilkova 1990), the plant-cover structure modifications shown in Table 3.16 will provide a decrease in this growth down to only a factor of 4. This threshold value may be further lowered by 60% by increasing the areas occupied by the ecosystems *F*, *D*, *G*, *Z*, and *Y*. In a hypothetical situation where coniferous forests occupy the zone of 35°N–55°N and evergreen tropical forests the zone of 35°N–35°S, the maximum CO_2 concentration excess will not be greater than 180% of the preindustrial level. Taking into account the goal of a 25% decrease in CO_2 emission adopted by many countries, the expected CO_2 concentration growth may be kept down to the 855-ppm level solely via reconstruction of soil-plant formations. Of course, such reconstructions should be brought into agreement with the climate fluctuations. So far, a purely hypothetical situation has been considered, and the feedback to the climatic system is parameterized by a simple diagram of the CLIMATE unit (see Fig. 3.6).

3.8.2 Diversion of Siberian Rivers to Central Asia

It is assumed that some water is taken from all Siberian rivers and evaporates over the central Asian territory uniformly. The average annual water volume removed from Siberian rivers is assumed to be 1%. The results of calculations performed using the GSM are the following. As may be seen, changes of 1–30% in climate and precipitation become significant on a global scale. For instance, whereas in Europe the temperature drops by 0.5°C and precipitation increases by 10%, the climate of the North Africa region becomes still hotter and more arid, and in Siberia an average 0.7°C drop in temperature is accompanied by a 15% precipitation decrease. In central Asia, a 0.3°C temperature drop and a 30% precipitation increase ensue. These changes are entailed by modifications made in the WATER unit of the hydrological cycle of the GSM.

3.8.3 Forecast for a Regional-Level Ecosystem Dynamics

To demonstrate the efficiency of the GSM in this case we consider the Sea of Okhotsk ecosystem described in the paper of Legendre and Krapivin (1992). The

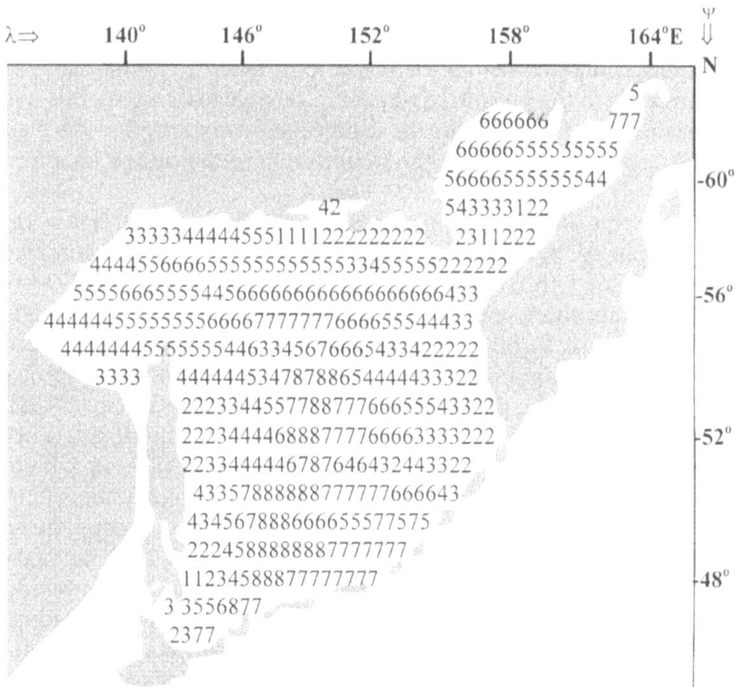

Fig. 3.16. Application of the GSM to the benthos biomass simulation in the Okhotsk Sea under climate conditions of February 1992 (g m^{-2}). Scale: *1* > 900; *2* 500–900; *3* 250–500; *4* 100–250; *5* 50–100; *6* 10–50; *7* 1–10; *8* < 1

data concerning the distribution of temperature, precipitation, wind direction and velocity, cloudiness, atmospheric humidity, and ice cover were kindly supplied by the Hokkaido Meteorological Research Institute. $\Delta\varphi = \Delta\lambda = 2°$ is used (Shinohara and Shikama 1988). Computational experiments were performed using the simulation unit of the GSM described in Chapter 5. Boundary conditions for the water and temperature regimes were obtained from the other units of the GSM.

Figure 3.16 shows forecast estimates for the ice conditions. Comparison with results obtained from the JERS-1 satellite revealed qualitative agreement between the predicted and measured ice edge location. According to the forecast made in November 1999, the sea near the northern coast of Hokkaido in February 2000 would have a stable ice cover, which was actually the case. The forecast given in Fig. 3.16 has proved false only in the western coastal zone of Kamchatka. According to satellite-survey data for 7 February 2000, there was an open-water strip about 25 km wide in this zone, which makes up 12% of the spatial resolution of the GSM for this region.

Another evaluation of forecast accuracy was made in 1990–1992. Comparison of pictures taken from on-board the Almaz-1 satellite revealed stability of the ice-condition predictions in the Sea of Okhotsk. It also became clear that improvement of forecast accuracy in detail requires a finer geographic grid in the database.

The data in Fig. 3.16 demonstrate the capability of the model for estimating the abiotic parameters of the Sea of Okhotsk ecosystem. The qualitative reliability of this estimate follows from comparison of the simulated results with the in situ measurements. According to this study, the oxygen regime in the Sea of Okhotsk is stable and characterized by a decrease in the oxygen concentration southwards in the near-bottom layer. For instance in the Southern Deep-Sea Depression the oxygen saturation is 20–28 %, which is close to the GSM estimated values (10–30 %). Whereas the oxygen saturation near the western Kamchatka littoral reaches 50–70 %, the GSM gives 40–70 % for its quantity.

As follows from the Sea of Okhotsk subsystem structure, the GSM permits the evaluation of the distribution of the benthos biomass, fish species, phytoplankton, and zooplankton, but comparison with measurement results is difficult because of inconsistency of the latter (see Chap. 7).

3.8.4
Other Global Simulation Model Applications

Below are some instances of simulation experiments using the GSM for prediction of natural ecosystems on the global scale. We limit our consideration to a brief review of the results obtained.

One of important tasks in the present-day global ecology is the evaluation of the carbon dioxide quantity absorbed by plants in various climatic zones. The GSM allows this for a $\Delta\varphi \times \Delta\lambda$ geographic grid and for soil-plant formation types specified in Fig. 3.8. A fragment of such a calculation for the Russian territory is given in Table 3.12.

Let us present the next model experiment results. It would be useful for cognitive purposes to follow up the effect of various parameters on the dynamics of the biosphere as regards the rates of land development in agriculture and the distribution of investments. Simulation experiments show that if by 2000 the area under agricultural land use has increased by 10 %, livestock production by 10 % and the intensity of generating pollutants has decreased by 10 %, then compared to the situation when there are no such changes, the population size will increase by 3.1 %, environmental pollution by 20.8 % and atmospheric cloudiness by 1.2 %, while the requirements for capital investment in agriculture will fall by 4.7 %. Food supplies will be improved only by 6.8 % and the level of well-being will rise by 5.8 %. However, if the areas of all the lands suitable for agricultural use have been developed by the year 2050, the average worldwide food supply of the population by 2200 will make up 94.7 % of the 1970 level. The environmental parameters will remain within the permissible range.

The GSM helps in formulating requirements to be placed on the database. An example is given in Table 3.17, showing the effect of errors in the GSM parameters on the stability of the simulation results. A comparison of acceptable parameter errors with maximum deviations from the simulated results which will follow after a span of 100 years is made in Table 3.17. An acceptable error is a maximum range of deviation in the GSM parameter within the limits of which the simulated results are stable in the qualitative ranges for each of the GSM parameters. However, this would be a tedious and time-consuming work requiring special investigation.

Table 3.17. Requirements placed on some GSM parameters

Area of soil-plant formation (see Table 3.8)	Allowed error of parameter (%)	Resulting error of the model (%)
$D + F + G$	35	28
$Z + Y$	30	32
$C + M + L$	68	18
$W + X + V$	54	21
$S + T + @$	41	38
Equatorial zone (20°N–20S)	36	29
Tropical summer-rainfall region	32	33
Intermediate zone with winter rainfall	29	35
Total land zone	24	27

Table 3.18. Results of the simulation experiment using the GSM for estimation of the global sulfur cycle

Parameter	Hemisphere	
	North	South
General velocity of sulfur sedimentation in the World Ocean ($mg \cdot m^{-2}\ day^{-1}$)	31.2	24.5
Sulfur moving from the ocean to land ($mg \cdot m^{-2}\ day^{-1}$)	74	21
Sulfur moving from land to the ocean ($mg \cdot m^{-2}\ day^{-1}$)	185	41
Sulphur falling on the surface ($mg \cdot m^{-2}\ day^{-1}$)		
Land (anthropogenic part, %)	3.5 (31)	2.7 (26)
Ocean (anthropogenic part, %)	1.4 (18)	0.9 (14)
Sulfur flow to the oceans from the atmosphere ($mg \cdot m^{-2}\ day^{-1}$)	0.2	0.2
Time interval for doubling of the reduction) in the atmospheric SO_2 (days	0.04	0.04
Anthropogenic sulfur part in the river flows (%)	48	27
Velocity of sulfur accumulation by soil-plant formations ($mg \cdot m^{-2}\ day^{-1}$)		
Forest	34.6	41.3
Bushes	2.7	3.4
Grass	4.1	5.6
Runoff of sulfur from continents to the World Ocean (10^6 t $year^{-1}$)		
North America	21.9	–
South America	2.3	15.4
Europe and Asia	45.5	4.1
Africa	14.2	5.4
Australia	–	2.8

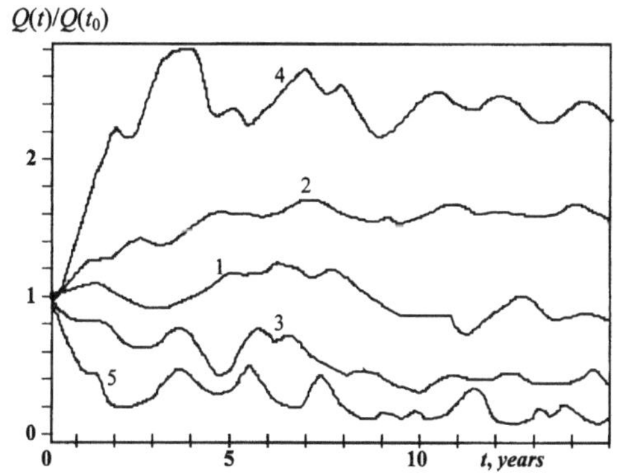

Fig. 3.17.
Dependence of sulfur concentrations on the initial data: *1* data corresponding to Table 3.12; *2* reduced by 50%; *3* increased by 50%; *4* reduced by 70%; *5* increased by 70%. Q is the total sulfur storage in the biosphere

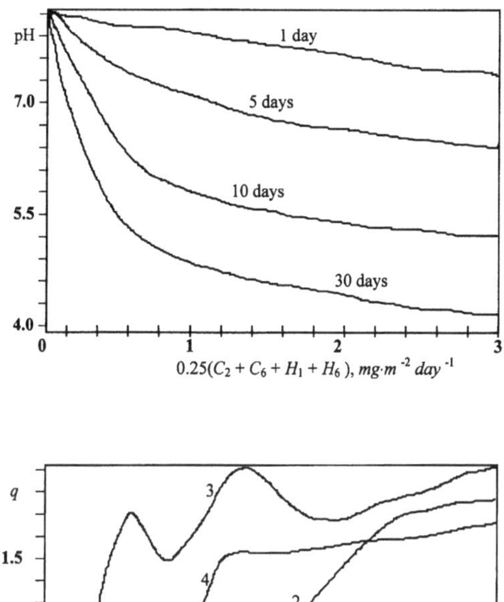

Fig. 3.18.
Dependence of average rain acidity on anthropogenic sulfur flow. The change in the anthropogenic activity is taken to be homogeneous in all territories. The time after the start of the experiment is shown on the curves

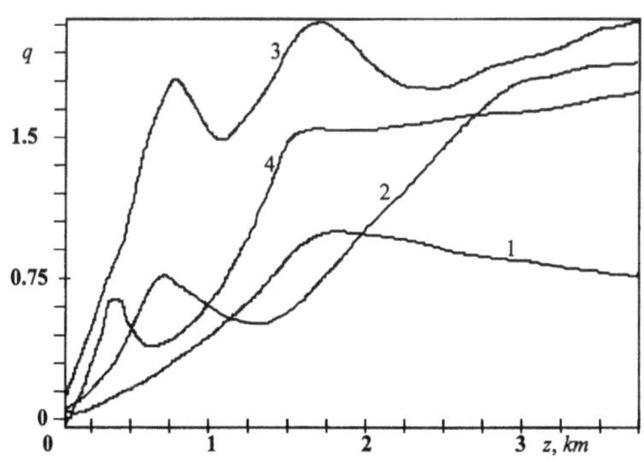

Fig. 3.19.
Average production of H_2S (mg m^{-3} day^{-1}) in the oceans: *1* Arctic; *2* Pacific; *3* Indian; *4* Atlantic. $q = H_{17} + H_{23}$ (mg m^{-2} day^{-1})

The results presented in Figs. 3.17–3.19 and Table 3.18 characterize the sulfur dynamics in the biosphere. Extensive variations in the initial data (± 70%) and in the model parameters (± 25%) do not influence the system dynamics. Increasing the initial sulfur storage brings variations to the system dynamics for only 2 years after a disturbance, while decreasing the initial sulfur storage prevents stability in the system for 5 years.

CHAPTER 4

Modeling of Ocean Ecosystem Dynamics 4

4.1 The World Ocean as a Complex Hierarchical System

Investigation of the structure and functioning of oceanic ecosystems has become an important and rapid developing area of sea biology and environmental studies. Different aspects of relevant problems have been studied in the framework of many international biological and environmental programs. One of the tasks is the substantiation of the possibilities to forecast the behavior of the system under the impact of its variable parameters. Because of the unique and enormous spatial extension of the World Ocean ecosystems, it is difficult to assess their parameters at different times and for the various aquatories. That is why the use of modeling technology is one effective approach to study and solve various problems.

The World Ocean, occupying about 71% of the Earth's surface, provides a substantial amount of food for the population. At the same time, the full volume of organic matter produced by the oceans approximately equals the productivity of the land ecosystems. The total biomass of the nekton is estimated by the value of 5.3 Gt. The extraction of commercial food elements from the World Ocean is about 70×10^6 tons/year. It includes approximately 20% of the protein consumed by humanity. For the time being, traditional fishing is near its acceptable limit ($\approx$ 90–100 million t/year). There is no limit for the commercial exploitation of the ocean ecosystems, in general, since there are many necessities in biological objects which are rarely used by people.

The disparity between the roles of the land and oceanic ecosystems in food production is explained first of all by the fact that on the land intensive agriculture is well developed, while in the oceans it is almost completely undeveloped. That is why the means of increasing the ocean ecosystem productivity is now clear. First, humanity uses mainly the upper trophic chains of the natural ecosystems. It is known that the oceanic ecosystems have long trophic chains with transfer coefficients between trophic levels equal to about 0.1. From the oceans and seas people extract the production of three to five upper levels (fish, crabs, whales, etc.). Between these levels and the phytoplankton there are more than ten trophic levels. Therefore, the problem of optimal exploitation of the oceanic ecosystems exists and its solution is possible in the framework of the complex simulation approach to the study of global change. Here, the question of the artificial improvement of ocean biological community arises.

Sea communities are complex biological systems comprised of populations of different species whose interaction results in the state of dynamic interaction within the community. The spatial structure of the community is largely dependent on the composition of numerous biotic and abiotic factors which depend on the set of oceanological fields. The latter are defined by the general ocean water circulation which includes ebbs and flows, convergence and divergence zones, wind and thermocline, etc.

During the last 10 years, the problem of forecasting the oceanic systems under growing anthropogenic influence has arisen acutely. Also, it is important to have estimations of the anthropogenic role in biosphere dynamics. In recent investigations of global change, it has been shown, for instance, that the role of the oceanic ecosystem in the formation of the greenhouse effect was underestimated, especially as related to the Arctic basin of the World Ocean (Kondratyev 1993).

The influence of oceanic ecosystems on the intensity of the biogeochemical cycles is realized at the boundary of *atmosphere–water* and usually is parameterized with the use of the data observed. However, the vertical structure of the oceanic processes is to be take into account in this context. The Arctic latitudes are characterized by the greatest correlation between the *atmosphere-water* boundary state and vertical hydrophysical processes. An investigation of this correlation is an important problem of global change studies.

The World Ocean has a spotty structure formed by the composition of the temperature, salinity, nutrients, insolation, and hydrophysical parameters. This spottiness of the water surface is reflected in the phytoplankton production. The spotty variability of the topology correlates with the vertical distribution of the phytoplankton biomass. Observations of these correlations give only fragmentary descriptions of the actual existing abiotic, biotic and hydrophysical processes. To obtain the global description of the World Ocean ecosystem dynamics it is necessary to develop elaborate mathematical models. This chapter describes a set of such models.

4.2 Common Principles for the Synthesis of Ocean Ecosystem Models

Each element of the oceanic ecosystem A can be described by means of a set of parameters: $x(t) = \{x_j(t), j = 1, \ldots, N\}$, where t is time. The ecosystem A is characterized by its structure $|A(t)|$ and behavior $A_B(t)$:

$$A(t) = \{|A(t)|, A_B(t\} = F(x(t)) \tag{4.1}$$

Consequently, according to Eq. (4.1), system A has some trajectory in $(N+1)$-space. The abstract formation

$$A_M(t) = F_M(x^M(t)) \tag{4.2}$$

is called the ecosystem A model, where $M \leq N$, $\{x^M\} \in \{x\}$. Parameter M defines the divergence level between the trajectories of the $A(t)$ and $A_M(t)$ ecosystems. Let us introduce some goal function to the ecosystem trajectory

$$V = Q(\{x_i(t)\}) \tag{4.3}$$

Fig. 4.1.
Typical block-diagram of the ocean ecosystem model. The *arrows* show the directions of the energy fluxes

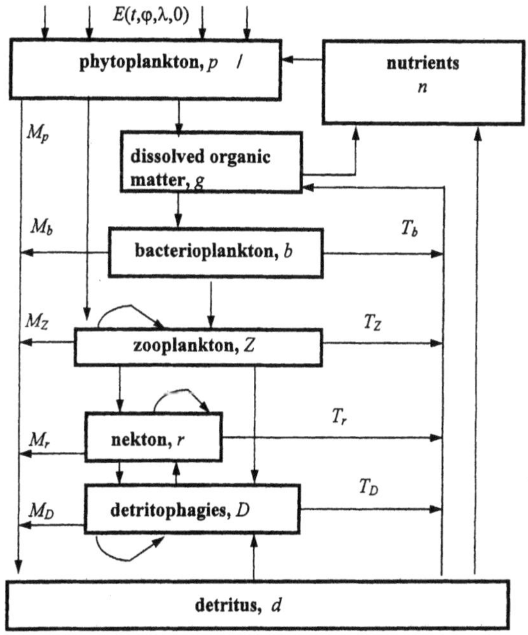

The form of Q is defined by the character of the system's goal. It is considered that the natural evolution process leads A to the optimal system A_{opt}. Therefore, model $A_{M,\text{opt}}$ giving an extreme value of Q is called the optimal model of ecosystem A. The value of divergence between the trajectories of A and $A_{M,\text{opt}}$ is a function of the correspondence of V to the real goal $\underline{A}$ of the ecosystem A. The set $\underline{G} = \{\underline{A}_1, \dots, \underline{A}_r, \dots, \underline{A}_m\}$ of the possible real goals $\{\underline{A}_r\}$ of the ecosystem A can be formed on the basis of the oceanological information. Then, defining

$$A_{M,\text{opt},r} = g_r(\underline{A}_r), \underline{A}_r \in \underline{G} \tag{4.4}$$

we proceed to the restricted set of possible optimal systems $A_{M,\text{opt},r}$ $(r = 1,\dots,m)$, the trajectories of which, together with the A trajectory, are in the space of potential trajectories. Defining $A_{M,\text{opt},r0}$ with minimal divergence from $A(t)$, we can obtain the value for the most probable ecosystem goal: $\underline{A}_{r0} = g_r^{-1}(A_{M,\text{opt},r0}(t))$.

According to the above principles, the construction of the mathematical model for the oceanic ecosystem A (Fig. 4.1) demands either the formation of a detailed description of all space for its states or the creation of a full set of mathematical models describing the occurrence in A of processes of energy exchange between trophic levels and of interactions having biotic, abiotic and hydrophysical origin. Of course, it is supposed that an appropriate set of assumptions about the nature of the balance correlations in the ecosystem exists.

The supposition that the solar energy E is the unique source of energy for all forms of life is assumed to be the basic criterion. According to many theoretical and experimental investigations, the law of insolation distribution across the ocean layers has the exponential form:

$$E(t,\varphi,\lambda,z) = uE_0 \exp[-\int_0^z U(x)dx - \alpha z] + (1-u)E_0 \exp(-\varsigma z) \tag{4.5}$$

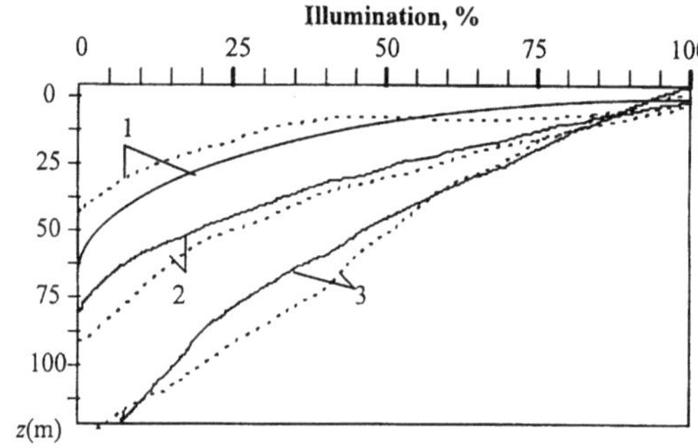

Fig. 4.2. Dependence of the illumination on the depth: theoretical (*continuous curves*) and experimental (*dashed curves*). *1* Ionian Sea, *2* western Pacific Ocean, *3* average for the tropical regions of the World Ocean

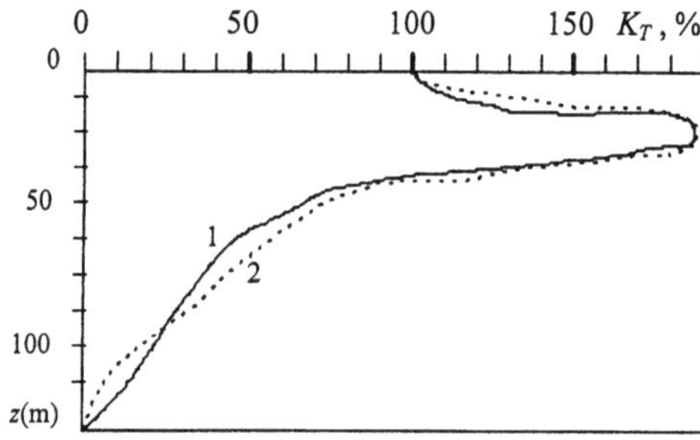

Fig. 4.3. Dependence of photosynthesis on underwater illumination: *1* experimental data, *2* theoretical curve calculated by Eq. (4.7) for $k_2 = 0.05$, $\alpha = 0.015$, $m = 0.4$, $E_{max} = 700$ cal m^{-2} day^{-1}, $E_0 = 3000$ cal m^{-2} day^{-1}

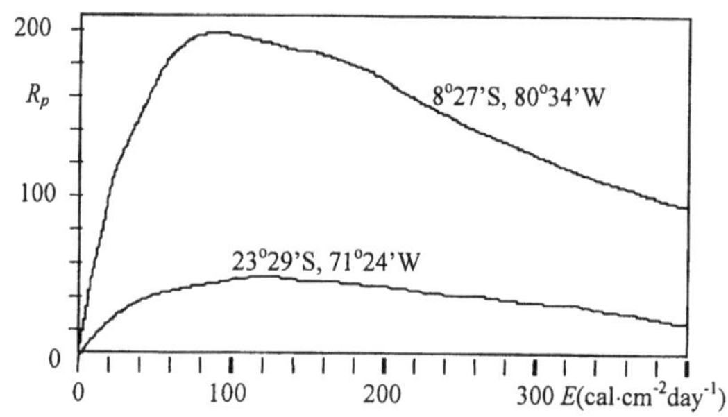

Fig. 4.4. Dependence of photosynthesis R_p (mg m^{-3} day^{-1}) on illumination for two geographical points in the Pacific Ocean off the coasts of Peru (*upper curve*) and Chile (*lower curve*)

where $U(x) = \delta p(t, \varphi,\lambda, x) + \beta d(t, \varphi,\lambda, x) + \nu Z(t, \varphi,\lambda, x)$, $E_0 = E(t, \varphi,\lambda, 0)$ is the surface illumination; α is the absorption coefficient for seawater; δ, β and ν are relaxation coefficients due to attenuation by the phytoplankton p, detritus d and zooplankton Z, respectively; u and ζ are the parameters selected in the concrete situation with the goal of the shortest distance between $E(t, \varphi,\lambda, z)$ and the real data.

Figures 4.2–4.4 give examples of such approximations for the different regions of the World Ocean. Note that the influence of other trophic levels on the water clarity are ignored.

An insolation level influences the photosynthesis R_p. It is known that R_p as a function of E has a maximum under some optimal value E_{max} decreasing with both the increase and decrease in the illumination from this value. The maximum R_p for the various latitudes φ is located at depths which change with the time of year. This variability of the maximal photosynthesis with the depth is the most substantial in the tropical zones. The average location of maximal photosynthesis is situated at 10 to 30 m, where $E_{max} = 65–85$ cal cm^{-2} day^{-1}. Beginning with the depths where $E = 20–25$ cal cm^{-2} day^{-1}, the photosynthesis decreases proportionally to E. The actual depression of the photosynthesis due to illumination is observed for $E > 100$ cal cm^{-2} day^{-1}. Such estimates are different for the northern latitudes where maximal photosynthesis is usually situated near the water surface.

The variation of the photosynthesis on the depth z depends on the water temperature T_W, nutrients n and phytoplankton biomass p, as well as on some other factors which are less important and are not taken into consideration. To simulate this dependence different equations are used to take the limiting role of the elements E, n and p into account. Considering that $\partial p/p\partial z \to 0$ when $n \to 0$ and $\partial p/p\partial z \to$ const with increasing n, the following function is used as the basic correlation to simulate the photosynthesis:

$$R_p\,(t,\varphi,\lambda,z) = k_0(T_W)K_T f_2(p) f_3(n) \tag{4.6}$$

where

$$K_T = Af_1(E),\, A = kA_{max}/E_{max},\, f_1(E) = E\cdot \exp[m(1-E/E_{max})],$$
$$f_2(p) = [1-\exp\{-\gamma_1 p\}],\, f_3(n) = [1-exp\{-\gamma_2 n\}]^{\theta}; \tag{4.7}$$

k is the proportionality coefficient; $k_0(T_W)$ is the function characterizing the dependence of the photosynthesis on the water temperature T_W; A_{max} is the assimilation number in the area of maximal photosynthesis (increase in the phytoplankton species per unit weight); γ_1, γ_2, θ and m are constants, the selection of which determines the specific characteristics of the phytoplankton type. For A_{max} the following estimate is valid: $A_{max} = 5.9\ E_{max}$ for the area of maximal photosynthesis and $A_{max} = 2.7\ E_{max}$ for the other areas. Corresponding to this value, the assimilation number of the tropical phytoplankton in the area of maximal photosynthesis equals 11–12 mg C h^{-1}. For instance, for the Peruvian upwelling $A_{max} = 6.25$ mg C h^{-1}. Thus, the light saturation of the photosynthesis in the equatorial latitudes takes place for $E_{max} = 9$ cal cm^{-2} day^{-1}.

As to the function of $k_0(T_W)$, it is known that the specific variation of the photosynthesis intensity with rising temperature at first increases until it achieves a maximal value for some optimal p temperature and then starts to fall with subsequent increase in the temperature. Many authors use the following approximation: $\ln k_0(T_W) = (T_W - T_{W,opt}) \ln \theta_0$, $0 < \theta_0 \le 2$.

The exponential dependence of the photosynthesis rate on the biogenic elements n (phosphorus, silicon, nitrogen, etc.) as expressed in the formula (4.6) by R_p (mg C m^{-3} day^{-1}) is in reality of course more complex. Nutrients are one of the most important elements of the ecosystem since they regulate the energy fluxes. Nutrients are consumed during the photosynthesis with the rate R_n usually approximated by the expression $R_n = \delta R_p$ where δ is the proportionality coefficient. The replenishment of the nutrients occurs at the expense of the deep waters where the stored nutrients are formed as a result of the chemical processes of decomposition of the dead organic matter. It is assumed that the nutrients stored in the deep ocean waters are unlimited.

The process of dead organic matter decomposition is controlled by the range of abiotic conditions which are specific for the different climatic zones of the World Ocean. The vertical flow of the nutrients is determined by the conditions of the water mixing. In the tropical zones where the vertical water structure has a pronounced three-layer configuration with the occurrence of a layer of sharp temperature change (the thermocline) the vertical movement of the nutrients is restricted by this layer. In those aquatories where the thermocline lies at a depth of 40–100 m, the upper layer is usually made poorer and the flow to this layer takes place only in the upwelling zones. In this case, the average speed of the vertical water transport under the thermocline fluctuates in the range of 10^{-3} – 10^{-2} cm s^{-1}. In the upwelling zones, the vertical water transport achieved is of the order of 0.1 cm s^{-1}.

4.3 Equations Describing the Ocean Ecosystem Dynamics

All depths of oceanic water are considered as unique regions of biogeocenosis in which the basic connecting factor is the organic matter flow produced in the upper layers and then spreading to maximal ocean depths. It is assumed that all model parameters are changed depending on the geographical coordinates and the season and that their parametrical description is realized by means of deterministic models. The food interactions between the trophic levels are proportional to their effective biomass.

A significant role in the trophic chains of the ocean ecosystem is played by the bacterioplankton b. According to many investigations, the bacterioplankton biomass can form condensed bodies which are consumed by the zooplankton Z. This fact is important since the bacterioplankton production of many ocean aquatories occurs at the same level as the phytoplankton production. The bacterial plankton has a changeable energy balance depending on the available food supply. The production of the bacterial plankton R_b is limited by the food abundance (detritus d and dissolved organic matter g excreted during photosynthesis):

$$R_b = k_b b[1 - \exp(-k_{1,d} d - k_{1,g} g)] \tag{4.8}$$

where k_b, $k_{1,d}$ and $k_{1,g}$ are empirical coefficients.

The equation describing the dynamic of the bacterioplankton biomass has the following form:

$$\partial b/\partial t - U_b + \sum_{s \in \Gamma_b} C_{bs} R_s = 0 \tag{4.9}$$

where

$$U_b + V_\varphi \partial b/\partial \varphi + V_\lambda \partial b/\partial \lambda + V_z \partial b/\partial z$$
$$= R_b - T_b - M_b + k_{2,\varphi} \partial^2 b/\partial \varphi^2 + k_{2,\lambda} \partial^2 b/\partial \lambda^2 + k_{2,z} \partial^2 b/\partial z^2;$$

T_b and M_b are the losses of the bacterioplankton biomass at the expense of the rate of exchange and mortality, respectively; Γ_b is the set of the trophic subordination of the bacterial plankton (in a typical case Γ_b consists of Z); C_{bs} is the consumption coefficient of the sth trophic level as regards to the bacterioplankton. Functions T_b and M_b are described by the following formulae (Nitu et al. 2000a):

$$T_b = t_b\, b \tag{4.10}$$

$$M_b = \max\{0, \mu_b\,(b - \underline{B}_{\,b})^{\xi}\} \tag{4.11}$$

where t_b is the specific loss on the exchange; μ_b is the mortality coefficient; $\underline{B}_b$ and ξ are the constants defining the dependence of the bacterial plankton mortality at the expense of its biomass. The coefficients $k_2 = (k_{2,\varphi}, k_{2,\lambda}, k_{2,z})$ determine the process of the ocean water turbulence.

The dynamic equation to describe the phytoplankton biomass balance is

$$\partial p / \partial t - U_p + \sum_{s \in \Gamma_p} C_{ps} R_s = 0 \tag{4.12}$$

where

$$U_p + V_\varphi\, \partial p/\partial \varphi + V_\lambda\, \partial p/\partial \lambda + V_z\, \partial p/\partial z$$
$$= R_p - T_p - M_p + k_{2,\varphi}\, \partial^2 p/\partial \varphi^2 + k_{2,\lambda}\, \partial^2 p/\partial \lambda^2 + k_{2,z}\, \partial^2 p/\partial z^2;$$

Γ_p is the set of the trophic subordination of the phytoplankton (in a typical case Γ_p consists of Z); C_{ps} is the part of the phytoplankton biomass consumed by the sth trophic level; T_p is the rate of exchange; and M_p is the mortality of the phytoplankton cells. Similarly to Eqs. (4.10) and (4.11), the functions T_p and M_p are represented by the following correlations:

$$M_p = max\{0, \mu_p(p - \underline{p})^{\theta}\} \tag{4.13}$$

$$T_p = t_p p \tag{4.14}$$

where t_p is the specific loss of phytoplankton biomass at the expense of the exchange with the environment; μ_p is the mortality coefficient; $\underline{p}$ and θ are coefficients characterizing the dependence of the mortality of the phytoplankton cells on their biomass.

The zooplankton is an important element of the ocean ecosystems. It is represented in Fig. 4.1 by the unique integrated level Z, the interior of which there are many sublevels: small herbivores, large herbivores, omnivores, carnivores, etc. The zooplankton consume phytoplankton and bacterial plankton. The trophic scheme of Fig. 4.1 reflects the set of the trophic dependence of the zooplankton (in a typical case Γ_Z consists of Z, r and D). The zooplankton production R_Z is approximated by an exponential law:

$$R_Z = k_Z\,(1 - exp[-vB_a]) \tag{4.15}$$

where k_Z is the maximal zooplankton production; v is the coefficient reflecting the hunger level; $B_a = \max\{0, B - B_{\min}\}$, and $B_{\min}$ is the minimal biomass of the food consumed by the zooplankton. The maximal production is defined by the relation: $k_Z = T_1\ u\ (1 - q_{2,\max})$, where $q_2 = P_1/(P_1 + T_1)$, $1/u$ is the efficiency of assimilation, P_1 is the maximum rate of production under the given rate of exchange T_1, and $q_{2,\max} = \max q_2$.

Formula (4.15) explains that under small food storage the zooplankton ration increases proportionally to B_a and then as the ration approaches the maximal value of k_Z it depends on B_a as little as possible. Expression (4.15) reflects the absence of the full extermination of the trophic levels.

Thus, the change of zooplankton biomass obeys the differential equation:

$$\partial Z / \partial t - U_Z + \sum_{s \in \Gamma_Z} C_{ps} R_s = 0 \tag{4.16}$$

where

$$U_Z + V_\varphi \partial Z/\partial \varphi + V_\lambda \partial Z/\partial \lambda + V_z \partial Z/\partial z$$
$$= R_Z - T_Z - M_Z - H_Z + k_{2,\varphi} \partial^2 Z/\partial \varphi^2 + k_{2,\lambda} \partial^2 Z/\partial \lambda^2 + k_{2,z} \partial^2 Z/\partial z^2,$$
$$H_Z = h_Z R_Z,\ T_Z = t_Z Z,\ M_Z = (\mu_Z + \mu_{Z,1} Z) Z \tag{4.17}$$

The coefficients h_Z, t_Z, μ_Z and $\mu_{Z,1}$ are determined empirically.

As follows from formula (4.16), the zooplankton is considered as a passive ecosystem element subjected to the physical processes of the water mixing. However, it is known that the zooplankton has vertical migrations. The effect of the vertical migration of the zooplankton on the trophic relationships is simulated in the model in such a way that the food requirements of the zooplankton inhabiting the $0 - z_0$ layer are supplemented by a certain portion of the total food requirements of the components of the same community occurring in deeper layers ($z_0 - H$).

The coefficients C_{as} ($a = p,Z$) in the formulae (4.12) and (4.16) will be determined on the basis of the assumption that the consumption of different food types by the sth trophic level is proportional to their effective biomasses:

$$C_{as} = k_{sa} B_a \Big/ \sum_{a \in S_s} k_{sa} B_a \tag{4.18}$$

where B_a is the effective biomass of the ath food; S_s is the food spectrum of the sth component; and k_{sa} is the index of satisfaction of the nutritive requirements of the sth component at the expense of the ath component biomass.

According to Fig. 4.1, the equations to describe the biomass dynamics of nekton, detritophages, detritus, dissolved organic matter and nutrient salts have the following form:

$$dr / dt = R_r - H_r - T_r - M_r - \sum_{s \in \Gamma_r} C_{rs} R_s \tag{4.19}$$

$$dD / dt = R_D - H_D - T_D - M_D - \sum_{s \in \Gamma_D} C_{Ds} R_s \tag{4.20}$$

$$\partial d/\partial t + V_\varphi \partial d/\partial \varphi + V_\lambda \partial d/\partial \lambda + V_z \partial d/\partial z$$
$$= M_b + M_D + M_r + M_p + M_Z + H_Z + H_r + H_D - \mu_d d -$$
$$C_{dD} R_D + k_{2,\varphi} \partial^2 d/\partial \varphi^2 + k_{2,\lambda} \partial^2 d/\partial \lambda^2 + k_{2,z} \partial^2 d/\partial z^2 \tag{4.21}$$

$$\partial n/\partial t + V_\varphi \partial n/\partial \varphi + V_\lambda \partial n/\partial \lambda + V_z \partial n/\partial z$$
$$= \mu_d d - \delta R_p + k_{2,\varphi} \partial^2 n/\partial \varphi^2 + k_{2,\lambda} \partial^2 n/\partial \lambda^2 + k_{2,z} \partial^2 n/\partial z^2 \tag{4.22}$$

$$\begin{aligned}\partial g/\partial t + V_\varphi \partial g/\partial \varphi + V_\lambda \partial g/\partial \lambda + V_z \partial g/\partial z \\ = T_p + T_b + T_r + T_D + T_Z - C_{gb}R_b + k_{2,\varphi}\partial^2 g/\partial \varphi^2 + k_{2,\lambda}\partial^2 g/\partial \lambda^2 \\ + k_{2,z}\partial^2 g/\partial z^2\end{aligned} \tag{4.23}$$

Here, $H_a = (1-h_a)R_a$ is the nonassimilated food for the ath component ($a = r, D$); $T_a = t_a\, a$ is the rate of exchange; $M_a = (\mu_a + \mu_{a,1}\, a)$ where a is the mortality; ρ_g is the rate indicator to replenish the nutrient salts at the expense of the dissolved organic matter; and δ is the coefficient of nutrients consumption by photosynthesis.

As follows from Eqs. (4.19) and (4.20), the model does not consider the possible spatial motion of the nekton and detritophages with the moving water. It is considered that these components migrate in the water space independently from the hydrophysical conditions. There are two possible versions of the migration process modeling. The first version consists in the addition of terms to the right parts of Eqs. (4.19) and (4.20) describing a turbulent mixing process with the coefficient $k_2{}^* > k_2$. In other words, this scheme of the migration process identifies the fish migration with intense turbulent diffusion. In this case, it is necessary to include stochastic elements in the model.

The investigations by many authors have revealed that the process of fish migration is accompanied by an external appearance of purposeful behavior. According to the biological principle of the adaptation, the fish migration process is subject to the composite maximization of the effective nutritive ration under the conservation of the environmental parameters within the limits of the living environment of the fish. Consequently, the traveling of migrating species takes place at characteristic velocities for them in the direction of the maximum gradient of effective food, given adherence to temperature and restrictions of other environmental parameters (salinity, dissolved oxygen, illumination, pollution, and other abiotic conditions). Schematically, this algorithm is described in Fig. 4.5.

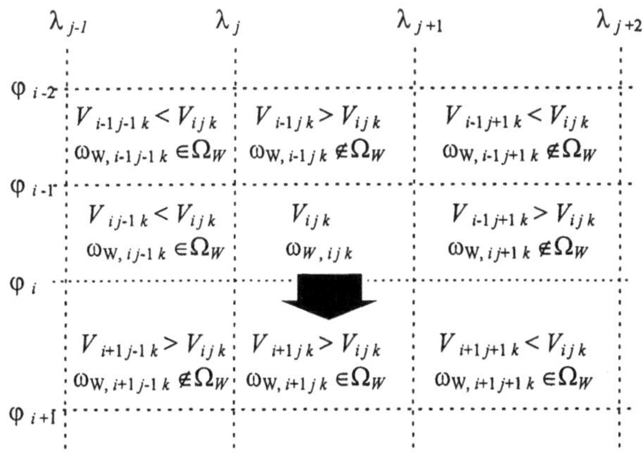

Fig. 4.5. Horizontal cross section of the conventional scheme describing the fish migration taken into account in the ocean ecosystem model: Ω_W is the area of available values for the abiotic parameters of the water environment; ω_W is the vector of current values of the abiotic parameters; $V_{ijk} = B(\varphi_i, \lambda_j, z_k, t)$ is the effective food ration; and $T_{W,ijk}$ is the water temperature

4.4
Analysis of the Vertical Structure of the Ocean Ecosystem

Equation (4.12) shows that the number s of maxima of the phytoplankton biomass function $p(t,\varphi,\lambda,z)$ under fixed time at any point (φ,λ) is defined by the number l of sign changes in the derivative $\partial p/\partial z$; moreover, $s \le 0.5(l+1)$. Information about the position of the phytoplankton biomass maxima as a function of depth is important for understanding all of the set of biogeocenotic processes within the ocean ecosystem. Certainly, the number and position of these maxima are determined by many environmental characteristics of the water, one of which is the hydrological heterogeneity between the different water layers. Numerous experimental investigations have shown that s can reach up to four and these maxima can occur from the surface down to 500 m. Naturally, $s \le 2$ in most cases with depths of 25 – 100 m.

For theoretical analysis, Eq. (4.12) is rewritten in the simplified form:

$$dp/dz = F(z)p - \nu N(z) - w(z) \tag{4.24}$$

where F, N and w are the velocity of photosynthesis, the zooplankton density and the sedimentation, respectively, versus the depth z. It is supposed that the concentration $n(z)$ of nutrient salts is accounted for by the function F.

Equation (4.24) is linear as to p. That is why it does not change when its right and left parts are multiplied by the following expression:

$$\exp\left\{-\int_0^z\left[F(\xi)-\nu N(\xi)\right]\right\}.$$

We have

$$\exp\left\{-\int_0^z\left[F(\xi)-\nu N(\xi)\right]\right\}\left\{dp/dz-[F(z)-\nu N(z)]p\right\}= \\ w(z)\exp\left\{-\int_0^z[F(\xi)-\nu N(\xi)]\right\} \tag{4.25}$$

Note that

$$\exp\left\{-\int_0^z\left[F(\xi)-\nu N(\xi)\right]\right\}\left\{dp/dz-F(z)-\nu N(z)]p\right\}= \\ (d/dz)\left[p\exp\left\{-\int_0^z[F(\xi)-\nu N(\xi)]\right\}\right] \tag{4.26}$$

Putting (4.26) into (4.25) and integrating by z, after simplifications, we will obtain:

$$p(z)\exp\left\{-\int_0^z\left[F(\xi)-\nu N(\xi)\right]\right\}= -\int_0^z w(y)\left\{\exp\left[-\int_0^y [F(x)-\nu N(x)]dx\right]\right\}dy+C \quad (4.27)$$

From (4.27) for $z = 0$, we find $c = p(0)$. Finally, Eq. (4.27) gives the formula for the vertical phytoplankton distribution:

$$p(z)=\left\{p(0)-\int_0^z w(y)\exp\left\{-\int_0^y\left[F(x)-\nu N(x)\right]dx\right\}dy\right\} \exp\left\{\int_0^z\left[F(\xi)-\nu N(\xi)\right]d\xi\right\}$$

Based on this dependence, we will study the distribution $p(z)$ as regards the number of maxima and their location. Let us introduce

$$g(h)=\int_0^h N(z)dz,\ \psi(h)=\int_0^h F(z)dz \quad (4.28)$$

then Eq. (4.2) can be rewritten in the form:

$$p(h)=\left\{p(0)-\int_0^h w(y)\exp\left[\nu g(y)-\psi(y)\right]dy\right\} \exp[\psi(h)-\nu g(h)] \quad (4.29)$$

Thus, $p(h)$ is a function of $g(h)$ and $\psi(h)$. The function g has a simple interpretation: if $h_2 > h_1$, then the difference $g(h_2) - g(h_1)$ is the zooplankton biomass of the layer $z \in [h_1, h_2]$. Functions $g(h)$ and $\psi(h)$ are monotonously undiminished since $N(h)$ and $F(h)$ cannot be negative.

Before starting the qualitative analysis of function $p(h)$ some auxiliary results will be derived. Introduce the following spaces $H^+ = \{h \geq 0\colon F(h)/N(h) = \nu/F'(h)/N'(h) > \nu\}$, $H^- = \{h \geq 0\colon F(h)/N(h) = \nu/F'(h)/N'(h) < \nu\}$ and the function $\varphi(h) = \nu g(h) - \psi(h)$.

Lemma 4.1. *The function $\varphi(h)$ is non-decreasing under $\forall h \in \Omega_1^+ \cup \Omega^+_2$, where $\Omega^+_1 = \{h \geq 0\colon h \in [0,h^+_1] / h^+_1 = \inf H^+, F(0)/N(0) < \nu\}$, $\Omega^+_2 = \{h \geq 0\colon h \in [h^-_i, h^+_j] / h^-_i \in H^-, h^+_j \in H^+, (h^-_i, h^+_j) \cap H^- = \varnothing\}$.*

Lemma 4.2. *The function $\varphi(h)$ is non-increasing under $\forall h \in \Omega_1^- \cup \Omega^-_2$, where $\Omega^-_1 = \{h \geq 0\colon h \in [0,h^-_1] / h^-_1 = \inf H^-, F(0)/N(0) < \nu\}$, $\Omega^-_2 = \{h \geq 0\colon h \in [h^+_i, h^-_j] / h^+_i \in H^+, h^-_j \in H^-, (h^+_i, h^-_j) \cap H^+ = \varnothing\}$.*

The correctness of these lemmas follows from the elementary characteristic of the function $\varphi(h)$. Specifically, we have $\varphi(0) = 0$, $\varphi'(0) = \nu N(0) - F(0), \ldots, \varphi^{(i)}(0) = \nu N^{(i-1)}(h) - F^{(i-1)}(h)$.

Comment 4.1. *Affirmation of lemma* 4.1 remains correct for $F(0)/N(0) = \nu$, if $(\exists s)[F^{(s)}(0)/N^{(s)}(0) < \nu, F^{(j)}(0)/N^{(j)}(0) = \nu, j = 0,1,\ldots, s1]$.

Comment 4.2. *Affirmation of lemma* 4.1 remains correct for $F(0)/N(0) = \nu$, if $(\exists s)[F^{(s)}(0)/N^{(s)}(0) > \nu, F^{(j)}(0)/N^{(j)}(0) = \nu, j = 0,1,\ldots, s1]$.

According to (4.29), the function $p(h)$ depends on $\varphi(h)$:

$$p(h) = \left\{ p(0) - \int_0^h w(y) ep[\varphi(y)] dy \right\} \exp[-\varphi(h)] \tag{4.30}$$

Based on lemmas 1 and 2, from Eq. (4.30) we obtain the monotonic characteristics for the function $p(h)$.

Since $\varphi(h)$ does not decrease in the space $\Omega^+ = \Omega^+_1 \cup \Omega^+_2$, then the function $\exp[-\varphi(h)]$ will be non-increasing under $\forall h \in \Omega^+$. For $w(h) \geq 0$ the function $p(0) - \int_0^h w(y)\exp[\varphi(y)]dy$ is non-increasing under any values $h \geq 0$. Therefore, $p(h)$ is a non-increasing function $\forall h \in \Omega^+$.

In the space $\Omega^- = \Omega^-_1 \cup \Omega^-_2$ $p(h)$ is the product of non-decreasing and non-increasing functions. Hence, in Ω^- the function $p(h)$ is either non-decreasing or non-increasing or has maxima. As for $\forall h \in \Omega^-$ $F(h)/N(h) > \nu$, then $p(h)$ will be a non-increasing function for $\forall h \in H_0 = \{h: h \geq h_0\}$, where

$$p(0) = \int_0^{h_0} w(y)\exp[\varphi(y)]dy$$

Let us designate by H_1 the space of values $h \geq 0$ for which the following condition is fulfilled

$$w(h)\left[F(h) - \nu N(h)\right]^{-1} \exp\left[\varphi(h)\right] + \int_0^h w(y)\exp\left[\varphi(y)\right]dy > p(0)$$

Then $p(h)$ will be a non-increasing function for $\forall h \in H_1 \cap \Omega^-$.

The results received are formulated as the following theorem.

Theorem. *The function $p(h)$ is non-increasing for $\forall h \in H^+ \cup (H_0 \cap H^-) \cup (H_1 \cap H^-)$ and non-decreasing for $\forall h \in H^- \cap (E \backslash H_0) \cap (E \backslash H_1)$, where $E = \{h: h \geq 0\}$.*

In the more common case, $\partial p/\partial z$ can be represented in the form:

$$\partial p/\partial z = (\partial p/\partial z)_{EB} + (\partial p/\partial z)_O + (\partial p/\partial z)_\tau + (\partial p/\partial z)_M + (\partial p/\partial z)_Z + (\partial p/\partial z)_T \tag{4.31}$$

where each item on the right reflects the change in the phytoplankton biomass at the expense of variations in the illumination and nutrient salts (EB), sedimentation (O), temperature (τ), mortality (M), consumption by the zooplankton (Z), and turbulence (T). From Eq. (4.30) we find:

$$(\partial p/\partial z)_{EB} \approx F(t,p,z) = Af_1(E)f_3(n)[1-10^{-\gamma p}].$$

For $\gamma p \ll 1$ we have

$$p^{-1}(\partial p/\partial z)_{EB} \approx Af_1(E)f_3(n).$$

As is known, the water temperature under stationary conditions is monotonically decreasing with depth, i.e., $(\partial p/\partial z)_\tau \leq 0$. Similarly, the term $(\partial p/\partial z)_M$ is a step-by-step continuous non-positive function. However, it is possible for the situation to arise when this monotony can be destroyed.

The values of the item $(\partial p/\partial z)_Z$ depend on the distribution $N(h)$. The item $(\partial p/\partial z)_T$ is fully determined by the hydrophysical conditions for all the water depths and its behavior can change in an arbitrary way. It is obvious that in the layer located above the thermocline, $(\partial p/\partial z)_O \leq 0$.

The investigation of all items from Eq. (4.31) will allow the determination of the number of changes in the derivative $(\partial p/\partial z)$ that defines the number of maxima in the vertical phytoplankton biomass distribution. Specifically, if $\partial p/\partial z$ changes its sign seven times under $(\partial p/\partial z)_{z=0} > 0$ and $(\partial p/\partial z)_{z\to\infty} < 0$, the function $p(z)$ has four maxima.

Finally, considering that the zooplankton consumes the phytoplankton uniformly throughout the whole water layer, and ignoring all terms in (4.31) except the first, on the basis of (4.28), we obtain the following expression to describe the specific rate of the phytoplankton biomass change:

$$p^{-1}(\partial p/\partial t) \approx b - ap$$

where

$$a = (k\gamma^2/2)E\,10^\mu,\ b = k\gamma E \cdot 10^\mu,\ \mu = m(1-E/E_{\max}),$$

$$m = (E_{\max}/E_0)\exp(\alpha H_{\max}\ln 10)$$

Equation (4.32) has the solution $p(t,h) = [\gamma/2 + \exp\{-tk\gamma E \cdot 10^L\}]^{-1}$, where $L = \exp[(E_{\max}/E_0)\exp\{\alpha H_{\max}(1 - E/E_{\max})\ln 10\}\ln 10]$, since when $h \to \infty$ and $E \to 0$, then $p(t,h) \approx [\gamma/2 + K\exp\{-tk\gamma E\}]^{-1}$, where $K = \exp\{-10^\beta\}$, $\beta = -(E_{max}/E_0)\exp(\alpha H_{max}\ln 10)$.

4.5 Mathematical Model of the Upwelling Ecosystem

It is known that the World Ocean zones characterized by vertical water motions (upwelling zones) have high productivity. An upwelling zone is a result of many phenomenon: water removal from the coast line by the wind, changes of ocean currents, etc. The water velocity and stability of an upwelling zone are determined by a set of synoptic parameters. The most specific value of the vertical speed of the water motion in the upwelling zone is 0.77×10^{-3} cm s^{-1}. The depths where the water flow motion starts vary within 200 m.

We shall proceed from the concept of successive development of a community from the time of its origin in the region of deep-water invasion up to its

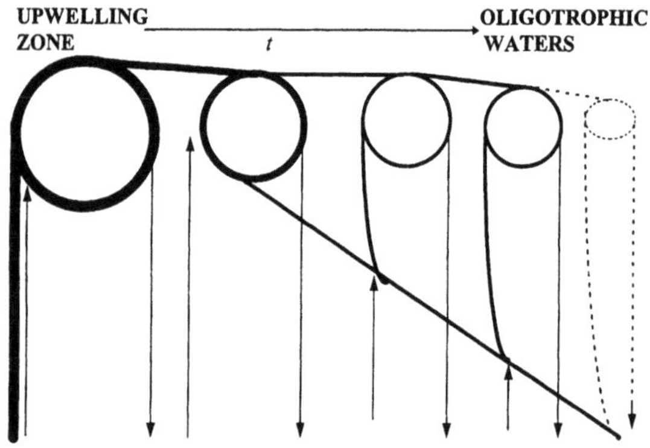

Fig. 4.6.
Scheme describing the cycle of nutrient salts and organic matter in the ocean waters. (Vinogradov et al. 1973)

climax in the oligotrophic convergence region. Within these times the system develops and moves along with the water flow. In addition, the total energy of the system and its structure (spatial, trophic and specific) are changed. The observed general characteristics of these changes occurring in time and space are now available, and one of the major criteria of adequacy of the model is its agreement with the actual picture observed in the oceans.

It is supposed that the ecosystem motion from the upwelling zone is homogeneous in the horizontal plane. The ecosystem state is characterized by depth z with step Δz ($\approx$ 10 m) and by time t with interval Δt (daily). The horizontal speed of the water current from the upwelling zone is $V = V_\varphi = V_\lambda$, so that the distance of the water volume from the upwelling zone equals $\Delta r = (\Delta\varphi^2 + \Delta\lambda^2) = V\,\Delta t$. Schematically, this situation is given in Fig. 4.6.

The ecosystem state on each layer $z = const$ is determined by the light intensity $E(z,t)$, by the nutrient salts concentration $n(z,t)$, and by the biomass of detritus $d(z,t)$, phytoplankton $p(z,t)$, bacterioplankton $b(z,t)$, protozoa $Z_1(z,t)$, microzoa $Z_2(z,t)$, small herbivores $Z_3(z,t)$, large herbivores $Z_4(z,t)$, small predators – cyclopoida $Z_5(z,t)$, predators – calanoida $Z_6(z,t)$, and large predators – chaetognatha and polychaeta $Z_7(z,t)$. The protozoa include infuzorii and radiolarii. The microzoa include the nauplii stages of copepods. Based on plankton feeding studies made by many authors (Vinogradov et al. 1972), the small herbivores are now believed to include, apart from the protozoa and nauplii, young copepod stages of calanoida and adult copepods whose size does not reach 1.0 mm, such as *Clausocalanus, Acrocalanus, Paracalanus, Calocalanus,* etc. The group of large herbivores consists of animals whose size exceeds 1.0 mm, such as *Undinula, Eucalanus, Rhincalanus, Neocalanus, Lucicutia,* juveniles of *Euphausiacea*, etc. The group of omnivores includes *Centropages, Pleuromamma, Scolecithrix, Undeuchaeta, Conchoecia*, etc., while *Chaetognatha, Candacia, Euchaeta, Cyclopodia,* etc. are grouped with the carnivores.

It is accepted that 30% of the bacterioplankton biomass is in natural clots more than 3–5 μm in size which can be consumed by the herbivores (Z_3 and Z_4). The microzoa (Z_2), protozoa (Z_1) and small herbivores (Z_3) can consume unclotted bacterioplankton also.

The trophic relations between the components are described by means of the energetic principle. The biomass, production, respiration, mortality and rations are measured by the energy scale in cal m^{-3} or cal m^{-2}. Similar to the case of the ocean ecosystem model, the source of energy and matter in the community is the primary phytoplankton production (R_p). The input source of energy is solar radiation (E) and nutrient salts (n). The vertical structure of the water environment is described by a three-layer model: the layer above the thermocline, the layer of the thermocline, and the layer below the thermocline. The thermocline boundaries are ($z_b \leq z \leq z_l$):

$z_b = 10 + 2.2\, t$ for $0 \leq t \leq 50$ days,

$z_b = 120 + 0.6\,(t-50)$ for $t>50$ days,

$z_l = 30 + 2.4\, t$ for $0 \leq t \leq 50$ days,

$z_l = 150 + 1.4\,(t - 50)$ for $t > 50$ days.

Thus, the thermocline at the initial time ($t = 0$) occupies the space from 10 to 30 m descending little by little to depths of 120–150 m during the next 50 days and to depths of 150–190 m on the 100th day.

The replenishment of nutrient salts in the layer 0–200 m is realized at the expense of the detritus decomposition and the rise from the deep layers ($z > 200$ m). The concentration of nutrient salts in the organic matter is 10% and is considered as constant.

In this case, Eq. (4.22) is rewritten in the form:

$$R_p = k_T\,[1 - 10^{-0.25p\gamma(t)}]\,[1 - 10^{-0.1n}]^{0.6},$$

where the function $\gamma(t)$ characterizes the time dependence of the phytoplankton P/B-coefficient. It is accepted that the maximal P/B value for the phytoplankton in the upwelling zone equals five. The value of P/B changes during succession as an exponential function and reaches the value of one after 15 days; then it is taken as constant.

Equation (4.12) is rewritten in the form:

$$\partial p / \partial t = A_p - \sum_{j \in \omega_0 \setminus p} C_{pj} R_j$$

where

$$A_p = R_p - t_p p - \mu_p p + k_2\, \partial^2 p/\partial z^2 + (V_z - w_p)\, \partial p/\partial z$$

k_2 is the turbulence diffusion coefficient, the coefficient w_p describes the gravity sedimentation process and is taken to be equal to the following value:

$$w_p = \begin{cases} w_{p1} & when \quad 0 \leq z \leq z_b \quad (\approx 5\text{cm} \cdot \text{day}^{-1}); \\ w_{p2} & when \quad z_b < z < z_l \quad (\approx 0.1\text{cm} \cdot \text{day}^{-1}); \\ w_{p3} & when \quad z \geq z_l \quad (\approx 3\text{cm} \cdot \text{day}^{-1}) \end{cases}$$

The coefficient k_2 has the following actual values: from the ocean surface to the upper boundary of the thermocline z_b, $k_2 \approx 200$ cm^2 s^{-1}; between depths z_b and z_l (a layer with a high density gradient), $k_2 \approx 50$ cm^2 s^{-1}; in the water layers below z_l, $k_2 \approx 150$ cm^2 s^{-1}.

In Eq. (4.23), it is considered that the following correlations hold:

$$T_p + T_b + T_Z + T_r + T_D = 0.3\,R_p,$$

$$\mu_b = 0.01,\ t_b = 0.75\,(1 - 10^Q),\ R_p = 3b(1-10^Q)$$

where $Q = -\,0.2\,d - 0.3\,R_p$.

The nutritive ration of the bacteria consists of detritus and dissolved organic matter extracted by the phytoplankton. The following restriction is made: it is supposed that during the day the bacteria can consume no more then 10% of the detritus biomass located in the same water layer.

Finally, Eq. (4.9) can be represented in the form:

$$\partial n/\partial t = A_n + 0.05 \sum_{i\in\omega_0} t_i B_i$$

where $A_n = -\,0.1\,R_p + 0.1d + k_2\,\partial^2 n/\partial z^2 + V_z\,\partial n/\partial z$, V_z is the velocity of vertical water motion, and B_i is the ith component of the biomass, $\omega_0 = \{p, b, Z_j\,(j = 1 \div 7)\}$.

Formula (4.5) may be simplified for the given case:

$$E(z_n, t) = E_0 \cdot 10^c,$$

where z_n is the depth of nth water layer (m),

$$c = -\alpha z_n,\ \alpha = 0.01 + 0.0001\sum_{k=0}^{n}\left[p(z_k,t) + d(z_k,t)\right] / z_n$$

An appreciable limiting of the photosynthesis by the light is observed when $E > 100$ cal cm^{-2} day^{-1}. Hence, formula (4.7) is rewritten in the form: $K_T = 0.041E\,10^B$, where

$$B = 0.25[1 - E/E_{max}],\ E_{\max} \approx 70 \text{ cal cm}^{-2}\text{ day}^{-1}.$$

For the case considered, the formula (4.6) is transformed to the state:

$$\partial b/\partial t = A_b - \sum_{j\in\omega_0\setminus(p,b)} C_{bj} R_j$$

where

$$A_b = R_b - \mu_b - t_b b + k_2\,\partial^2 b/\partial z^2 + (V_z - w_p)\,\partial b/\partial z.$$

The rate $\partial Z_i/\partial t$ ($i = 1 \div 7$) of biomass change for each zooplankton component, by analogy with Eq. (4.16), is determined by the ration $R_{Z,i}$, the assimilability rate $1/u_{Z,i}$, the respiration rate $T_{Z,i} = t_{Z,i}\,Z_i$, the mortality $\mu_{Z,i} = 0.01Z_i$, the consumption of ith component by the jth component with the coefficient C_{ij}, and by the age transition of Z_2 component to the Z_j ($j = 3$–6) component. Thus, we have:

$$\partial Z_i/\partial t = U_i - \sum_{j\geq i}^{7} C_{ij} R_{Z,j},\ (i = 1 \div 7);$$

where

$$h_{Z,i} = 1 - 1/u_{Z,i},\ U_i = \Omega_{m,i} + (1 - h_{Z,i})R_{Z,i} - (t_{Z,i} + \mu_{Z,i})Z_i$$

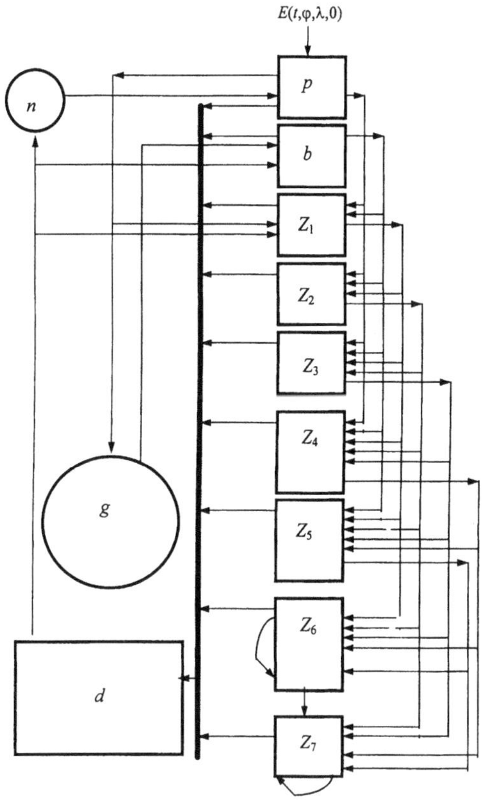

Fig. 4.7. Block diagram of energy flow (cal m^{-2} day^{-1}) through the community inhabiting the upper 0–200 m layer of the mesotrophic tropical areas of the ocean. The *symbols* for the ecosystem components are defined in Table 4.1

and the function $\Omega_{m,i}$ characterizes the age transition of small herbivorous plankters to the other stages. Let us define

$$\Omega_{m,3} = Z_2Z_3\{15(Z_3 + Z_4 + Z_5 + Z_6)\}^{-1},$$

$$\Omega_{m,4} = Z_2Z_4\,\{20(Z_3 + Z_4 + Z_5 + Z_6)\}^{-1},$$

$$\Omega_{m,4} = Z_2Z_5\,\{15(Z_3 + Z_4 + Z_5 + Z_6)\}^{-1},$$

$$\Omega_{m,6} = Z_2Z_6\,\{20(Z_3 + Z_4 + Z_5 + Z_6)\}^{-1}.$$

The coefficient ν in the formula (4.15) is determined as $\nu = 0.01/B_{min}$. The trophic structure and food chains are represented by Figs. 4.7 and 4.8. Under the valuation of the coefficients C_{ij}, it is supposed that the food requirements are related to the various fodder objects in proportion to their biomass with consideration of the nutrition specification.

The effect of vertical migrations of zooplankton on trophic relationships is simulated in the model in that the food requirements of large herbivores, omnivores and carnivores inhabiting the 0–50 m layer are supplemented by a certain portion (k_Z) of the total food requirements of the components of the same community occurring in the deeper layers (50–200 m). In doing so, it is supposed that these food requirements are distributed homogeneously within the 0–50 m

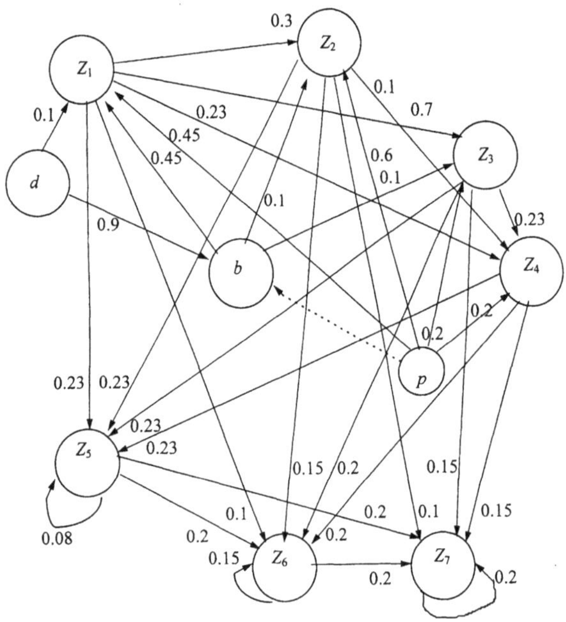

Fig. 4.8.
Scheme of the trophic interactions between the upwelling ecosystem components. The *symbols* are the same as in Fig. 4.7. The values of the coefficients C_{ij} are shown on the *arrows*

layer. This coefficient is taken to be dependent on time according to Vinogradov et al. (1972):

$$k_Z = 0.02 + 0.0016\, t \text{ at } t \le 50 \text{ days},$$

$$k_Z = 0.1 \text{ at } t > 50 \text{ days}.$$

Here, the existence of different scenarios is possible to describe the migration and consumption processes for the zooplankton components.

The equation describing the detritus dynamics can be written in the following form:

$$\partial d/\partial t = \sum_{i\in\omega_0}\left(H_{Z,i} + M_{Z,i}\right) - \sum_{j\in\omega_0\setminus p} C_{dj}R_j + k_2\, \partial^2 d/\partial z^2 + \left(V_z - w_d\right)\partial d/\partial z - \mu_d d$$

where the coefficient of gravity sedimentation is given in (cm day^{-1}):

$$w_d = 25, \text{ when } z_b \le z \le z_l,$$

$$w_d = 50, \text{ when } z < z_b, z > z_l.$$

To accomplish these calculations it will be necessary to assume that all ecosystem components at the upwelling point ($t = 0$) have a uniform distribution by depth within the water column: $n(z,0) = 250$ mg m^{-3}, $d(z,0) = 0$, $b(z,0) = 1$, $p(z,0) = 0.5$, $Z_i(z,0) = 0.01 \div 0.5$ ($i = 2 \div 7$), $Z_1(z,0) = 0.0001$ cal m^{-3}. Let us note that variations of these estimations that decrease or increase by a factor of 50 practically do not change the character of all the system dynamics in the future. Really the system during 50 days of its development "forgets" the variations of the initial concentrations.

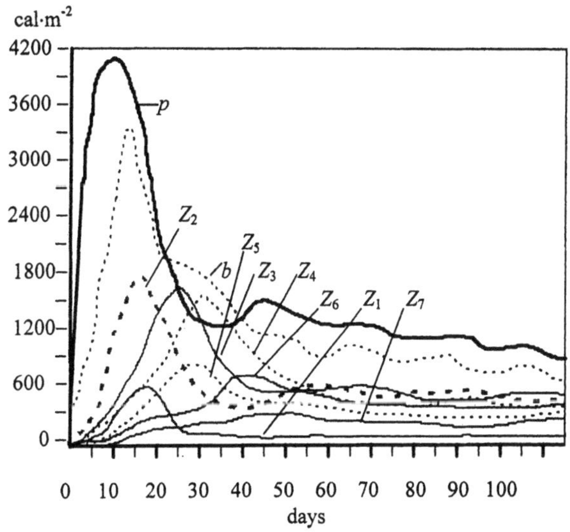

Fig. 4.9.
Time dependence of the total biomasses of living components of the upwelling ocean ecosystem in the 0–200 m water layer

Zero gradients for all components excepting n and d are given on the upper ($z = 0$) and lower ($z = 200$ m) boundaries. The boundary conditions for n and d components have the values:

$$\partial n/\partial z|_{z=0} = 0; \quad n(200,t) = 250 \text{ mg} \cdot \text{m}^{-3};$$

$$\partial d/\partial z|_{z=0} = 0; \quad \partial d/\partial z|_{z=200} = -0.5.$$

When estimating changes of the system over time, it was assumed that water takes more than 60 days to cover the distance from the upwelling zone to the oligotrophic zone of the planetary convergence. Figure 4.9 gives the representation of the change of the biomasses as a function of time. We see that the phytoplankton biomass increases most rapidly here, reaching its maximum ($\approx$ 4500 cal m^{-2}) on the 5–10th day of existence of the system. After this, the phytoplankton biomass decreases. The maximum of the bacterioplankton biomass is reached on the 10–15th day. The small herbivores lag somewhat behind the phytoplankton in development, and their biomass reaches its maximum only on the 30th day. Nevertheless, its joint influence with the decrease in the nutrient salts leads to a sharp drop in the phytoplankton and bacterioplankton biomasses. Namely, inverse chains of the community give weak contributions to R_p and R_b. After the 40th day the phytoplankton mainly functions at the expense of the biogenic elements rising in the eutrophic zone across the thermocline from the deeper layers. Subsequently, the phytoplankton biomass decreases sufficiently slowly. At this time period the stability factors begin to influence the community at the expense of the exterior energy flows.

The carnivores prove to be still more inertial than the herbivores as their biomass attains its maximum only on the 35–50th day. At this time, namely the 50–60th day, the system reaches its quasi-stationary state characterized by low concentration of all living components.

Table 4.1. A comparison of the model results with the experimental assessment of the upwelling ecosystem components

Ecosystem component	Community in the middle maturity stage (30–40 days)			Mature community (60–80 days)	
	Model		Observation	Model	Observation
	Day 30	Day 40	Days 30–40	Day 70	Days 60–70
Phytoplankton, p	1319	1092	2000	827	900
Bacteria, b	1673	864	4100	564	2180
Microzoa, Z_2	394	303	321	300	–
Small herbivores, Z_3	1338	612	525	290	74
Large herbivores, Z_4	1416	726	420	252	164
$Z_3 + Z_4$	**2754**	**1338**	**945**	**542**	**238**
Small predators, Z_5	624	491	495	203	236
Predators – calanoida, Z_6	288	600	610	191	175
Large predators, Z_7	184	183	15	102	51
$Z_5 + Z_6 + Z_7$	**796**	**1274**	**1110**	**496**	**462**

Thus, the present model demonstrates the spatial disconnection and alternation of the biomass maxima of the phytoplankton, herbivores and carnivores as the water moves farther away from the upwelling zone. A system degradation observed 60 days after its development is reported by many authors.

It is very interesting to compare the quantitative estimations of any values obtained from the model with expedition investigations. Table 4.1 gives an example of such a comparison. The visible discrepancy between the model and experimental estimations of the bacterioplankton biomass is a result of the inexact description of its food chain.

The vertical distribution of the biomass component is also subject to essential change in time (Fig. 4.10). It is seen that at times when the system has only just formed ($t \approx 5$ days), and the total phytoplankton biomass is near its maximum, the phytoplankton biomass has a uniform vertical distribution in the layer 0–50 m. All other living components have more or less pronounced maxima narrow enough in depth to be connected to the thermocline. However, after 10 days, the biogenic salt storage in the upper layer is almost exhausted. In this state it is possible to form the phytoplankton biomass maximum only at a depth of 10–20 m. A second maximum is formed deeper, at the expense of biogenic salts moving across the thermocline. A similar structure having two maxima is formed in the vertical distribution of the other components of the ecosystem. It arises at the 20–30th day and is particularly clear when both maxima are well expressed and, owing to the thermocline descending by up to 55–75 m, they are clearly separated by depth.

The upper maximum is formed mainly due to the horizontal transport of nutrient salts in the production–destruction cycle of the community. As can be seen from Fig. 4.10, the upper maximum is at first more distinctly pronounced than the lower one. As the nutrient salts of the surface layer are consumed, the vertical transport across the thermocline begins to play a significant role, and

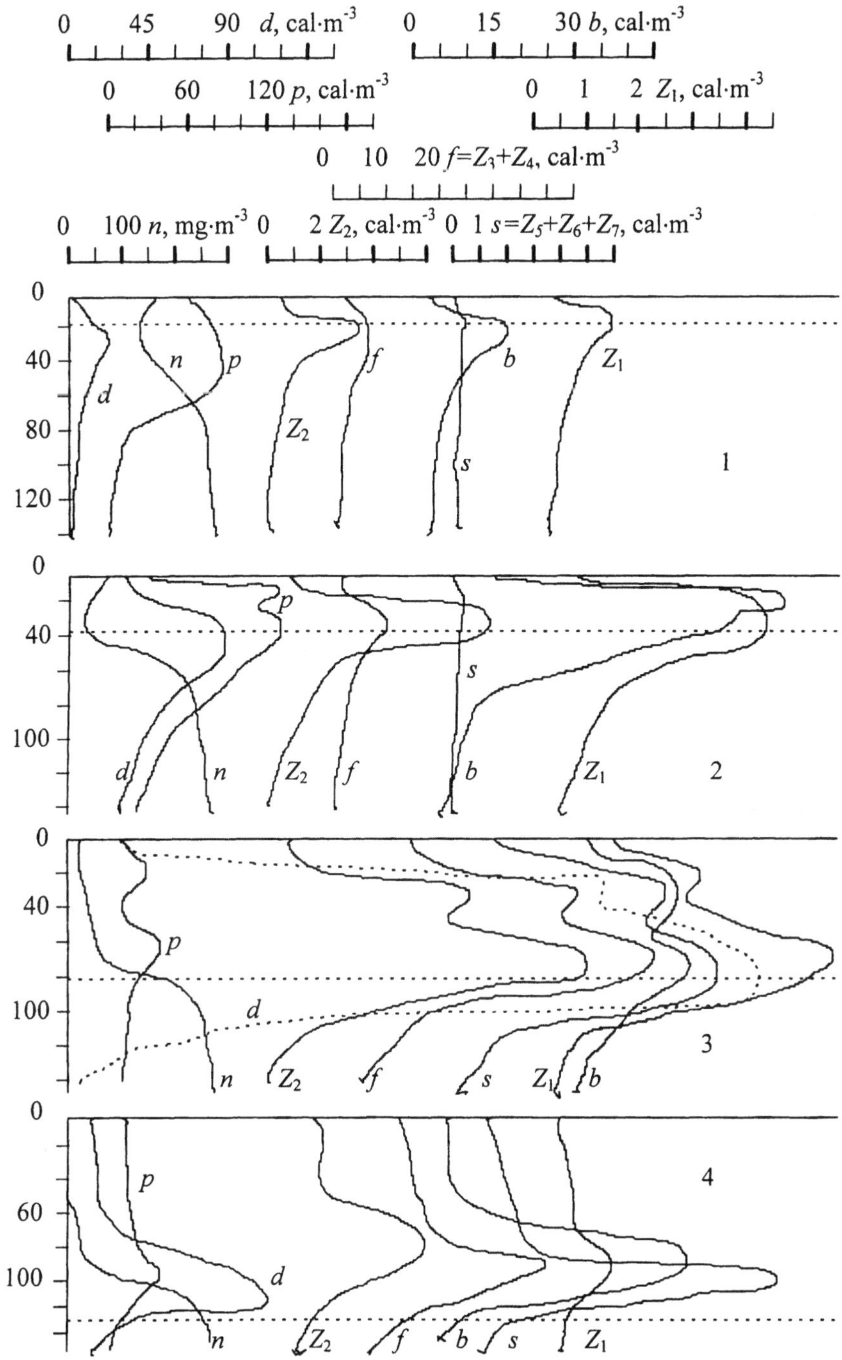

Fig. 4.10. Changes in time of the vertical distribution of elements of the ecosystem. The *horizontal dashed line* corresponds to the upper boundary of the thermocline. Comparison made at intervals of 1–5 days, 2–10 days, 3–30 days, 4–60 days

owing to this, the lower maximum is now absolutely larger than the upper one. Subsequently, under the thermocline descending by up to 80–100 m and more, illumination becomes the limiting factor for photosynthesis. A situation appears when the lower maximum moves away from the thermocline and its position is defined by the illumination limit and by the flow of nutrient salts from the deeper layers. This picture of the vertical distribution of the phytoplankton is consistent with the patterns obtained from observations in the World Ocean.

Under a more mature state of the system (t > 50–60 days), the complete disappearance of the upper maximum is typical for the oligotrophic and ultra-oligotrophic waters. This picture exists mainly for the vertical distribution of the phytoplankton biomass. Other living components of the ecosystem can have a small upper maximum due to environmental processes that the model does not taken into account.

The results given in Figs. 4.9, 4.10 and Table 4.1 do not contradict the existing notion about the structure and behavior of the upwelling ocean ecosystems. It is impossible to demand exact numerical correspondence between the model results and the real state of the ecosystem. We can consider in this context the more general characteristics such as the stability of the ecosystem. The ecosystem is considered stable in the time interval (0,T) if its state variables (in this case B_i, $i = 1 - n$) are within the interval:

$$B_{i,\min} \le B_i \le B_{i,\max} \; (i = 1 - n) \tag{4.32}$$

Actually, the right part of this condition is unnecessary. Namely, when only one of the ecosystem components has $B_i \to \infty$, another component always exists with $B_j \to 0$, as follows from the data in Fig. 4.7. Therefore, the stability condition for the upwelling ocean ecosystem can be rewritten in the form:

$$B_{\min} \le \sum_{i=1}^{n} B_i \tag{4.33}$$

where $B_{\min}$ is the total minimal value of the ecosystem biomass (specifically, $B_{\min} = B_{1,\min} + \cdots + B_{n,\min}$).

The criterion (4.33) was used to estimate the model stability under alteration of the various parameters. Some results of the calculations are given in Tables 4.2–4.5. From Table 4.2 we see that a deviation in the food assimilation (u^{-1}) by ± 20% leads to the variation of $Z_3 + Z_4$ by 48% and of Z_2 by 70%. The variations of the respiration coefficients t_Z are also not very strong. From Table 4.3, the same variations in the rate of energy assimilation T and the non-assimilated food H cause equivalent changes in the vertical distributions of the zooplankton components. The small zooplankton and herbivores are more sensitive to variations in T and H than are the predators.

It follows from Table 4.5 that large variations in the initial data do not significantly influence the ecosystem dynamics. More detailed calculations show that the ecosystem relaxes these variations during 20–25 days when its interior energetic interactions are balanced. Stable correlations between the trophic levels are set during the beginning period near the upwelling area.

Table 4.2. Changeability of the distribution of microzoa and herbivores under variation of the model parameters: $\hat{Z}_2, \hat{Z}_3, \hat{Z}_4$ are expedition estimations by Vinogradov et al. (1977); K_2 is the coefficient of the consumed food utilized by the growth; α is the absorption coefficient by the screening seawater [see Eq. (4.5)]; $\Delta_1 = [(1/200)\int_0^{200} \left(Z_3(z,40) + Z_4(z,40) - \hat{Z}_3 - \hat{Z}_4\right)dz]^{1/2}$, $\Delta_2 - [(1/200)\int_0^{200} \left(Z_2(z,40) - \hat{Z}_2\right)dz]^{1/2}$

Scenario	Simulation results	
	Δ_1	Δ_2
Model	0.29	0.75
$u^{-1} - 20\%$	0.35	0.91
$u^{-1} + 20\%$	0.43	1.27
$t_Z - 2\sigma$	0.76	1.02
$t_Z + 2\sigma$	3.36	1.71
$\alpha + 50\%$	7.9	1.78
$K_2 = 0.3$	1.44	3.5
$K_2 = 0.2$	2.6	6.0

Table 4.3. Standard deviation Δ of the trophic levels of biomass under variations of the rate of energy exchange (T) and the nonassimilated food (H). Z_i, $Z_{i,\,exp}$ are the model values when parameter variations are absent or present, respectively; $a = \Delta\sum_{i=2}^{4} Z_i$, $b = \Delta\sum_{i=5}^{7} Z_i$, $\Delta_i = \left\{\frac{1}{200}\int_0^{200}\left[Z_i(z,40) - Z_{i,\mathrm{exp}}(z,40)\right]dz\right\}^{1/2}$

Δ_i	Parameter variations							
	ΔT (%)				ΔH (%)			
	− 20	− 10	+ 10	+ 20	− 20	− 10	+ 10	+ 20
Δ_2	1.7	1.2	0.6	1.9	2.4	0.7	1.9	2.6
Δ_3	0.3	0.2	0.3	1.0	0.5	0.5	0.9	1.3
Δ_4	0.5	0.4	0.6	1.3	0.9	0.4	0.6	1.5
a	0.6	0.5	0.6	1.2	0.9	0.6	1.0	1.8
Δ_5	1.4	0.9	0.7	1.2	0.9	0.6	1.2	1.1
Δ_6	0.1	0.1	0.1	0.1	0.1	0.1	0.1	0.2
Δ_7	0.8	0.4	0.1	0.2	0.5	0.2	0.1	0.2
b	1.8	1.2	1.2	1.3	1.2	0.6	1.1	1.3

Table 4.4. Standard deviations Δ of the vertical distribution of various components of the ecosystem on the 30th day of its functioning under the change of the initial concentration of the nutrient salts. Designations are the same as in Table 4.3

Δ_i	Change of $n(z, t_0)$		
	- 50%	- 20%	+ 20%
Δ_2	5.0	0.6	0.9
Δ_3	0.4	0.4	0.3
Δ_4	1.3	0.1	0.1
a	1.5	0.4	0.4
Δ_5	1.1	0.5	0.4
Δ_6	1.1	0.04	0.03
Δ_7	1.1	0.03	0.02
b	1.2	0.5	0.4

Table 4.5. Ecosystem structure dynamics when there is a change in the initial biomass of the predators (cal m^{-2}). The basic initial biomasses are $Z_5(z,t_0) = Z_6(z,t_0) = Z_7(z,t_0) = 0.1$ cal m^{-3};

$$Z_0 = \int_0^{200} \{Z_5(z,t_0) + Z_6(z,t_0) + Z_7(z,t_0)\} dz$$

Component	5th day			30th day		
	Z_0	$100Z_0$	10^3Z_0	Z_0	$100Z_0$	10^3Z_0
b	1789	2263	2263	2269	2159	1995
p	3264	3462	3794	3971	2086	2614
Z_1	31	12	3	18	3	2
Z_2	178	106	101	109	265	465
Z_3	10	85	80	80	201	415
Z_4	67	14	74	145	156	191
Z_5	18	14	9	114	119	53
Z_6	1	25	41	14	25	22
Z_7	36	234	183	213	437	821

Figure 4.11 shows how the variations in the illumination influence the vertical ecosystem structure. We see that an increase in illumination on the ocean surface leads to an insignificant lowering of the lower maximum and to a more rapid decrease in the biogenic salt concentration above the thermocline. When E_0 = 2000 kcal m^{-2} day^{-1}, the maximal photosynthesis is at the surface. Variations in E_0 from 2000 to 7000 kcal m^{-2} day^{-1} without biogenic limitation lead to a change in the vertical distribution of the ecosystem biomass.

Table 4.6 shows how a change in the initial biogenic concentration can influence the phytoplankton biomass dynamics. From which it follows that variations of the nutritive salts above 100 mg m^3 at the moment t_0 have practically no influence on the ecosystem state in distant days but only at the initial period of the ecosystem development.

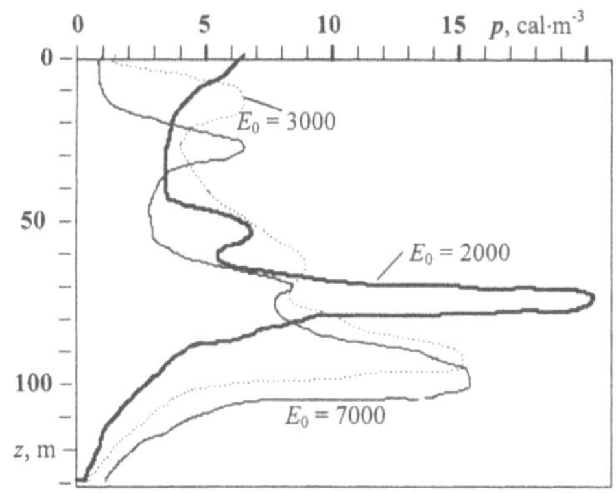

Fig. 4.11. Dependence of the ecosystem vertical structure on the illumination E_0 (kcal m^{-2} day^{-1}) at the 30th day after the upwelling

Table 4.6. Phytoplankton biomass (cal m^{-3}) as a function of the variation of the nutritive salts at the initial time

Depth (m)	$n(z,t_0)$ = 100 mg m^{-3}			$n(z,t_0)$ = 400 mg m^{-3}		
	Interval of ecosystem functioning (days)					
	10	50	100	10	50	100
0–10	0.3	0.1	0	0.3	0.1	0
10–20	31.7	8.7	10.9	37.4	16.1	15.4
20–30	39.6	9.4	10.7	289.7	7.8	10.4
30–40	39.4	9.2	10.6	207.7	7.8	10.4
40–50	36.6	9.7	10.9	295.1	7.9	10.7
50–60	36.7	9.5	10.9	153.7	7.1	10.5
60–70	41.7	11.4	10.9	62.1	18.9	8.8
70–80	13.4	15.7	11.2	15.6	10.3	17.9
80–90	1.9	31.4	17.0	2.1	39.0	17.9
90–100	0.6	3.6	12.8	0.6	4.2	1.4
100–150	3.1	2.1	2.1	3.1	2.2	2.4
150–200	0.9	0.4	0.5	0.8	0.7	0.7

The P/B coefficients of the phytoplankton and other ecosystem components are important model parameters. For example, if we suppose that the P/B coefficient of the phytoplankton decreases during 15 days by up to 0.6, the ecosystem practically becomes extinct by the 35th day. An increase in the P/B coefficient of the phytoplankton leads to a rapid impoverishment of the biogenic reserve in the upper layer. This precipitates disappearance of the upper maximum. As follows from Fig. 4.12, under $R_p/p \equiv 3$ the phytoplankton biomass fluctuates with high amplitude and an unreal situation has arisen. This says that in reality the dependence of the phytoplankton P/B coefficient on time is defined by a diminishing function. The investigation of this function is an important stage for the modeling procedure.

Considering the model of the upwelling ecosystem here raises many questions concerning the interactions between trophic levels as well as the question

Fig. 4.12.
Dynamics of the biomass of the ecosystem components when the P/B coefficient is constant and equal to 3

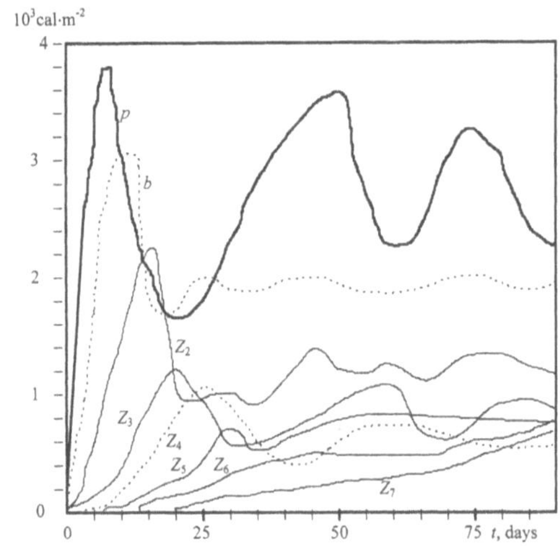

Fig. 4.13.
Hypothetical dynamics of the ecosystem when predator zooplankters are absent

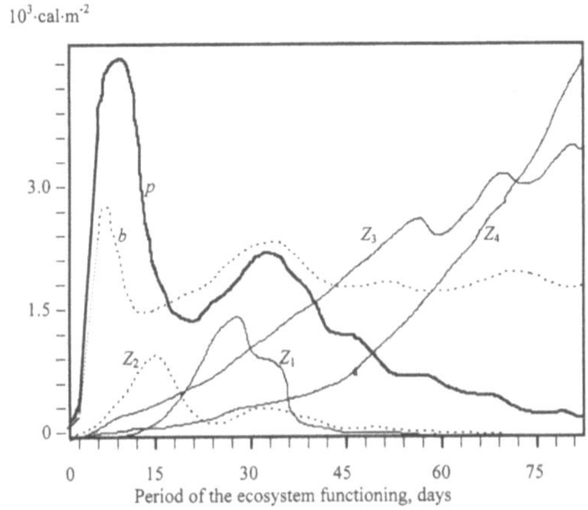

of the aggregation class for the trophic levels. Figure 4.13 gives an example of some variations in the trophic pyramid of the ecosystem. We see that in the absence of predators the biomass of herbivores increases without limit and the ecosystem loses its stability by the 50th day. This says that the relations between the trophic levels and the disaggregation of the ecosystem are certainly important to achieve the best correlation of model results with the field observations. Undoubtedly, further refining of the simulation techniques and incorporation of components of higher trophic levels – primarily of fishes – will make the model more complete and representative.

The set of model parameters, the variations of which are important for the ecosystem stability, includes the vertical speed v_z of the motion of water. Model calculations show that the ecosystem is stable in its dynamics for $v_z \in [10^{-4}, 10^{-1}]$ cm s^{-1}. When v_z is faster or slower, the vertical structure of the ecosystem is changed to a great extent and its stability is broken.

4.6 Probabilistic Model of the Interaction Between Ocean Ecosystem Components

A knowledge of the interaction laws between the ecosystem components is one of the main conditions for adequacy of the mathematical methods used to model its dynamics. This task is complex because the variety of trophic interactions in the ocean ecosystems have not been adequately studied. In addition, the mathematical description of these interactions needs improvement. Therefore, we continue the consideration of this task by introducing probabilistic elements.

Let us study an ecosystem with three basic components: s, f and m. Population s is the predator for population f, which in turn is the predator for population m. The consumption of population m by the species of population s is absent. The existence of possible interactions between these components with other ecosystem trophic levels is taken into account in the mathematical equations by means of the introduction of mortality and reproducing factors.

We shall take into account that the interaction between species of the s and f populations is realized at random time intervals; moreover, the interval τ_f during which the population f is the free state independent of the population s is distributed with the density function $p_f(\tau_f)$. In addition, let us suppose that the interval τ_{sf} of the interaction between the f and s populations is distributed with the density function $p_{sf}(\tau_{sf})$. Finally, let us assume that the distribution of the interactions between the species of the f and m populations is determined by Poisson law; moreover, the probability of real interaction between species under the contacting state of populations s and f within the interval $[t, t+h]$ equals λh, where $h > 0$.

Let us designate: $P_{nF}(\tau_f, t)h$ is the probability that until moment t the n interactions between the species of the f and m populations will be realized under the condition that the time passed since the population f is in the free state lies between τ_f and $\tau_f + h$; $P_{nc}(\tau_{sf}, t)h$ is the probability that until t the n interactions between species of the f and m populations are realized under the condition that time passed from the moment when population f interacts with population s lies between τ_{sf} and $\tau_{sf} + h$.

Thus, according to Eq. (4.16), we can write for the each ecosystem component the equation:

$$dB_i/dt = R_i - M - \Psi_i \tag{4.34}$$

where B_i is the biomass of the ith component (f,m,s); R_i is its ration; M_i is the total loss of biomass for the ith component at the expense of mortality, respiration and other causes; Ψ_i is the reduction of the ith component biomass at the expense of consumption by other components.

It is assumed that the external conditions under which the ecosystem exists do not change in time. This means that the functions R_i, M_i and Ψ_i depend only upon the biomasses $\{B_i\}$. These functions are defined proceeding from the following suppositions. It is accepted that the rate of reproduction of the m population is proportional to its volume, but that the biomass accumulation is restricted by its consumption by the population f and by other causes. The accumulation rate of the biomass for the population f is considered to be proportional to the food volume consumed by the species of population f during unit time:

$$f = f_{\max}\left[1 - 10^{-\beta_0 f_n}\right]$$

where $f_{\max}$ is the maximal ration under sufficient food volume, when nutritive demands are fully satisfied; f_n is the average efficient food biomass which is proportional to the difference between the biomass of the m and its minimal value, $m_{\min}$, under which the consumption is absent.

The decrease in the biomass s also takes place in accordance with Eq.(4.15) during those time intervals when the populations s and f are interacting. Finally, it is considered that accumulation of the biomasses for populations s, f and m is restricted by the mortality, respiration, and other causes of species death. In addition, mortality rate of such losses for the populations f, s and m equals μ_1, μ_2 and μ_3, respectively. In addition, it is considered that these losses take place under the biomasses of populations f, s and m exceeding f_0, s_0 and m_0, respectively. The assumptions accepted above reflect indirectly in the model the real situations where, as a rule, utter destruction of one trophic level by another one and absolute mortality are not realized.

Thus, Eq. (4.34) is rewritten for the populations f, s and m in the following forms:

$$ds/dt = \beta_{12}\,\delta_Y(t)s(t)[1 - \exp\{-\beta_{21}f(t)\}] - \mu_2 s^*(t) \tag{4.35}$$

$$\begin{aligned} df/dt = {} & \beta_1 f(t)[1 - \exp\{-\beta_0 n^*(t)m^*(t)\}] \\ & - \beta_{12}\,\delta_Y(t)s(t)[1 - \exp\{-\beta_{21}f^*(t)\}] - \mu_1 f^*(t) \end{aligned} \tag{4.36}$$

$$dm/dt = Am(t) - \beta_1 f(t)[1 - \exp\{-\beta_0 n^*(t)m^*(t)\}] - \mu_3 m^0(t) \tag{4.37}$$

where

$$\delta_Y(t) = \begin{cases} 1 & for \quad t \le Y, \\ 0 & for \quad t > Y, \end{cases} \qquad Y = \int_0^\infty \tau p_{sf}(\tau)d\tau$$

$$P_n(t) = \bar{P}_{nF}(t) + \bar{P}_{nC}(t),\ n^*(t) = n_0\bar{n}(t) = n_0\int_0^\infty nP_n(t)dn$$

$$\bar{P}_{nF}(t) = \int_0^\infty P_{nF}(\tau,t)d\tau, \qquad \bar{P}_{nC}(t) = \int_0^\infty P_{nC}(\tau,t)d\tau\ ,$$

$$m^*(t) = \max\{0, m(t) - m_{min}\};\ s^*(t) = \max\{0, s(t) - s_0\};$$

$$f^*(t) = \max\{0, f(t) - f_0\};\ m^0(t) = \max\{0, m(t) - m_0\}.$$

Here, the following designations are made: $f_{\min}$ and $m_{\min}$ are the minimal unconsumed biomass for the populations f and m, respectively; &ymacr; is the mean time interval during which the populations s and f interact; &nmacr; (t) is the mean number of interactions between the species of populations f and s up until the moment t; n_0 is the proportionality coefficient; β_{12}, β_l and A are coefficients characterizing the reproduction rate; β_{21} and β_0 are coefficients defining the saturation rate; s_0, m_0, f_0 and μ_i ($i = 1,2,3$) are constants setting the nature of mortality and other transitions of the living biomass to the detritus; $P_n(t)$ is the probability that until time t the n interactions between the species of populations f and m are realized independently from the state of populations s and f.

Equations (4.35) – (4.37) give some average pictures of the reproduction and destruction of the populations s, f and m. The model described by these equations is one of the selection of integrated models of the ecosystem represented in Fig. 4.7.

Under the given distributions p_f and p_{sf} and for the assumption that interactions between populations f and m have Poisson form, the distribution $P_n(t)$ can be calculated:

$$P_{nF}(\tau_f + h, t + h) = P_{nF}(\tau_f, t)[1 - p_f(\tau_f)h]; \tag{4.38}$$

$$P_{nC}(\tau_{sf} + h, t + h) = P_{nC}(\tau_{sf}, t)[1 - p_{sf}(\tau_{sf})h]\,[1 - \lambda h] + P_{n-1\,C}(\tau_{sf}, t)\,\lambda h. \tag{4.39}$$

Given the boundary conditions for P_{nF} and P_{nC} from Eqs. (4.38) and (4.39) we can obtain these probabilities. Specifically, let we consider that the distributions p_f and p_{sf} have the following forms:

$$p_f(\tau) = l(\tau)\exp\left[-\int_0^{\tau} l(u)du\right], \qquad p_{sf}(\tau) = r(\tau)\exp\left[-\int_0^{\tau} r(v)dv\right] \tag{4.40}$$

where $l(u)$ and $r(v)$ are arbitrary non-negative functions.

Putting the expression (4.40) into Eqs. (4.38) and (4.39) and substituting for $p_f(\tau_f)h$ and $p_{sf}(\tau_{sf})h$ by their first approximations $l(\tau_f)h$ and $r(\tau_{sf})h$, respectively, under $h \to 0$ we obtain the following equations:

$$\partial P_{nF}(\tau_f, t)/\partial \tau_f + \partial P_{nF}(\tau_f, t)/\partial t = -l(\tau_f)\, P_{nF}(\tau_f, t); \tag{4.41}$$

$$\partial P_{nC}(\tau_{sf}, t)/\partial \tau_{sf} + \partial P_{nC}(\tau_{sf}, t)/\partial t =$$
$$-\,[r(\tau_{sf}) + \lambda]\, P_{nC}(\tau_{sf}, t) + \lambda\, P_{n-1\,C}(\tau_{sf}, t). \tag{4.42}$$

To these equations it is necessary to add natural boundary conditions:

$$P_{nF}(0,t) = \int_0^{\infty} P_{nC}(\tau,t) r(\tau) d\tau, \qquad P_{nF}(\tau_f, 0) = \delta(\tau_f)\delta_{ni} \tag{4.43}$$

$$P_{nC}(0,t) = \int_0^{\infty} P_{nF}(\tau,t) l(\tau) d\tau, \qquad P_{nC}(\tau_{sf}, 0) = 0 \tag{4.44}$$

where $\delta(\tau_f)$ is the Dirac delta function, δ_{ni} is the Kronneker delta function, i is the number of interactions between the species of populations f and m until the moment $t = 0$ under the condition that at the moment $t = 0$ the populations s and f go to the free state.

Let us consider the case $l(\tau) = l_0$, $r(\tau) = r_0$. Then, using the method by Arora (1962), we obtain:

$$P_n(t) = \begin{cases} a_p(n) \sum_{m=0}^{n} (-1)^{n-m} C_n^m (2n-m)! \sum_{k=0}^{m} b_p(k,m) & for \quad u > u_0, \\ a_q(n) \sum_{m=n+2}^{2n+1} C_{2n+1}^m (n-1)! b_q(m) / (m-n-2)! & for \quad u \le u_0, \end{cases}$$

where

$$a_p(n) = \lambda^n l_0 / (2n \cdot n!) e^{-\varphi t/2}, a_q(n) = \lambda^n l_0 \, e^{\mu t},$$

$$u = t_0 + r_0 + \lambda, b_p(k,m)$$

$$= [t^{m-k} \rho^{m-1-2n} / (n-k)!] \{ \gamma^{n-k-1} e^{\rho t/2} [(n-2k) r_0 + \rho + n(l_0 - \lambda)]$$

$$+ (-1)^{m-1} G^{n-k-1} e^{-\rho t/2} [(n-2k) r_0 - \rho + n(l_0 - 1)],$$

$$b_q(m) = (\mu_c + l_0)^{m-n-2} t^m [\mu_c + l_0 + r_0 + \{(2n-m+1)/(m-n+1)\}(\mu_c + l_0)]$$

$$+ t^{n+1} (2n+1)!/(n+1)!,$$

$$\mu_c = -\varphi/2, \varphi = l_0 + \lambda + r_0, r_c = \lambda l_0, \gamma = -(r_0 - l_0 + \lambda + \rho)/2,$$

$$\rho = (\varphi^2 - 4 r_c)^{1/2}, G = -(r_0 - l_0 + \lambda - \rho)/2, u_0 = 2(\lambda l_0)^{1/2}.$$

Finally, note that the dependence (4.40) permits us to assess $P_n(t)$ and likewise for the other classical distributions. For example, if $l(\tau) = l_0\tau$ and $r(\tau) = r_0\tau$ then Eq.(4.40) transforms to the Rayleigh distribution.

First, we consider a specific case when the interacting interval of the state of the populations s and f is negligibly small compared with the studied interval of their interaction, i.e., &ymacr;≈0. In addition, it is considered that $m_0 = f_0 = m_{\min} = f_{\min} = 0$. Then, from Eqs. (4.35)–(4.37) we have:

$$ds/dt = -\mu_2 s(t),$$

$$df/dt = -\mu_1 f(t) + \beta_1 f(t)[1 - \exp\{-\beta_0 n^*(t) m(t)\}],$$

$$dm/dt = (A - \mu_3) m(t) - \beta_1 f(t)[1 - \exp\{-\beta_0 n^*(t) m(t)\}]. \tag{4.46}$$

From the first Eq.(4.46), it follows that population s dies out during the time $t_s = -(1/\mu_2)\ln[s_{\min}/s(0)]$, where $s_{\min}$ is the minimal critical biomass of the population s.

Further, let we suppose that $n^*(t)$ is small. Then, from the last two Eqs. (4.46) we have:

$$df/dt = f(t)[\beta_0\beta_1 n^*(t) m(t) - \mu_1],$$

$$dm/dt = m(t)[A - \mu_3 - \beta_0\beta_1 n^*(t) f(t)]. \tag{4.47}$$

Let us approximate $n^*(t)$ by the step-by-step function: $n^*(t) \cong n^*(t_i) = n_i$ for $t_i \le t \le t_{i+1}$. The precision of such an approximation depends on the max $|t_{i+1} - t_i|$. The function $n(t)$ in Eq.(4.47) is substituted by the above step-by-step function. Finally, we have:

$$dm/df = m(A - \mu_3 - \beta_0\beta_1 n_i f)/(\beta_0\beta_1 n_i m - \mu_1). \tag{4.48}$$

Integrating Eq.(4.48) gives:

$$F(m,f) = const = F_0 = -\mu_1 lg m + \beta_0\beta_1 n_i m - (A - \mu_3) \, lg f + \beta_0\beta_1 n_i f.$$

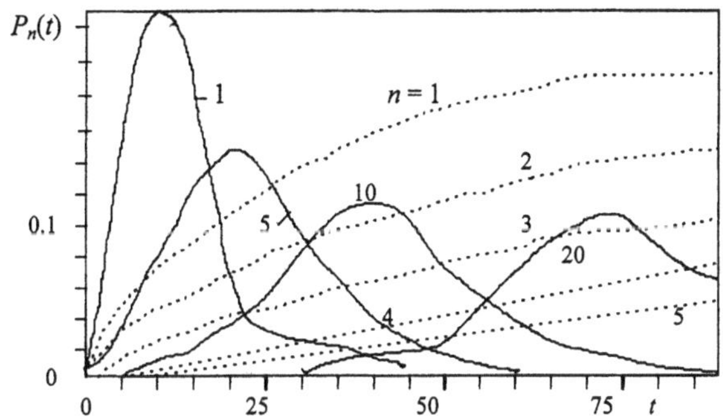

Fig. 4.14.
Time dependence of $P_n(t)$: $l_0 = r_0 = \lambda = 0.5$ (*solid curves*); $l_0 = 0.02, r_0 = \lambda = 0.5$ (*dashed curves*)

From this we see that the point $M(\mu_1/[\beta_0\beta_1 n_i], (A - \mu_3)/[\beta_0\beta_1 n_i])$ of the state plane (m,f) is the point of ecosystem stability. Moving from other points to point M is realized without fading. The ecosystem behavior in the small neighborhood of this point has a periodic law with the period equaled to $T = 2\pi[(A - \mu_3)\mu_1]^{-1/2}$.

Analytical analysis of the integrated model is a difficult task. Therefore, to study the general case of the interactions between the populations s, f and m, we consider s as the predator zooplankton, f as the herbivores, and m as the phytoplankton. Table 4.7 gives the value of the coefficients for Eqs. (4.35) – (4.37). Other coefficients were determined with the following estimations:

$$s_0 = 0.9, \beta_1 = 1.043, \beta_0 = 0.2, f_{min} = 14.5, m_{min} = 0.3, f_0 = 1.2,$$

$$A = 0.56, m_0 = 6, \mu_1 = 0.2626, s(0) = 9, f(0) = 12, m(0) = 60.$$

Figure 4.14 demonstrates an example of calculations using Eq. (4.4). We see that the function $P_n(t)$ has a unique maximum value which decreases with increasing n, and its position shifts with the increase in time. An increase in the time interval when the populations s and f are in a free state leads to the expansion of curves $P_n(t)$ under conditions of the conservation of their behavior. Under such conditions $P_n(t)$ is close to the uniform distribution for a sufficiently wide time interval. This is natural, since the distribution of interactions between the populations f and m approaches the uniform distribution when the average time of free state of populations s and f is $1/l_0$ for r_0 = constant.

Table 4.7 demonstrates the calculation results for the time moments t^* when the population f reaches f_{min} for $r_0 = 0.01$. The value of f^* is the solution of

Table 4.7. The model parameters estimation used for the solution of Eqs. (4.35) – (4.37), and the results of the calculation for the time t^* when the population f reaches the level f_{min}

Predator, s	β_{12}	β_{21}	μ_2	t^* (days)
Average, s_{mid}	0.71	0.83	0.3	8.1
Herbivores, Z_4	0.41	200	0.25	23.2
Microzoa, Z_2	0.32	100	0.25	56.6
Small, Z_5	0.71	0.83	0.52	18.2
Large, Z_7	0.35	0.83	0.23	33.3
Fish larvae, r	0.13	0.68	0.07	74.7

the equation $f(t^*) = f_{\min}$ when $l_0 = \lambda = 0.5$. From Eq.(4.35) it follows that if the population s until the moment $t = t^*$ has a biomass larger than $s_{\min}$ it dies out at the moment of time determined by the following expression: $t^{**} = t^* - (1/\mu_0)ln|(s_{min} - s_0)/(s(t^*) - s_0)|$. After the moment t^* the development of population m will be mainly determined by the correlation between the coefficients A and μ_3.

The solutions to Eqs. (4.35) – (4.37) show that the time when the ecosystem is in the living state depends on the parameters of population s and $P_n(t)$. For example, if &ymacr; ≈ 100 days then the ecosystem has time to go one cycle of its development. Just the other way, about &ymacr; ≈ 10 days, the ecosystem has rapidly ceased to exist as a unique formation.

Chapter 5

Application of a Global Model to the Study of Arctic Basin Pollution 5

5.1 Introduction

The introduction of man-made radionuclides, heavy metals and oil hydrocarbons into the Arctic basin has been of international concern since the 1992 release of information on Soviet dumping of nuclear reactor and solid industrial waste. The purpose of this chapter is to develop and to investigate a simulation model of the pollution dynamics in the Arctic basin. There are many observational and theoretical results giving estimates of the growing dependencies between the pollution dynamics in the World Ocean and the state of the continental environment.

The problem of pollution of the Arctic basin causes perhaps the most anxiety to investigators (Krapivin 1995; Krapivin et al. 1997b; Yudakhin 2000; Krapivin and Phillips 2001a; Alexeeva et al. 2001; Miguel 2001; Yablokov 2001). It is known that the ecosystems of the Arctic seas are vulnerable to a considerable extent in comparison with the ecosystems of other seas. The processes that clean the Arctic ocean are slower and marine organisms of the Arctic ecosystem live in a polar climate where the vegetation period is restricted. Some feedback mechanisms have significant time delays and the capacity of the ecosystem to neutralize the effects of human activity is feeble. Apart from these reasons, the Arctic ecosystem has specific boundary conditions connected with the sea-ice ergocline which reduce its survivability level (Demers et al. 1986; Kelley et al. 1999).

In connection with this circumstance, the Arctic basin has been the object for investigations in the frameworks of many national and international environmental programs, such as the international Geosphere-Biosphere Program, the US Global Change Research program, the international program "Arctic System Science" (ARCSS) (McCauley and Meier 1991), the US Arctic Nuclear Waste Assessment Program (ANWAP) and the international Arctic Monitoring and Assessment Program (AMAP). The research strategy of these programs includes the theoretical and experimental study of the tundra ecosystems, Siberian rivers, and near-shore and open arctic waters.

An understanding of the environmental processes in the Arctic regions, which is a prerequisite for finding scientific solutions to the problems arising there, can be found only by combining many disciplines, including ecology, oceanography, mathematical modeling, and system analysis. This chapter synthesizes data from many sources and knowledge from various scientific fields in the form of a

Spatial Simulation Model of the Arctic Ecosystem (SSMAE). Separate units of the SSMAE were created earlier by many authors (Riedlinger and Preller 1991; Muller and Peter 1992; Legendre and Krapivin 1992; Krapivin 1995; Shabas and Chikin 2001). The sequence of these units in the SSMAE structure and the adaptation of it to the GSM provide a technology for computer experiments (Krapivin 1993).

This investigation realizes the idea of the ARCSS Program, which was initiated at the US National Science Foundation as part of their contribution to the Global Change Research Program (McCauley and Meier 1991). The present chapter describes a simulation system based on sets of computer algorithms for processing data from the monitoring of arctic regions and applying mathematical models of natural and anthropogenic processes.

The basic units of the SSMAE are oriented on the simulation of the dynamics of any pollutant. For the consideration of a specific pollutant it is necessary to include in the SSMAE an additional unit with the description of its physical and chemical characteristics. This procedure is demonstrated by examples of units that simulate the characteristics of radionuclides, heavy metals and oil hydrocarbons. The consideration of these pollutants is restricted to elements with averaged properties.

The Arctic basin aquatory Ω studied in this chapter has boundaries which include the peripheral Arctic seas as well as the coast line and southern boundaries of the Norwegian and Bering Seas.

5.2 The Spatial Simulation Model of the Arctic Ecosystem Structure

A conceptual diagram and a unit content of the SSMAE are shown in Fig. 5.1 and Table 5.1. The functioning of the SSMAE is supported by the GSM and by the climate model (CM). The inputs to the SSMAE are data about the pollutant sources of the near-shore Arctic basin, the ice areas and the current maps. The SSMAE contains three types of units: mathematical models of the natural ecological and hydrophysical processes, service software and the scenarios generator. The marine biota unit MMB describes the dynamics of the energy flows in the trophic chains of the Arctic basin ecosystem. The hydrological unit HB describes the water circulation in the Arctic. The pollution simulation model PSM contains the anthropogenic scenarios and the service control unit SB provides for the control of the simulation experiment.

Let us designate the Arctic basin aquatory as $\Omega = \{(\varphi,\lambda)\}$, where φ and λ are latitude and longitude, respectively. Spatial inhomogeneity of the Arctic basin model is provided for by the spatial discretization of Ω on the set of cells Ω_{ij} with latitude and longitude steps of $\Delta\varphi$ and $\Delta\lambda$, respectively. These cells are the basic spatial structure of Ω for the realization of the computer algorithms. The cells Ω_{ij} are heterogeneous as to their parameters and functioning. There are a set of cells which are adjacent to the river mouths (Ω_R) and to the ports (Ω_P), bordering on the land (Ω_Γ), in the Bering Strait (Ω_B) and on the southern boundary of the Norwegian Sea (Ω_N). The aquatory Ω is divided in depth z by step Δz. The distribution of depths is given as the matrix $||h_{ij}||$ where $h_{ij} = h(\varphi_i, \lambda_j), (\varphi_i, \lambda_j) \in \Omega_{ij}$.

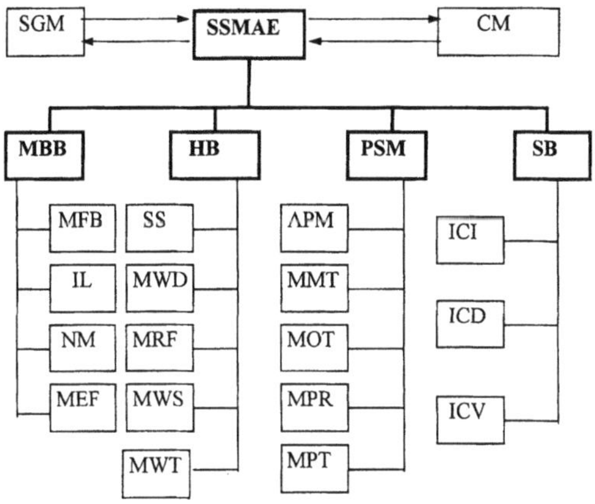

Fig. 5.1. Block diagram of the spatial simulation model of the arctic ecosystem (*SSMAE*). Coupled to the SSMAE are the global spatial model (*GSM*) and the climate model (*CM*) or the climate scenario. *MBB* is the marine biota unit, *HB* is the hydrological unit, *PSM* is the polar simulation model and *SB* is the service unit Descriptions of the units and their subunits are given in Table 5.1

As a result, the full water volume of Ω is divided on the volumetric compartments

$$\Xi_{ijk} = \{(\varphi,\lambda,z)/\varphi_i \le \varphi \le \varphi_{i+1};\lambda_j \le \lambda \le \lambda_{j+1};z_k \le z \le z_{k+1}\}$$

with volumes $\sigma_{ijk} = \Delta\varphi_i\Delta\lambda_j\Delta z_k$. Within the Ξ_{ijk} boxes the water body is considered as a homogeneous structure. The water temperature, salinity, density, and biomass of Ξ_{ijk} are described by box models. The anthropogenic processes on the aquatory Ω, are described for the four seasons: τ_w winter, τ_s spring, τ_u summer, and τ_a autumn.

The procedure of spatial discretization is provided for via the ICI unit of the SSMAE database, including the set of identifiers $A_k = \left\|a_{ij}^k\right\|$, where a_{ij}^k is a specific symbol to identify a real element of Ω_{ij} in the computer memory. Identifier A_1 reflects the spatial structure of the Arctic basin and adjoining territories – for $(\varphi_i, \lambda_j) \notin \Omega\ a_{ij}^1 = 0$; for $(\varphi_i, \lambda_j) \in \Omega, a_{ij}^1 = 1$ when (φ_i, λ_j) belongs to the land (islands); and a_{ij}^1 equals the aquatory identifier symbol from the second column of Table 5.4 when (φ_i, λ_j) belongs to a given sea. Identifier A_2 shows the position of the cells Ω_R, Ω_P, Ω_N, Ω_S, Ω_Γ and describes the spatial distribution of the pollutant sources. Other identifiers A_k $(k \ge 3)$ are used to describe ice fields $(k = 3)$, the spatial distribution of solar radiation $(k = 4)$, and the dislocation of upwelling zones $(k = 5)$.

The user of the SSMAE in free-running mode may choose different ways to describe the many input parameters. Units ICI and ICD realize the on-line entry to A_k and to the database. For example, if the user has data about the spatial distribution of ice fields in Ω, he can form the identifier A_3 with $a_{ij}^3 = 0$ for the ice-free water surface, $a_{ij}^3 = 1$ when cell Ω_{ij} corresponds to new ice, and $a_{ij}^3 = 2$ when cell Ω_{ij} corresponds to old ice. In this case unit SS enables the input of data of the climate model concerning the ice fields.

Table 5.1. Description of the SSMAE units of Fig. 1

Unit	Unit description
MBB	Marine Biota Unit containing the set of models for energy flows in the trophic chains of the Arctic basin ecosystem (Harrison and Cota 1991; Legendre and Krapivin 1992; Krapivin 1995; Legendre and Legendre 1998; Sazykina et al. 2000)
MFB	Model of the Functioning of the Biota under the conditions of energy exchange in the trophic chain of the Arctic basin ecosystem (Legendre and Krapivin 1992)
IM	The illumination model (see Sect. 4.2)
NM	The nutrients model (Krapivin 1996)
MEF	Model for energy flow transport in the Arctic basin ecosystem (Harrison and Cota 1991; Hrol 1993)
HB	Hydrological unit describing the water circulation in the Arctic seas and the movement of ecological elements (Krapivin 1995, 1996)
SS	Simulator of scenarios describing the ice fields, and the synoptical situations and change in the hydrological regimes (Porubaev 2000)
MWD	Model for the water dynamics of the Arctic basin (Riedlinger and Preller 1991; Rudels et al. 1991)
MRF	Model of river flow to the Arctic basin (Krapivin et al. 1998a)
MWS	Model of water salinity dynamics (Berdnikov et al. 1989)
MWT	Model for calculating the water temperature (Krapivin 1993)
PSM	Arctic basin pollution simulation model including the set of anthropogenic scenarios (Krapivin 1993, 1995)
APM	Air pollution transport model (Muller and Peter 1992)
MMT	Model for heavy metals transport through food chains (Krapivin et al. 1998a)
MOT	Model for the process of oil hydrocarbons transport to the food chains (Payne et al. 1991)
MPR	Model for the process of radionuclide transport to the food chains (Thiessen et al. 1999)
MPT	Model for pollution transport through water-exchange between the Arctic basin and the Atlantic and Pacific Oceans (Bourke et al. 1992)
SB	Service unit to control the simulation experiment
ICI	Interface for control of the identifiers
ICD	Interface for control of the database
ICV	Interface for control of the visualization

The unit structure of the SSMAE is provided with a C++ program. Each of the units from Table 5.1 is C++ function. The main function provides the interactions between the SSMAE, SGM, and CM. This functional specification supports overlapping output and input streams of the SSMAE units. With the conversational mode the user can toggle the data streams between the slave units.

The calculation procedure is used on the subdivision of the Arctic basin into boxes $\{\Xi_{ijk}\}$. This is realized by means of a quasi-linearization method (Nitu et al. 2000b). All differential equations of the SSMAE are substituted in each box Ξ_{ijk} by easily integrable ordinary differential equations with constant coefficients. The water motion and turbulent mixing are realized in conformity with current velocity fields which are defined on the same coordinate grid as the $\{\Xi_{ijk}\}$ (Krapivin et al. 1998b).

5.3 Marine Biota Unit

The ice-water ergocline plays an important role in the biological productivity of the north seas. According to the hypothesis of Legendre and Legendre (1998), energetic ergoclines are the preferential sites for biological production in the Arctic Ocean. The primary production in the food chains of the Arctic basin ecosystems is determined by the phytoplankton productivity. This is associated with complex variations in the meteorological, hydrodynamic, geochemical and energetic parameters of the sea environment. The problem of parameterization of the phytoplankton production in the north seas was studied by Legendre and Legendre (1998). Table 5.2 gives the structure for the seasonal composition of conditions affecting the primary production in Ω. This scheme is applied for each of the Ξ_{ijk}.

A unit MWF calculates the water temperature T_w by means of the averaging of temperatures for the mixed water volumes. In addition, the following correlations are applied (Krapivin and Phillips 2001a):

$$T_g = T_r = T_f = \begin{cases} a_1 & when \quad g+r+f \le 50cm, \\ a_2 & when \quad g+r+f > 50cm \end{cases}$$

where $a_1 = -\,0.024\,(g+r+f)+0.76T_0 + 8.38$, $a_2 = -\,0.042(g+r+f) + 0.391T_0 - 0.549$, T_0 is the surface temperature, g is the snow depth, r is the thickness of floating ice and f is the depth of the ice submerged below the water surface. If we designate by ρ_g, ρ_r and ζ the density of snow, ice and seawater, respectively, we obtain for the depth of ice beneath the surface:

$$f = (g\rho_g + r\rho_r)/(\zeta - \rho_r).$$

Figure 5.2 shows a conceptual flow chart of the energy in an ecological system. The energy input during time interval t is provided by solar radiation $E_A(t,\varphi,\lambda,z)$ and the upward transport of nutrients from the deep-sea layers. The concentration of nutrients $B_{6,A}(t,\varphi,\lambda,z)$ at the depth z is determined by photosynthesis R_{pA}, advection and destruction of suspended dead organic matter or detritus B_7. The role of hydrodynamic conditions manifests itself in maintenance of the concen-

Table 5.2. Vertical structure of the Arctic basin aquatic system

Layer (A)	Parameters					
	Layer thickness	Temperature	Irradiance	Turbulence coefficient	Irradiance relaxation coefficient	Reflection coefficient
Surface		T_0	E_0			
Snow	g	T_g	E_g		α_g	β_g
Floating ice	r	T_r	E_r		α_r	β_r
Submerged ice	f	T_f	E_f		α_f	β_f
Water	z–f	T_w	E_w	k_w	α_w	β_w

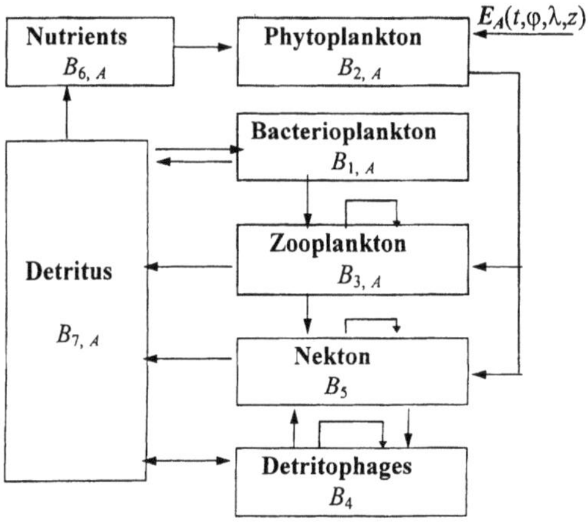

Fig. 5.2. Block diagram of the energy flows (cal m^{-3} day^{-1}) in the trophic pyramid of the Arctic basin ecosystem. This is realized as unit MEF. The *boxes with elements* denote the generalized trophic levels of the Arctic ecosystem. All of the elements are described by means of averaged parameters for the biological community of the north seas. It is supposed that this trophic pyramid takes place in each of the Arctic basin seas. The trophic relations between the elements of the model are described on the basis of the energetic principle. Biomasses, the rates of production, and exchange (respiration), and the food rations are expressed in energy units. Total nitrogen serves as a "nutrient salts" prototype in the model

tration of nutrients required for photosynthesis which occurs via transport from other layers or aquatories of the sea where the concentration is sufficiently high. Taking into account the designations of Table 5.2, we have:

$$E(t,\varphi,\lambda,z)=\begin{cases} E_0 & when \quad z\le-(g+r), \\ E_g & when \quad z\in[-(g+r),-r], \\ E_r & when \quad z\in[-r,0], \\ E_f & when \quad z\in[0,f], \\ E_w & when \quad z>f, \end{cases} \tag{5.1}$$

where

$$E_g = (1-\beta_g)E_o\exp\{-\alpha_g z\}, E_r = (1-\beta_r)E_g(t,\varphi,\lambda,-r)\exp\{-\alpha_r z\},$$
$$E_f = (1-\beta_f)E_r(t,\varphi,\lambda,0)\exp\{-\alpha_f z\}, E_w = (1-\beta_w)E_f(t,\varphi,\lambda,f)\exp\{-\alpha_w z\},$$

and the values of α_A $(A = g,r,f,w)$ depend on the optical properties of the Ath medium. The irradiance E_0 arrives at the surface of Ω. The estimate of E_0 is obtained from monitoring or is calculated from the climatic model. The flow of E_0 is attenuated by snow, ice and water according to the scheme of Table 5.2. In each cell Ω_{ij} the structure of these layers is changed corresponding to the time of year. Within each layer, the attenuation of the irradiance with the depth is described by an exponential model (Legendre and Krapivin 1992). The parameters α_A and β_A are functions of the salinity, turbidity, temperature, and biomass. The form of this dependence is given as a scenario or else the standard functions are used (unit IL).

As a basic scheme for the flow of nutrients in the water, the scheme proposed by Vinogradov et al. (1972, 1973, 1977) is accepted as corrected for the conditions of the Arctic basin by Legendre and Legendre (1998). It is supposed that the spatial distribution of the upwelling zones is given with seasonal variations. Unit NUM realizes this scheme in reference to the current structure of the upwelling regions.

The dynamic equation for the nutrients $B_{6,A}$ in the environment, where $A = \{S\text{–snow}, I\text{–ice}, W\text{–water}\}$, is given by

$$\partial B_{6,A} / \partial t + v_{\varphi}^{A} \partial B_{6,A} / \partial \varphi + v_{\lambda}^{A} \partial B_{6,A} \partial \lambda + v_{z}^{A} \partial B_{6,A} / \partial z = Q_A + k_2^w \partial^2 B_{6,A} / \partial z^2 + \beta_v \partial B_{6,A} / \partial z + \rho_1 \sum_{i=1}^{5} T_i - \delta_1 R_{pA} + \varepsilon_1^A H_1^\varepsilon \quad (5.2)$$

where $v_{\varphi}^{A}, v_{\lambda}^{A}, v_{z}^{A}$ are the velocity projections of motion in the environment; Q_A is the input of biogenic elements to A resulting from the decomposition of detritus ($Q_A = \delta_n R_D^A$) with $R_D^A = \mu_A B_7$; δ_n is the content of nutrients in dead organic matter; μ_A is the rate of decomposition of detritus into the environment A; k_2^W is the kinematic coefficient of vertical diffusion; δ_1 is the velocity of nutrient assimilation by the photosynthetic process per unit of phytoplankton production; ε_1^A is the proportional part of the εth radionuclide which is chemically analogous to $B_{6,A}$ on substrate A; H_I^ε is the rate of input flow of the εth radionuclide; T_i is the rate of exchange with the environment; ρ_1 is the part of the biomass losses due to exchange which transform into nutrients; and β_v is the upwelling velocity. Equation (5.2) is the basic element of unit NM.

The phytoplankton production R_{pA} in the environment A is a function of the solar radiation E_A, the concentration of nutrients $B_{6,A}$, the temperature T_A, the phytoplankton biomass $B_{2,A}$, and the concentration of pollutants ξ_A. There are many models for the description of the photosynthesis process (Vinogradov et al. 1977; Legendre and Krapivin 1992; Dilao and Domingos 2000). For the description of this function in the present study, the equation of the Michaelis-Menten type is used (unit MFB):

$$R_{pA} = a_A \, k_I^A p_{A,max} / (E_A + k_I^A) \quad (5.3)$$

where k_I^A is the irradiance level at which $R_{pA} = 0.5 \cdot R_{pA,\max}$, and $p_{A,\max}$ is the maximum quantum yield. The coefficient a_A reflects the dependence of the phytoplankton production on the environmental temperature T and the concentration of nutrients $B_{6,A}$. A unit MFB realizes the following equation for the calculation of a_A:

$$a_A = a_1 K_0 (T,t) / [1 + B_{2,A} / (a_2 B_{6,A})] \quad (5.4)$$

Here, a_1 is the maximal rate of absorption of nutrients by the phytoplankton (day^{-1}), a_2 is the index of the rate of saturation of photosynthesis, and

$$K_0(T,t) = a_3 \max\left\{0, \frac{T_c - T}{T_c - T_{opt}} \exp\left[1 - \frac{T_c - T}{T_c - T_{opt}}\right]\right\} \quad (5.5)$$

where a_3 is the weight coefficient, and T_c and T_{opt} are the critical and optimal temperatures for photosynthesis, respectively (°C).

Kiefer and Mitchel (1983) have shown that Eq. (5.3) adequately fits the laboratory data. The relationships (5.4) and (5.5) make the description of the phytoplankton production more accurate for critical environmental conditions when the concentration of nutrients and the temperature have high fluctuations. The coefficients of these relationships are defined on the basis of estimates given by Legendre and Legendre (1998).

The dynamic equation for the phytoplankton biomass in the environment A has the following form:

$$\partial B_{2,A} / \partial t + v_\varphi^A \partial B_{2,A} / \partial \varphi + v_\lambda^A \partial B_{2,A} / \partial \lambda + v_z^A \partial B_{2,A} / \partial z =$$
$$R_{pA} - \theta_p^A - M_p^A + k_2^A \partial^2 B_{2,A} / \partial z^2$$
$$-\left[k_{Zp}^A R_{ZA} / \xi_Z^A + k_{Fp}^A R_{FA} / \xi_F^A\right] B_{2,A} \tag{5.6}$$

where R_{ZA}, R_{FA} are the production and ξ_Z^A, ξ_F^A are the food spectrum of zooplankton B_3 and nekton B_5, respectively; M_p^A is the mortality; and θ_p^A is the rate of exchange. The balance equations for the other ecological elements of Fig. 5.2 are given by Krapivin (1995, 1996).

The energy source for the entire system is the solar radiation energy $E_A(t,\varphi,\lambda,z)$, the intensity of which depends on time t, latitude φ, longitude λ, and depth z. The equations that describe the biomass dynamics of the living elements are:

$$\frac{\partial B_i}{\partial t} + \xi_i [V_\varphi \frac{\partial B_i}{\partial \varphi} + V_\lambda \frac{\partial B_i}{\partial \lambda} + V_z \frac{\partial B_i}{\partial z}] = R_i - T_i - M_i - H_i$$
$$- \sum_{j \in \Gamma_i} C_{ij} R_j + \xi_i [\frac{\partial}{\partial \varphi}\left(\Delta_\varphi \frac{\partial B_i}{\partial \varphi}\right)$$
$$+ \frac{\partial}{\partial \lambda}\left(\Delta_\lambda \frac{\partial B_i}{\partial \lambda}\right) + \frac{\partial}{\partial z}\left(\Delta_z \frac{\partial B_i}{\partial z}\right) + \beta_V \frac{\partial B_i}{\partial z}], (i = 1,3,4,5); \tag{5.7}$$

where $V(V_\varphi, V_\lambda, V_z)$ are the components of the water current velocity ($V_\varphi = v_\varphi^W$, $V_\lambda = v_\lambda^W$, $V_z = v_z^W$; R_i is the production; M_i is the mortality; H_i is nonassimilated food; and Γ_i is the set of the trophic subordination of the ith component: $C_{ji} = k_{ji} F_i / \sum_{m \in S_i} k_{jm} F_m$; S_i is the food spectrum of the jth component; k_{jm} is the index of satisfaction of the nutritive requirements of the jth component at the expense of the mth component biomass; $F_i = \max\{0, B_i - B_{i,min}\}$, $B_{i,min}$ is the minimal biomass of the ith component consumed by other trophic levels; $\Delta(\Delta_\varphi, \Delta_\lambda, \Delta_z)$ are components of the turbulent mixing coefficient (on the assumption of isotrophism of vertical mixing in the horizontal plane $\Delta_\varphi = \Delta_\lambda = v_H$); and β_V is the upwelling velocity. Functions R_i, M_i, H_i, and T_i are parameterized according to the models by Legendre and Krapivin (1992) and Krapivin (1996). The equations describing the dynamics of the abiotic elements are represented in conformity with Berdnikov et al. (1989). Functions M_4 and M_5 include the biomass losses at the expense of fishing. Parameter ξ_i characterizes the subjection of the ith component relative to the current. It is supposed that $\xi_i = 1$ for $i = 1,2,3$ and $\xi_i = 0$ for $i = 4, 5$.

Fig. 5.3.
Block diagram of the energy flows (cal m^{-3} day^{-1}) in the *snow-ice-water* interface

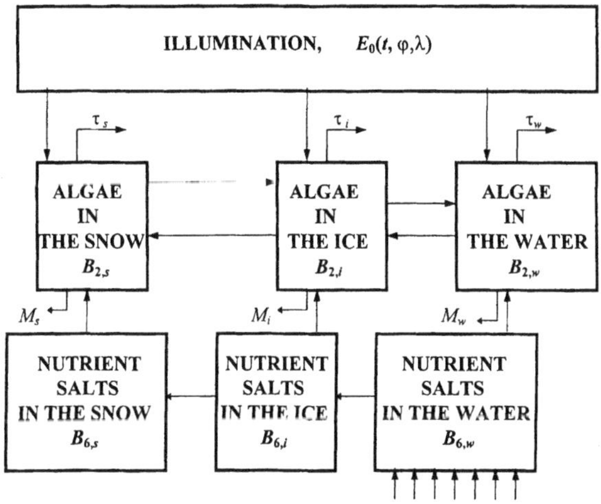

The inert components are described by the following equations (Krapivin 1996):

$$\frac{\partial B_7}{\partial t}+V_\varphi\frac{\partial B_7}{\partial \varphi}+V_\lambda\frac{\partial B_7}{\partial \lambda}+V_z\frac{\partial B_7}{\partial z}=\sum_{i=1}^{5}(M_i+H_i)-\mu_W$$

$$-(v^*-\beta_v)\partial B_7/\partial z-(k_{1,7}R_1/P_1+k_{3,7}R_3/P_3+k_{4,7}R_4/P_4$$

$$+k_{5,7}R_5/P_5)B_{7,\min}+\frac{\partial}{\partial\varphi}\left(\Delta_\varphi\frac{\partial B_7}{\partial\varphi}\right)+\frac{\partial}{\partial\lambda}\left(\Delta_\lambda\frac{\partial B_7}{\partial\lambda}\right)$$

$$+\frac{\partial}{\partial z}\left(\Delta_z\frac{\partial B_7}{\partial z}\right) \tag{5.8}$$

where $\frac{\partial e^*}{\partial t}=H_1^e-\alpha_1(H_L^{e\psi}+H_a^{e\psi})$; μ_W is the velocity of decomposition of detritus per unit of biomass; v_* is the velocity of settling due to gravity; and k_{ij} is a coefficient of the relation of the *i*th element with respect to the *j*th element of the ecosystem.

Equations (5.2) – (5.8) are used in the complete volume only when $(\varphi,\lambda,z) \in W$. In the other cases (i.e., in the layers of ice or snow), these equations are automatically reduced in accordance with the scheme represented in Fig. 5.3.

5.4
The Hydrological Unit

The circulation of the waters in the Arctic basin is a complex system of cycles and currents with different scales. Unit HB simulates the dynamics of the Arctic basin waters by a system of subunits presented in Fig. 5.1. The water dynamics in Ω is presented by flows between the compartments Ξ_{ijk}. The directions of water exchanges are represented on every level $z_k = z_0 + (k-1)\Delta z_k$ according to Aota et al. (1992) in conformity with the current maps assigned as the SSMAE input. The external boundary of Ω is determined by the coast line, the bottom, the Bering Strait, the southern boundary of the Norwegian Sea, and the *water-atmosphere* boundary.

The hydrological data are synthesized via a four-level structure according to the seasons (unit MWD). The velocity of current in the Bering Strait is estimated by the following binary function:

$$V(t)=\begin{cases} V_1 & for \quad t\in\tau_u\cup\tau_a; \\ V_2 & for \quad t\in\tau_w\cup\tau_s \end{cases}$$

The water exchange through the southern boundary of the Norwegian Sea is V_3. The water temperature $T^W{}_{ijk}$ in Ξ_{ijk} (unit MWT) is a function of the evaporation, precipitation, river flows, and inflows of water from the Atlantic and Pacific Oceans. Its change with time in Ξ_{ijk} is described by the equation of heat balance:

$$\zeta C\sigma_{ijk}\frac{\partial T^W_{ijk}}{\partial t}=\sum_{s,l.m}\left(W^{ijk}_{slm}+f^{ijk}_{slm}\right)-W_{ijk} \tag{5.9}$$

where ζ is the seawater density (g cm^{-3}); C is the water thermal capacity, (cal g^{-1} grad^{-1}); σ_{ijk} is the volume of Ξ_{ijk}; W^{ijk}_{slm} is the heat inflow to Ξ_{ijk} from Ξ_{slm}; f^{ijk}_{slm} is the heat exchange between Ξ_{slm} and Ξ_{ijk} caused by turbulent mixing; and W_{ijk} is the total heat outflow from Ξ_{ijk} to the bordered boxes. Heat exchange with the atmosphere is calculated with the consideration of empirical Eq. (5.1).

It is considered that the dissipation of moving kinetic energy, geothermal flows on the ocean bed, heat effects of chemical processes in the ocean ecosystem, and freezing and melting of the ice are not global determinants of the water temperature fields. The SSMAE does not consider these effects.

The dynamics of the water salinity $S(t,\varphi,\lambda,z)$ during the time interval t are described by the balance equation of Berdnikov et al. (1989) as unit MWS. The ice salinity is defined by a two-step scale: s_1 – old, s_2 – new. It is supposed that $S(t,\varphi,\lambda,z) = s_0$ for $z > 100$ m, $s_2 = k_s S(t,\varphi,\lambda,f)$ for $r+f > H_{\max}$ and $s_1 = k_r s_2 H_{\max}/(r+f)$ for $r+f < H_{\max}$, where the coefficients k_s and k_r are determined empirically and $H_{\max}$ is maximal thickness of new ice. In accordance with the estimations by Riedlinger and Preller (1991), the simulation experiments are realized for $H_{\max} = 50$ cm, $k_s = k_r = 1$.

The river flows, ice fields, and synoptical situations are described by scenarios given in the MRF and SS units and formed by the user of the SSMAE (Trushkov et al. 1992; Rovinsky et al. 1995).

The snow-layer thickness $g(t,\varphi,\lambda)$ may be described via statistical data with given dispersion characteristics: $g = \bar{g} + g_0$ where the value $\bar{g}$ is defined as the mean characteristic for the chosen time interval and the function $g_0(t, \varphi, \lambda)$ gives the variation of g for the given time interval.

An alternative description is the parameterization of the snow-layer dynamics process in the framework of the atmospheric process simulation algorithm (unit APM) relating the thickness of the growth and melting of the snow layer to the temperature and precipitation:

$$g(t+\Delta t, \varphi, \lambda) = g(t, \varphi, \lambda)+S_F-S_M,$$

where S_F is the part of the snow precipitated at temperatures close to freezing ($265K \leq T_0 \leq 275K$) and S_M is the snow ablation (i.e., evaporation + melting).

Unit SS gives the user the possibility to select between these algorithms. When statistical data on the snow-layer thickness exist, the function $g(t,\varphi,\lambda)$ is reconstructed for $(\varphi,\lambda) \in \Omega$ by means of the approximation algorithm with time and polynomial interpolation in space.

5.5 The Pollution Unit

A unit PSM simulates the pollution processes of the territory Ω as a result of the atmosphere transport, river and surface coastal outflow, navigation, and other human activity (Straub 1989; Mohler and Arnold 1992; Muller and Peter 1992; Krapivin et al. 1996a,b, 1997a). The variety of pollutants is described by three components: oil hydrocarbons O, heavy metals (e – solid particles, ψ – dissolved fraction) and radionuclides ε. It is supposed that the pollutants get into the living organisms only through the food chains.

The rivers bring a considerable contribution to the level of pollution of the arctic waters. The concentration of pollutant κ in the river γ is γ_κ. The pollutant κ enters compartment $\Omega_{ijk} \in \Omega_R$ with the velocity c_y reflecting the mass flow per unit time. Subsequently, the spreading of the pollutant in Ω is described by other subunits.

The influence of water exchanges between the Arctic basin and the Pacific and Atlantic Oceans on the pollution level in Ω is described by unit MPT. It is supposed that the watersheds of the Norwegian Sea Ω_N and the Bering Strait Ω_B are characterized by currents with varying directions given as a scenario.

The atmospheric transport of heavy metals, oil hydrocarbons and radionuclides is described by many models (Kudo et al. 1999; Mackay and Fraser 2000; Sportisse 2000). The application of these models to the retrieval of the pollution distribution over Ω makes it possible to estimate optimal values of $\Delta\varphi$, $\Delta\lambda$ and the step in time Δt. The present level of the database for the Arctic basin provides for the use of a one-level Euler model with Δt = 10 days and $\Delta\varphi = \Delta\lambda = 1°$ (unit APM). It is supposed that pollution sources can be located on the Arctic basin boundary. Their detailed distributions are given as the SSMAE input. The transport of pollutants to the Arctic basin and the formation of their spatial distribution are realized in conformity with the wind velocity field, which is considered as given.

It is postulated that the oil hydrocarbons $O(t,\varphi,\lambda,z)$ are transformed by the following processes (Payne et al. 1991): dissolving H_O^1, evaporation H_O^2, sedimentation H_O^3, oxidation H_O^4, biological adsorption H_O^5, bio-sedimentation H_O^6, and bacterial decomposition H_O^7. The kinetics equation for the description of the dynamics of the oil hydrocarbons in the Arctic basin is given by

$$\partial O / \partial t + v_\varphi^w \partial O / \partial\varphi + v_\lambda^w \partial O / \partial\lambda + v_z^w \partial O / \partial z =$$

$$Q_O + k_2^w \partial^2 O / \partial z^2 - \sum_{i=1}^{7} H_O^i \qquad (5.10)$$

where Q_O is the anthropogenic source of the oil hydrocarbons.

The process of diffusion of the heavy metals in the seawater depends on their state. The dissolved fraction of the heavy metals (ψ) takes part in the biogeochemical processes more intensively than the suspended particles (e). However, like suspended particles, the heavy metals fall more rapidly to the sediment. A description of the entire spectrum of these processes in the framework of this study is impossible. Therefore, unit MMT describes those processes which have estimates. The transport process of heavy metals in the sea water includes absorption of the dissolved fraction ψ by plankton (H_Z^ψ) and by nekton (H_F^ψ), sedimentation of the solid fraction (H_l^e), deposition with the detritus (H_D^ψ), adsorption by detritophages from bottom sediments (He_L^ψ), and release from bottom sediments owing to diffusion (He_a^ψ). As a result, the dynamic equations for the heavy metals become:

$$\frac{\partial e_W}{\partial t} + v_\varphi^W \frac{\partial e_W}{\partial \varphi} + v_\gamma^W \frac{\partial e_W}{\partial \lambda} + v_z^W \frac{\partial e_W}{\partial z} =$$

$$\sum_{i=1}^{3} \alpha_2^1 Q_{e\psi}^1 - H_1^e + \alpha_1 H_a^{e\psi} \tag{5.11}$$

$$\frac{\partial \psi_W}{\partial t} + v_\varphi^W \frac{\partial \psi_W}{\partial \varphi} + v_\lambda^W \frac{\partial \psi_W}{\partial \lambda} + v_z^W \frac{\partial \psi_W}{\partial z} =$$

$$\left(1-\alpha_1\right) H_a^{e\psi} + k_2^W \frac{\partial^2 \psi_W}{\partial z^2} - H_Z^\psi - H_F^\psi - H_D^\psi - H_a^\psi \tag{5.12}$$

$$\frac{\partial e^*}{\partial t} = H_1^e - \alpha_1 (H_L^{e\psi} + H_a^{e\psi}) \tag{5.13}$$

$$\frac{\partial \psi^*}{\partial t} = H_D^\psi - (1-\alpha_1)(H_L^{e\psi} + H_a^{e\psi}) \tag{5.14}$$

where $e_{w,}$ ψ_w and e^*, ψ^* are the concentrations of heavy metals in the water and in the bottom sediments as solid and dissolved phases, respectively; H_a^ψ is the output of heavy metals from the sea to the atmosphere by evaporation and spray; $Q_{e\psi}^i$ is the input of heavy metals to the sea with river water ($i = 1$), atmospheric deposition ($i = 2$) and ship's wastes ($i = 3$); α_2^i is the part of the suspended particles in the ith flow of heavy metals; and α_l is the part of the solid fraction of heavy metals in the bottom sediments.

Each radionuclide of εth type is characterized by the half-life τ^ε, the rate H_1^ε of input flow to the aquatory Ω, the accumulation rate H_α^ε in the living organisms $\alpha(p_A, B_A, Z, F, L)$ and the removal rate H_D^ε with the dead elements of the ecosystem. As a result, the concentration Q^ε of the radionuclide ε in Ω_{ijk} is described by the following system of equations:

$$\frac{\partial Q^\varepsilon}{\partial t} + v_\varphi^W \frac{\partial Q^\varepsilon}{\partial \varphi} + v_\gamma^W \frac{\partial Q^\varepsilon}{\partial z} = \frac{\sigma_{ijk}}{\sigma} H_1^\varepsilon + k_2^W \frac{\partial^2 Q^\varepsilon}{\partial z^2} - H_\alpha^\varepsilon - H_D^\varepsilon - \frac{\ln 2}{\tau^\varepsilon} Q^\varepsilon + H_*^\varepsilon \tag{5.15}$$

$$\frac{\partial Q_*^\varepsilon}{\partial t} = H_D^\varepsilon - H_*^\varepsilon - \frac{\ln 2}{\tau^\varepsilon} Q_*^\varepsilon \tag{5.16}$$

where Q_*^ε is the concentration of εth radionuclide in the bottom sediments and H_*^φ is the rate of the output flow of the εth radionuclide from the bottom sediments via desorption. The exchange of radionuclides between the water layers by migration of living elements is ignored as it has a small value in comparison with the flow H_D^ε.

5.6 Simulation Results

5.6.1 The Assumptions

The SSMAE allows for the estimation of the pollution dynamics of the Arctic basin under various a priori suppositions about the intensities of the flows of pollutants and about other anthropogenic impacts on the ecosystems of this region. Here, some of the possible situations are considered. The thermal regime of the Arctic basin is given by a normal distribution with average temperatures and with dispersions on the aquatories which are calculated by the World Resource Institute (Rosen 2000). The scheme of transport of pollutants in the atmosphere is adopted from Champ et al. (2000). The estimates of parameters for the units of Table 5.1 are given by literature sources or personal recommendations (see Table 5.3).

The vertical distribution of pollutants in the initial moment t_0 is taken as homogeneous. The average diameter of the solid particles are estimated in the range from 0.12 to 1000 μm and the vertical velocity of sedimentation is 0.003 m s^{-1}. The concentration of nutrients in the ice and snow equals 0. It is also supposed that the deep water temperature $Y(t,\varphi,\lambda) = 0$°C and the surface ice temperature $f_I(t, \varphi, \lambda)$ for $(\varphi,\lambda) \in \Omega$. It is supposed that $\varepsilon_1^A = 0$ and the phytoplankton productivity in the ice layer is 2.5% of the primary production in the water column $[(R_{p,r}+R_{p,f})/R_{p,w} = 0.025]$.

Let the ratio between solid and dissolved phases of heavy metals at the moment $t = t_0$ equal 1:2, i.e., $e(t_0,\varphi,\lambda,z)/\psi(t_0,\varphi,\lambda,z) = 0.5$. The flows of heavy metals, H_Z^φ, H_F^φ, and H_D^φ and H_L^φ, are described by linear models, $H_1^e = 0.01e_w$, $H_a^{e\varphi} = 0$. The boundaries of the Norwegian and Bering Seas are approximated by lines with $\varphi_N = 62$°N and $\varphi_B = 51$°N, respectively. Values for the other parameters are defined by Rudels et al. (1991), Muller and Peter (1992) and Wania et al. (1998). The initial data are defined in Tables 5.4 and 5.5.

5.6.2 The Dynamics of Arctic Basin Radionuclear Pollution

The intensity of external flows through the boundaries of the Arctic basin and the internal flows due to dead organisms H_D^ε, sediment H_*^ε and living organisms H_α^ε are described by linear models in accordance with Preller and Cheng (1999) and Preller and Edson (1995). Some results of the simulation experiment are given in Figs. 5.4 – 5.6. Figure 5.4 shows the tendency versus time of the average content of the radionuclear pollution on the whole Arctic aquatory. The distribution with depth is represented by a three layer model, upper waters (z < 1 km),

Table 5.3. The values of some parameters in the framework of the simulation experiments using the SSMAE

Parameter	Defined value
Step by	
Latitude, $\Delta\varphi$	1°
Longitude, $\Delta\lambda$	1°
Depth, Δz:	
$z \leq 100$ m	1 m
$z > 100$ m	$h - 1$ m
The ice thermal conductivity coefficient, λ_1	2.21 W m^{-1} $degree^{-1}$
The water thermal conductivity coefficient, λ_2	0.551 W m^{-1} $degree^{-1}$
The coefficient of vertical turbulence, Δ_z	10^4 m^2 s^{-1}
The coefficient of vertical diffusion, k_2^W	
For the free water surface	0.5×10^{-4} m^2 s^{-1}
When the water surface is covered by ice	5×10^{-6} m^2 s^{-1}
The specific heat of ice-thawing, q	334 kJ kg^{-1}
The rate coefficient of detritus decomposition, δ_n	0.1
The intensity of detritus decomposition, μ_A:	
$A = g, r, f$	0
$A = W$	0.01
The velocity of the water current in the Bering strait, V_i:	
i = 1	0.2 m s^{-1}
i = 2	0.05 m s^{-1}
The water thermal capacity, C	4.18 kJ kg^{-1} K^{-1}
The ice salinity, s_i:	
i = 1	5‰
i = 2	1‰
The water salinity for $z > 100$ m, s_0	34.95‰
The area of the Arctic basin, σ	16,795,000 km^2
Half-life of radionuclides, τ_ε	
$\varepsilon = {}^{60}C$	5.271 years
$\varepsilon = {}^{137}Cs$	30.17 years
Critical temperature for photosynthesis, T_c	- 0.5°C
The rate of detritus decomposition, μ_A:	
$A = g, r, f$	0
$A = W$	1% day^{-1}
The content of nutrients in the detritus, δ_n	10%
The density of	
Ice, ρ_r	0.9 g cm^{-3}
Snow, ρ_g	0.1 g cm^{-3}
Seawater, ζ	
$z \leq 200$ m	1.024 g cm^{-3}
$z > 200$ m	1.028 g cm^{-3}
Reflection coefficient for	
The snow layer, β_g	0.8
The ice layer, β_r	0.65
The submerged ice layer, β_f	0.6
The water layer, β_w	0.47
The part of exchange biomass losses transforming nutrients, ρ_1	0.1
The nutrient assimilation velocity coefficient, δ_1	0.1
The vertical velocity in the upwelling zone, β_V	0.1 m s^{-1}

Table 5.4. Initial data for the distribution of pollutants in the Arctic aquatories at the moment t_0

Aquatory	Identifier	Concentration: Radionuclides, Bq l^{-1} $^{137}C_s$	^{60}Co	Heavy metals mg l^{-1}	Oil hydrocarbons mg l^{-1}
Greenland Sea	Γ	0.05	0.05	0.5	0.2
Norwegian Sea	N	0.05	0.05	0.7	0.4
Barents Sea	B	0.07	0.07	0.8	0.6
Kara Sea	K	0.1	0.1	1.0	0.4
White Sea	∇	0.1	0.1	1.1	0.4
Laptev Sea	Λ	0.05	0.05	0.9	0.5
E. Siberian Sea	E	0.01	0.01	0.9	0.5
Bering Sea	S	0.02	0.02	0.8	0.7
Chukchi Sea	X	0.01	0.01	0.8	0.6
Beaufort Sea	Φ	0.05	0.05	0.7	0.2
Central Basin	U	0	0	0.1	0.1

Table 5.5. The input flows of radionuclides, heavy metals (suspended particles e and dissolved fraction ψ) and oil hydrocarbons O to Ω by water flows taken into account in the SSMAE

Source	Flow to Arctic basin (km^3 year^{-1})	Pollutant concentration: ^{137}Cs (Bq l^{-1})	^{60}Co (Bq l^{-1})	e (mg l^{-1})	ψ (mg l^{-1})	O (mg l^{-1})
Rivers						
Yenisey	600	0.3	0.5	0.3	5.1	2.3
Ob	400	0.1	0.1	0.4	6.9	4.7
Lena	500	0	0	1.1	8.8	6.9
Pechora	130	0.1	0.1	0.3	1.5	3.0
North Dvina	100	0	0	0.2	1.1	4.0
Other Siberian	200	0.1	0.1	0.1	0.5	2.3
North American	600	0.2	0.2	0.1	1.0	1.0
Evaporation	3500	0	0	0	0	0
Precipitation	5300	0	0	0.1	0.1	0
Southern boundary of						
Norwegian Sea	12,000	0.1	0.1	0.6	2.2	2.4
Bering strait	10,560	0	0	0.5	1.9	1.9

deep waters ($z > 1$ km), and sediments. The bottom depth is taken as 1.5 km. A more realistic depth representation of the shallow seas and the deeper Arctic basin will be considered in a future refinement of the model. The curves describe the vertical distribution with time of the radionuclide content in the two water layers and in the sediments. The transfer of radionuclides from the upper waters to the deep waters occurs with a speed which results in the reduction of radionuclear pollution in the upper waters by 43.3% over 20 years. These distributions for each of the Arctic seas are given in Fig. 5.5 (top) after 30 years and in Fig. 5.5 (bottom) after 50 years.

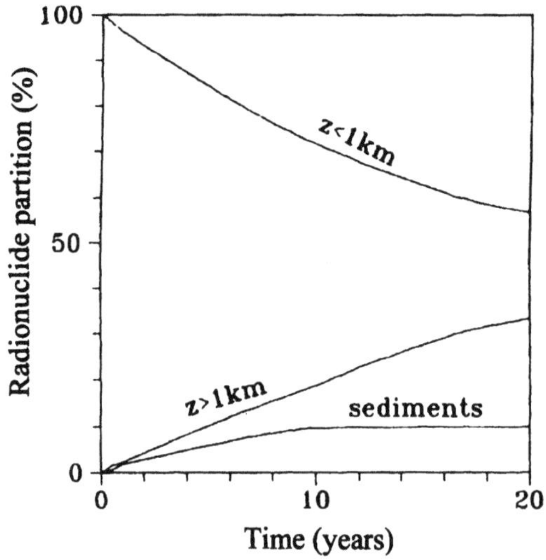

Fig. 5.4. Dynamics of the radionuclide distribution in the Arctic basin. It is supposed that at the moment $t_0 = 0$ radionuclear pollutants (^{137}Cs, ^{60}Co; see Table 5.4) are in the upper water layer $z \leq 1$ km only. The curves show the partition of the radionuclear pollutant distribution with time between the two water layers and the sediments, obtained by averaging the simulation results for all of the north seas

Local variations in the vertical distribution of radionuclides are determined by both hydrological and ecological conditions. The correlation between these conditions is a function of the season. Table 5.6 gives estimates of the role of the ecological processes in the formation of the vertical distribution of the radionuclear pollution of the arctic seas. These estimates show that the biological community plays a minor role in the radionuclide transport from the upper layers to the deep ocean.

The aquatories of the White, Laptev, East-Siberian, and Chukchi Seas are subject to visible variations in the radionuclear pollution. An accumulation of radionuclides is observed in the central aquatory of the Arctic basin. The aquageosystems of the Greenland and Kara Seas have some conservative character leading to buildup of radionuclear pollution, but in the Norwegian Sea there is even a decrease in the pollution level.

A somewhat stable situation is observed in the vertical distribution of the radionuclides. It is generally achieved 5–7 years following the initial moment t_0 with the exception of the East-Siberian, Laptev and Kara Seas where the stabilization processes of the vertical distribution are delayed by 10–12 years compared to the other aquatories of the Arctic basin.

The results of the simulation experiments show that variations of the initial data by ± 100% change the stabilization time by no more than 30%, so that the distributions take shape in 4–8 years. One unstable parameter is the river flow into the Arctic basin. Figure 5.6 shows the variations in the simulation results under a change in the river flow to the Arctic basin. The radionuclear pollution is reduced by 80% when the river flow decreases by 50%. When the river flow

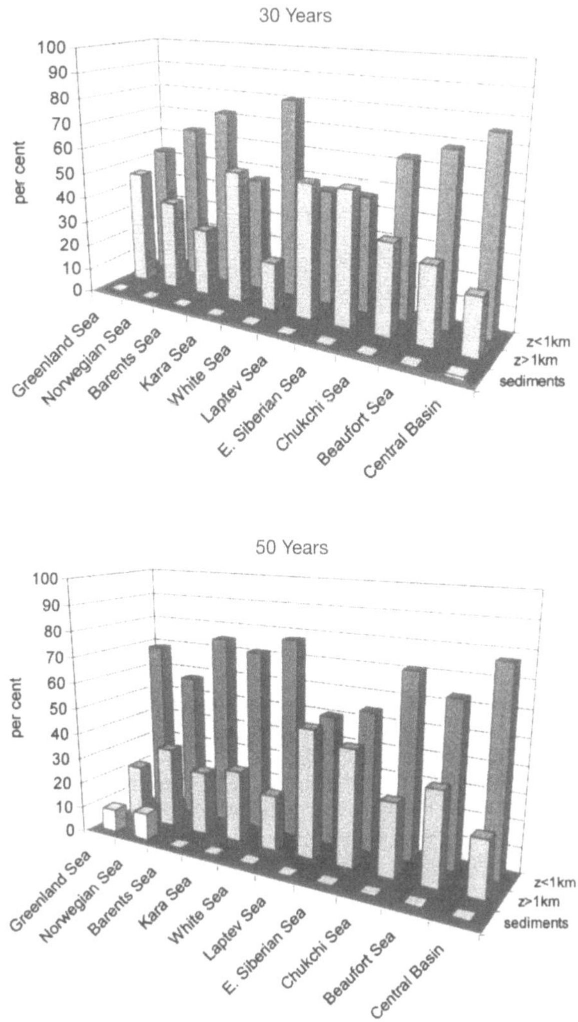

Fig. 5.5. *Top* Distribution of radionuclear pollution in the arctic aquatories 30 years after t_0. *Bottom* Distribution of radionuclear pollution in the arctic aquatories 50 years after t_0

increases by 50 % the radionuclear pollution of the Arctic basin increases by only 58 %. Hence, an error in the river flow estimate by ± 50 % can cause a deviation of the simulation results of less than 100 % for radionuclear pollutants. As follows from the other curves of Fig. 5.6, such deviations are even less for heavy metals and oil hydrocarbons.

The SSMAE allows for the estimation of a wide spectrum of radionuclear pollution parameters. Thus, this study shows a dependence of the biological transformation mechanism on the initial prerequisites. The biological transport of radionuclides downward out of the mixed layer varies in a wide interval from months to scores of years. The percent of vertical transport by living elements is divided between 11 % for the migration and 89 % for the dead organisms. There is a stable result in that the lower trophic levels of the arctic ecosystem have a greater concentration of radionuclides than the higher trophic levels. However,

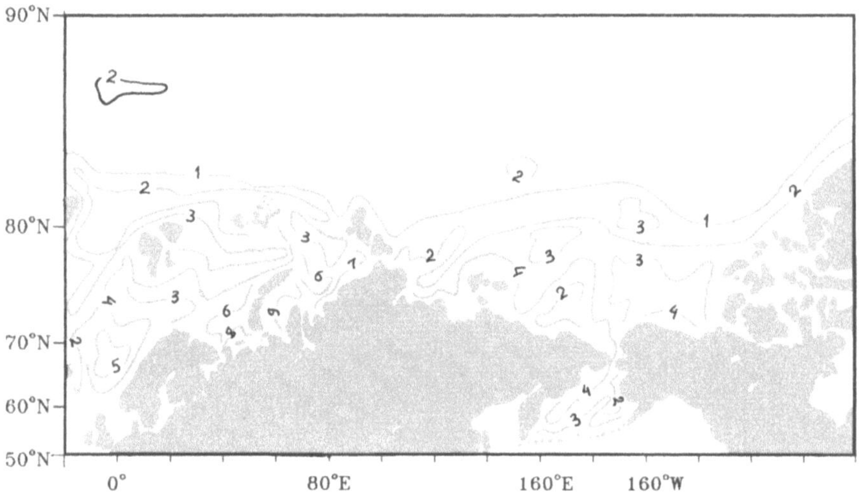

Fig. 5.6. Influence of variations in the river flows on the Arctic basin pollution level. Here, Δ_1 is the percentage variation in the river flow to Ω with respect to the value averaged on Ω_R in the last 3 years and Δ_2 is the content of the pollutant averaged on all rivers in Ω_R and normalized to the initial data (such that $\Delta_2 = 1$ for $\Delta_1 = 0$)

Table 5.6. A portion of the simulation experiment results using the SSMAE for the estimation of the vertical distribution of radionuclides in the Arctic basin. The contribution of ecological processes to the formation of the vertical distribution in the radionuclide content of the water is represented by the parameter ξ (%). The average content of the phytoplankton biomass is represented by the parameter p_w (g m^{-2})

Aquatory	Seasons							
	Winter		Spring		Summer		Fall	
	τ_w		τ_s		τ_u		τ_a	
	p_w	ξ	p_w	ξ	p_w	ξ	p_w	ξ
Greenland Sea	3.2	2	8.4	10	5.7	5	6.3	5
Norwegian Sea	2.9	2	7.8	9	5.9	5	6.7	6
Barents Sea	2.1	1	8.9	11	6.8	6	7.1	6
Kara Sea	2.4	1	9.2	12	5.3	5	6.0	5
White Sea	2.2	1	7.6	9	6.3	6	6.4	5
Laptev Sea	0.9	1	2.4	4	1.3	2	1.4	2
E. Siberian Sea	1.3	1	2.7	4	1.9	3	2.1	3
Bering Sea	2.5	2	7.1	9	3.9	4	5.3	4
Chukchi Sea	2.3	2	6.9	8	4.1	4	5.1	4
Beaufort Sea	1.9	2	5.7	7	4.8	4	4.9	4
Central Basin	1.0	1	1.7	2	1.5	2	1.6	2
Average value	2.1	1.5	6.2	7.7	4.3	4.2	4.8	4.0

it is evident that for a higher precision of unit MPR the model of the biological processes must consider each radioactive element separately and its interaction with the plankton. The variant of MPR realized here considers the physical processes as the major forces.

5.6.3 The Dynamics of Arctic Basin Pollution by Heavy Metals

The results of the simulation experiment are given in Table 5.7. We see that the stabilization of the average content of heavy metals on the full aquatory of the Arctic basin is achieved in 3–5 years. Under this stable regime, the concentration of heavy metals in the compartments $\Omega_R \cup \Omega_P$ (river mouths and ports) is six times higher than in the central aquatory and two times higher than in $\Omega_T \cup \Omega_B \cup \Omega_N$ (the near-shore waters, the Bering Strait, and the southern boundary of the Norwegian Sea). The concentration of heavy metals in the phytoplankton is 18 % lower than in the zooplankton and 29 % lower than in the nekton. A process of accumulation of heavy metals in the upper trophic levels is observed; moreover, the relation of the concentration of heavy metals in the phytoplankton to their concentration in the other trophic levels varies from 0.3 in $\Omega_R \cup \Omega_P$ to 0.5 in the open part of the Arctic basin.

Table 5.7. Results of the simulation experiment on the estimates of the dynamical characteristic parameters for pollution of the Arctic waters by heavy metals

Parameter	Estimate of the parameter after Δt (years)					
	$\Delta t = 1$	$\Delta t = 3$	$\Delta t = 5$	$\Delta t = 10$	$\Delta t = 15$	$\Delta t = 20$
Average concentration of heavy metals in the biomass (ppm):						
Phytoplankton	0.011	0.012	0.016	0.024	0.036	0.037
Zooplankton	0.013	0.014	0.019	0.028	0.041	0.043
Nekton	0.015	0.017	0.022	0.04	0.07	0.07
Detritophagies	0.033	0.037	0.048	0.088	0.15	0.16
Average content of heavy metals in the arctic waters, (ppm):						
	0.022	0.027	0.036	0.037	0.038	0.038
Flow of heavy metals from the upper layer to the deep waters (mg m^{-2} day^{-1}):						
Norwegian Sea	0.71	1.07	1.14	1.17	1.19	1.22
Barents Sea	0.72	1.08	1.25	1.19	1.24	1.16
Greenland Sea	0.26	0.62	0.71	0.82	0.76	0.89
White Sea	0.11	0.23	0.24	0.21	0.19	0.2
Kara Sea	0.34	0.47	0.57	0.61	0.63	0.64
Laptev Sea	0.55	0.78	0.81	0.89	0.74	0.77
East-Siberian Sea	0.59	0.79	0.95	0.97	1.02	1.07
Chukchi Sea	0.88	0.83	1.54	1.49	1.31	1.44
Beaufort Sea	0.34	0.67	0.66	0.81	0.74	0.69

Table 5.8. Estimates of heavy metal flows to and from the atmosphere

Heavy metal	Parameter	
	Atmospheric deposition, $Q^2_{e\psi}$ (mg m^{-2} h^{-1})	Evaporation and spray, H^{ψ}_{α} (t year)
Ag	0.7	7
Cd	1.1	58
Co	0.3	5
Cr	1.8	188
Cu	15	169
Fe	599	894
Hg	0.6	3
Mn	4.2	283
Ni	5.5	60
Pb	48	5
Sb	0.05	123
Zn	109	4471

The spatial features of the distribution of heavy metals in the sea water is characterized by more rapid accumulation into the aquatories adjoining the west coastline of Novaya Zemlya and situated on the boundary between the Jan-Mayen and East-Iceland currents. The central aquatory of the Arctic basin has a quasi-uniform distribution of heavy metals concentration. The vertical gradients of heavy metals vary in the interval from 0.11 to 1.54 mg m^{-2} day^{-1}. The forms of the vertical distributions of the dissolved fraction (ψ_W) and the suspended particles (e_W) of the heavy metals are not obviously expressed. The average relation of the concentration of heavy metals in the sediments and the water ($[\psi^* + e^*]/[\psi_W + e_W]$) varies on the aquatory of the Arctic basin from 1.9 to 5.7. For example, this relation for the Bering Sea is 3.3. The contribution of the biosedimentation process in the vertical distribution of the heavy metals is defined by values which vary from 0.23 to 1.24 mg m^{-2} day^{-1}.

The SSMAE provides the possibility of estimating the characteristics for separate types of heavy metals. An example of such calculations is given in Table 5.8.

5.6.4 The Dynamics of Arctic Basin Pollution by Oil Hydrocarbons

The spatial distribution of the forecast of the oil hydrocarbons in the arctic aquatories for 5 years after t_0 was calculated with the scale step of 0.003 mg l^{-1} under the following restrictions: $O(t_0,\varphi,\lambda,z) = 0$, $H_O^1 = 0.1$ mg m^{-3} day^{-1}, $H_O^2 = 0$ for $g(t) > 0$ and $H_O^2 = 0.01$ mg m^{-3} day^{-1} for $g(t) = 0$ and $z \geq \Delta z_1$, $H_O^3 =$ 0.01 mg m^{-3} day^{-1}, $H_O^4 = 0.02$ mg m^{-3} day^{-1}, $H_O^5 = k_D D_A^{1/3}$ (k_D the adsorption coefficient equals zero for $A = g, r, f$ and 0.005 day^{-1} when $A = W$), $H_O^6 = k_Z Z^{1/4}$ (k_Z the biosedimentation coefficient equals zero when $A = g, r, f$ and 0.004 day^{-1} for $A = W$), $H_O^7 = k_B B_A$ (k_B the bacterial destruction coefficient equals 0.01 day^{-1} for $A = g, r, f$ and 0.05 when $A = W$).

The simulation experiments show that the intensity of the anthropogenic sources of the oil hydrocarbons (Q_O) estimated by Bard (1999) and by Mcintyre (1999) is transformed to other forms by 56% in the environments of the surface snow, floating ice and submerged ice and by 72% in the water. The stabilization of the distribution of the oil hydrocarbons is realized in 3 years after t_0. The average level of oil pollution in the Arctic basin reaches the value of 0.005 mg l^{-1}. This is lower than the natural level of World Ocean pollution. In the influence zone of the Gulf-Stream current and in the Pacific waters an insignificant excess of this level is observed. The Barents and Kara Seas are the most polluted. Here, the concentration of oil hydrocarbons reaches the value 0.03 mg l^{-1}. An average summary content of oil hydrocarbons in the Arctic basin is 65331 t (metric tons) with a dispersion of 32%. The hierarchy of flows H_O^i (i = 1, ..., 7) (see Eq. 5.10) is estimated by the set of $H_O^2 > H_O^4 > H_O^1 > H^5 > H_O^6 > H_O^7$. This set is changed for each of the Arctic Seas. The order of preponderance of the destruction processes H_O^i of the oil hydrocarbons is defined by the seasonal conditions. The oxidation process at the expense of evaporation of the oil carbohydrates (H_O^2) prevails over the other processes in the summer.

In reality, the oil hydrocarbons evaporated from the surface of the Arctic Seas return to the Arctic basin with the atmospheric precipitation. These processes are simulated in the units APM and PSM. The maximal destruction of oil hydrocarbons is 0.0028 g m^{-2} day^{-1}. The flow H_O^7 due to bacterial decomposition averages 27 t $year^{-1}$. It has unequal values for different seas (t $year^{-1}$): Bering 3.7, Greenland 11, Norwegian 2.2, Barents 3.4, Kara 2.3, White 2.3, Laptev 2.8, East Siberian 2.8, Chukchi 3.4, Beaufort 2.5, Central Basin 0.5.

The total estimate of the role of the Arctic basin ecosystem in the dynamics of the oil hydrocarbons is traced for each of the Arctic Seas. For example, Fig. 5.7 gives such results for the Barents Sea. The discrepancy between the simulation results (solid curve 1) and the data of curves 2 and 3 (Terziev 1992) is explained by the assumption that the trophic structures of the different Arctic Seas are described by means of the general scheme represented in Fig. 5.2 and the discrimination between the ecological elements in each of the Arctic Seas is not taken into consideration in the SSMAE.

Figure 5.7 indicates that the vegetative period for the phytoplankton in the Barents Sea equals 4.9 months as shown by the effects on the ecosystem contribution to the self-clearing of the oil hydrocarbons (dashed curve 2). In the case considered, the Barents Sea ecosystem neutralizes about 25% of oil hydrocarbons during the vegetative period. The rest of the time this value oscillates near 3%. The dispersion of these estimates with latitude reaches 53%. For example, in the northern part of the Barents Sea the vegetative period varies from 2.6 to 3.1 months, while in the southern latitudes the variation is 5.3–5.8 months. Consequently, the role of the ecosystem in the sea cleaning of the oil hydrocarbons is 8 and 36% for the northern and southern aquatories, respectively. Such estimates can be calculated for each cell $\Delta\varphi \times \Delta\lambda \times \Delta z$ of the Arctic basin.

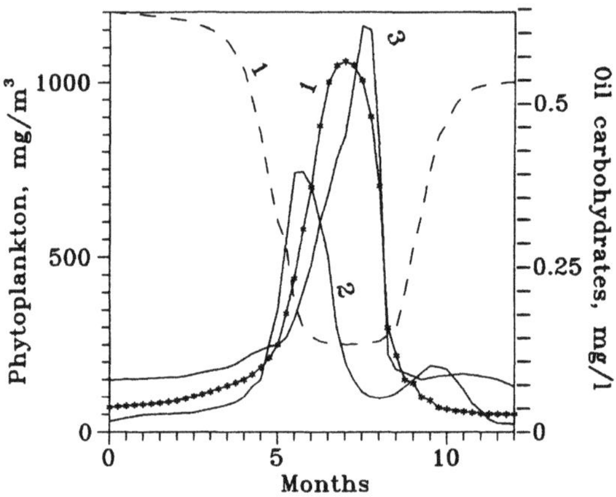

Fig. 5.7. Influence of the Barents Sea ecosystem upon the dynamics of the oil hydrocarbons in the seawater. *Curves 1* and *2* show the simulation results over a year's time for the phytoplankton and oil hydrocarbons, respectively. As the phytoplankton grow during the summer months, the oil hydrocarbons decline. *Curves 3* and *4* show the yearly distribution of phytoplankton in the southwestern and in the northern and northeastern aquatories of the Barents Sea, respectively. (Terziev 1992)

5.6.5 The Dynamics of the Pollutants in the Arctic Basin

An important problem is determining and understanding the role of the various pollutant sources in forming the pollution levels for the different Arctic basin areas (McCauley and Meier 1991; Crane et al. 2000). One major function of the SSMAE is to estimate the pollution dynamics in each of the cells $\Omega_{ij} \subseteq \Omega$ as a function of time. The influence of the pollutant sources on the Arctic ecosystem occurs through the boundary area $\Omega_R \cup \Omega_P \cup \Omega_B \cup \Omega_N \cup \Omega_\Gamma$ and through the atmosphere. The total picture of the spatial distribution of pollutants is formed from the local dynamic processes. The incompleteness of the arctic database forces the consideration of some scenarios in the framework of which assumptions necessary to make concrete the indefinite arctic system parameters are admitted. In Chapter 8 of this volume, there is the estimation of the flow of pollutants to the Kara Sea based on the experimental measurements of radionuclear pollution and heavy metals in the Angara-Yenisey river system. Therefore, the SSMAE is used to estimate the flow of pollutants from the Kara Sea to the other Arctic basin aquatories.

The Ob and Yenisey Rivers are considered as the main sources of radionuclear pollution, heavy metals and oil hydrocarbons for the Kara Sea (Table 5.5). Figure 5.8 shows the influence of the river flow on the volume of pollutants transported by the Kara Sea aquageosystem to the Central Basin. As shown by curves 1 and 2, the transfer of heavy metals and radionuclides from the Ob and Yenisey Rivers through the Kara Sea aquageosystem to the central aquatories of the Arctic basin amounts to 2.1% when the river flow is varied from 500 to 1000 km^3 $year^{-1}$ and after that it begins to grow linearly up to 7.6% for 2000 km^3 $year^{-1}$. Hence, there is a critical level of pollution for the Kara Sea

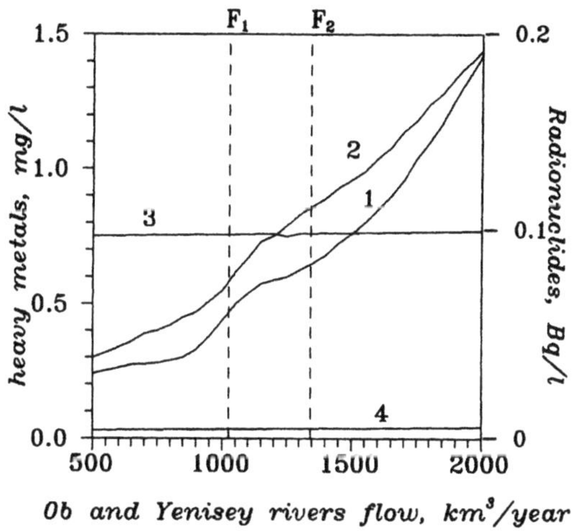

Fig. 5.8. Dependence of the concentrations of heavy metals (dissolved ψ + solids e) and radionuclides ($\varepsilon = {}^{137}Cs + {}^{60}Co$) at different geographical points as a function of the flow (F) of the Ob and Yenisey Rivers to the Kara Sea. The interval $[F_1, F_2]$ between the *dashed lines* corresponds to the range of variations in the flow in the real world. *Curves 1* and *2* show concentrations of the heavy metals and radionuclides, respectively, at the point with φ, λ = 75°N, 65°E in the northwestern part of the Kara Sea. *Curves 3* and *4* show the concentrations of heavy metals and radionuclides, respectively, in the Beaufort Sea above the Alaskan North Slope (φ, λ = 72°N, 150°W)

ecosystem beyond which it does not have time to dilute the flow of heavy metals and radionuclear pollutants. Similar estimations can be obtained for the other elements of set Ω_R using rivers referred to in Table 5.5.

The SSMAE allows the estimation of the flow of pollutants between different aquatories of the Arctic basin. For example, the transport of heavy metals and oil hydrocarbons from the Barents Sea to the Kara Sea is 631 and 473 kg year^{-1}, respectively. The total flow of pollutants from the Russian coast line to Alaska varies in the framework of Table 5.5 between 0.3 and 0.9% of the initial flow. As follows from curves 3 and 4 of Fig. 5.8, the flow of the Ob and Yenisey Rivers have practically no influence on the pollution level of the arctic waters in the Beaufort Sea near Alaska. This effect does not change over time.

5.7 Summary and Conclusion

We will discuss here three very important aspects of the SSMAE. The first concerns the incorporation of ecological, hydrophysical, climatic, and biogeochemical relationships in a model simulating the dynamics of the Arctic basin pollution. The main problem here is how to parameterize these relationships to achieve the satisfactory precision. The second concerns the key problem of database conformity to the model. In this case, the task is in the adaptation of the spatial-time scale to the database. The third concerns the user's ability to run the SSMAE in the scenarios space.

This investigation shows that the simulation experiment provides the possibility of studying both the temporal and spatial distributions of the pollutant dynamics in the Arctic basin. The precision of the results is a function both of the scenarios and the forms of the parameterization of the elements in Eqs. (5.1)–(5.16). It is obvious that the SSMAE is not effective when the climate conditions are varied to a critical state or when the anthropogenic impacts are grown to a critical value. However, in the SSMAE, the Arctic basin acts as a stabilizing subsystem of the biosphere. When the atmospheric temperature is reduced by 1 °C, an inverse connection occurs in the water balance of the *atmosphere-land-sea* system, which acts to stabilize the estimates. The parameterization of such variations in the framework of the SSMAE is not convenient. Nevertheless, the connection of the SSMAE to a global database having estimates of such parameters as ice area, temperature and albedo distributions, will allow the use of the SSMAE in the present form. The effectiveness of the SSMAE will increase with the use of models such as the coupled ice-ocean model described in the paper of Riedlinger and Preller (1991). In general, many different modifications of the SSMAE are possible. However, it is obvious that movement to the optimal SSMAE structure depends upon greater accuracy of the pollutant types, ecosystem structure, water cycle, ice movement, and climate model. The main difficulty is to realize the optimal modification at the same time for each of the units of the SSMAE.

The pollution level of the Arctic basin is formed mainly by inflow from the rivers. Because of this, the unit MRF plays a very important role in the SSMAE. A regular monitoring of the water flows and pollutant inputs by the rivers to the SSMAE is practically impossible. Consequently, the study and measurement of these flows during scientific expeditions and the modeling of the results are significant steps in the investigation of the Arctic basin pollution.

One example of such a step is the US/Russian expedition of 1995 to obtain on-site measurements of the pollution levels in the Yenisey and its tributary the Angara, in order to investigate the likely origins of land-based sources contributing to the pollution levels in the Yenisey estuary (Krapivin et al. 1997b; Phillips et al. 1997; Krapivin and Phillips 2001a).

The problem of verification of the SSMAE is important. However, it will be possible to realize this after essential modifications to the SSMAE, using models of greater precision to account for the hydrological, biogeochemical, ecological and climatic processes. The present SSMAE structure leads to a new technology of Arctic basin pollution monitoring. Greater, or at least better, accuracy in the SSMAE may be realized by means of simulation experiments where the model parameters will be varied over wide intervals. That is beyond the scope of this study.

We draw attention to several results of this investigation. In the framework of other scenarios, it follows that variations in the velocity of vertical advection from 0.004 to 0.05 cm s^{-1} does not affect the Arctic environmental state. An error of 32% in the ice area estimate leads to a variation of the simulation results by 36%. When this error is more than 32%, the simulation results become less stable and can vary by several times. The problem exists of the proper criterion to estimate the SSMAE sensitivity to variations in the model parameters. As is shown in Chapter 2, a survivability function $J(t)$ reflecting the dynamics of the total biomass of living elements enables this sensitivity to be estimated. In this instance

$$J(t)=\frac{\sum\limits_{i=1}^{5}\iint\limits_{(\varphi,\lambda)\in\Omega}\int\limits_{0}^{H(\varphi,\lambda)}B_i(t,\varphi,\lambda,z)d\varphi d\lambda dz}{\sum\limits_{i=1}^{5}\iint\limits_{(\varphi,\lambda)\in\Omega}\int\limits_{0}^{H(\varphi,\lambda)}B_i(t_0,\varphi,\lambda,z)d\varphi d\lambda dz}\,.$$

The index $J(t)$ provides an estimation of the uncertainty associated with the SSMAE parameters. Although a complete investigation of the influence of the SSMAE parameter variations on model results is an independent task, various estimations are given here. Preliminary simulation results show that the SSMAE permits variations of the initial data in the interval –70 to +150%. In this case, the model "forgets" these variations during ≈ 40 days. In addition, a large uncertainty (± 50%) is permitted in the value of such parameters as μ_A, δ_n, V_i, T_c, T_{opt}, ρ_1, k_{ij}. The correlation between variations of these parameters and the model results is linear. However, high model sensitivity is observed under variations of β_A, α_A, E_0, T_0. In general terms, the acceptable variation of these parameters is ±20%. Moreover, the deviation in the model results due to variations of these parameters is nonlinear. For example, fluctuations of the surface temperature T_0 within ± 5 K turn out to be not hazardous to the system, causing small variations of $J(t)$ by ±10%, but fluctuations of T_0 by ±7 K cause much larger variations in the value of $J(t)$ by ± 30%. Under this the temporal dependence of the system dynamics to variations in the parameters is diverse.

The SSMAE structure and its realization do not completely describe the processes taking place in the Arctic basin. An optimal extension of the SSMAE functions is possible by use of environmental monitoring to control the parametrical and functional model inputs. In this framework, the prognosis of the Arctic aquageosystem state is realized on the basis of the SSMAE and by processing of the observed data.

CHAPTER 6

Estimation of the Peruvian Current Ecosystem 6

6.1 Introduction

The Peruvian Current moves northward along the coast of South America where it causes an upwelling of cold, nutrient- and oxygen-rich water. This current belongs to those areas of the World Ocean where complex interrelated physical, biological, and chemical processes are taking place within a relatively small area, characterized by a high concentration of nutrients and large quantities of matter for the production of phytoplankton and other living elements (Sorokin 1977).

In this region negative environmental conditions occur caused by an *El Niño* event. During an El Niño, warm water appears at the surface of the ocean off South America and kills off marine life. In addition, there are many other global impacts owing to the resulting climate variability, e.g., on World Ocean pollution and fishing. Anthropogenic pressure on the Peruvian Current ecosystem (PCE) is the result of increasing fish production.

Off the coasts of Peru and Chile the production of organic matter is so large that there is not even enough oxygen for its oxidation. This results in the emergence of an oxygen-free layer at 100–800 m depth. The occurrence of such a layer is accompanied by perishing of the fauna. At the coast of Peru occurs the globe's largest upwelling of abyssal waters enriching the photic layer of the adjoining ocean area with nutrients. For a description of this process we will use the model described in Chapter 4. We will also consider the nutrients as a common element of the PCE and follow the hypothesis that in the PCE area the correlation between nitrogen, phosphate, and silicate is stable with nitrogen as the limiting nutrient.

The investigation of bio-producing processes in the area of the Peruvian Current was carried out during specialized cruises of research vessels from various countries (Vinogradov et al. 1977). The results of these investigations were taken into account in elaborating a simulation unit of the GSM which describes the bio-oceanographic fields in the Peruvian Current (Krapivin 1996) and in guiding its further improvement and investigations.

This study considers the PCE area Ω as a part of the East Pacific with latitudes $0 \le \varphi \le 45\,°S$ and longitudes $\lambda \le 90\,°W$. Contrary to earlier created models of the PCE this unit simulates the trophic pyramid in detail and gives the spatial distributions of the PCE elements over a wide area Ω.

6.2 Block Diagram and Principal Equations of the Peruvian Current Ecosystem (PCE) Model

In the hierarchical composition of the PCE trophic pyramid, it is possible to identify components as shown in Fig. 6.1. The characteristic feature of the given block-diagram is that it takes into account the element of interaction between the land and oceanic biocenoses, as expected by the presence of trophic relationships between E_5, E_7, E_9 and E_8. Introduction of this relationship appears to be necessary because seabirds produce a perceptible trophic pressure on the above components. Anchovy accounts for 80–90% of the birds' food spectrum. In Fig. 6.1 a single relationship describes the consumption of anchovy by other fishes. It is assumed that E_7 mainly refers to the Pacific bonito (*Sarda chiliensis*), though anchovy enters in the food spectra of other animals as well.

The block-diagram reflects the interaction elements between the atmosphere and the ocean, reduced to the reciprocal oxygen exchange at current incoming and outgoing rates. The input to the entire system is solar radiation energy $E(t,\varphi,\lambda,z)$ the intensity of which depends on time t, latitude φ, longitude λ and depth z:

$$E(t,\varphi,\lambda,z) = E(t,\varphi,\lambda,0)\exp(-\alpha z - \beta\int_0^z [E_2(t,\varphi,\lambda,u) + E_{10}(t,\varphi,\lambda,u) + E_{13}(t,\varphi,\lambda,u)]du) \tag{6.1}$$

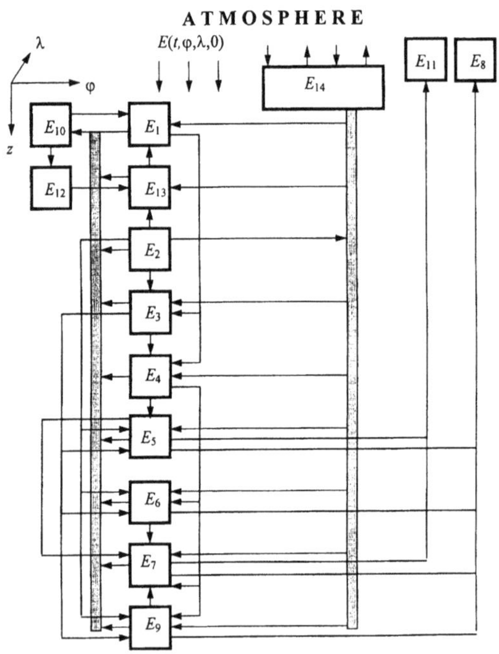

Fig. 6.1. Block diagram of the PCE. Notations are given in Table 6.1; φ latitude; λ longitude; t time; z depth (m)

where α (≈ 0.01) is the vertical light attenuation coefficient for clear seawater and β (≈ 0.0001) is the coefficient of light attenuation due to the detritus and dissolved organic matter. Commonly, formula (6.1) is generalized to the following equation:

$$E(t,\varphi,\lambda,z) = E(t,\varphi,\lambda,0)\exp(-\alpha z - \int_0^z \sum_{i=1}^m \beta_i E_i(t,\varphi,\lambda,u)du)$$

The dynamics of the biomass of the ith component ($i = 1$–9) is described by the balance equation:

$$\partial E_i / \partial t - U_{Ei} - \sum_{j\in\Gamma_i} C_{ij} R_j = 0 \tag{6.2}$$

where

$$U_{Ei} + v_\varphi \partial E_i/\partial \varphi + v_\lambda \partial E_i/\partial\lambda^+ v_z \partial E_i/\partial z$$
$$= R_i - T_i - M_i - H_i + \partial/\partial \varphi (\Delta_\varphi \partial E_i/\partial\varphi) + \partial/\partial \lambda (\Delta_\lambda \partial E_i/\partial \lambda)$$
$$+ \partial/\partial z(\Delta_z \partial E_i /\partial z),$$

$v(v_\varphi, v_\lambda, v_z)$ are components of the current velocity, R_i is the production, T_i the rate of exchange, M_i the mortality, and H_i the nonassimilated food. Γ_i is the set of the trophic subordination of the ith component:

$$C_{ji} = k_{ji} \overline{E}_i / \sum_{m\in S_j} k_{jm} \overline{E}_m \tag{6.3}$$

where S_j is the food spectrum of the jth component and k_{jm} is the index of satisfaction of the nutritive requirements of the jth component at the expense of the mth component biomass. $\overline{E}_i = \max\{0, E_i - E_{i,\min}\}$, where $E_{i,\min}$ is the minimal biomass of the jth component consumed by other trophic levels. $\Delta(\Delta_\varphi, \Delta_\lambda, \Delta_z)$ are components of the turbulent diffusion coefficient (on the assumption of isotrophism of vertical diffusion in the horizontal plane $\Delta_\varphi = \Delta_\lambda = v_H$). The coefficient of v_H is estimated by the value $\approx 10^2$ m^2 s^{-1} or is calculated by Ozmidov's formula $v_H = \varepsilon t_l^2$ where $\varepsilon \in [10^{-10}, 10^{-5}]$ m^2 s^{-3}, t_l^2 is the specific scale of time of the turbulent diffusion. The estimation of Δ_z is given by the formula: $= \Delta_z \Delta_z$ $(1 + 5\, r_i)^{-1}$ where r_i is Richardson's number. The coefficient of Δ_z is estimated by a value from 3 to 578.1 cm^2 s^{-1}. The values of Δ_z vary from ≈ 100 cm^2 s^{-1} over the thermocline to 1–10 cm^2 s^{-1} in the thermocline and 10–100 cm^2 s^{-1} below the thermocline.

The photosynthesis rate R_2 as a function of illumination E, nutrients E_{12}, water temperature σ and atmosphere pollutant concentration θ is simulated by the model

$$R_2 = \min\{\rho_2 E_2, R_2^*\}, \tag{6.4}$$

where

$$R_2^* = k_2 E_2[1 - \exp(-k_3 E_{12})]E \cdot \exp[k_5(1 - E/E_{opt})] \cdot \exp(-k_1\theta)(1 + k_4\sigma), \tag{6.5}$$

where E_{opt} is the optimal illumination for the photosynthesis and k_i ($i = 1$–5) is the adaptation coefficient reflecting the effect of the above-mentioned factors on photosynthesis. The factor k_2 is a function of the phytoplankton biomass and the water saturation by oxygen:

$$k_2 = k_2^* \exp(-b_2 E_2)[1 - k_6 \exp(-k_7 E_{14})]$$

or

$$k_2 = A_{max}(k_o E_{opt})^{-1} \theta^{\Delta T}$$

where $\Delta T = T - T^*$, T^* is the optimal temperature for photosynthesis, $A_{\max}$ is maximal diurnal assimilation coefficient, the k_o coefficient defines the quantity of chlorophyll into the algae cells and θ is the indicator of the dependence of photosynthesis on the temperature. The coefficient $A_{\max}$ is estimated by the expression: $A_{\max} = 5.9\ E_{\text{opt}}$ in the area of maximal photosynthesis and $A_{\max} = 2.7\ E_{\text{opt}}$ for "shady" phytoplankton.

For the Peruvian Current the values of $A_{\max}$, k_o, E_{opt} and θ are estimated as follows: $A_{\max} = 150$ mg C mg^{-1} of chlorophyll "a" per day, $k_o = 30$–100 mg C mg^{-1} of Chl. "a", $E_{\text{opt}} = 160$ kcal m^{-2} day^{-1} and $\theta = 2$.

The production of R_i at the other trophic levels is computed from the formula:

$$R_i = k_8[1 - \exp(-\sum_{j \in S_i} k_{ij} E_j)](1 + k_4 \sigma) E_i \tag{6.6}$$

where the factor k_8 is the function of maximal productivity $R_{i,\max}$ and optimal temperature σ_{opt} per unit of biomass:

$$k_8 = R_{i,\max}/(1 + k_4 \sigma_{\text{opt}}) \cdot \text{d}$$

Functions T_i ($i = 1$–9) in Eq. (6.2) are postulated in accordance with the investigations carried out by Nitu et al. (2000a) as a dependence:

$$T_i = \tau_i E_i^{\omega_i} \tag{6.7}$$

where

$$\tau_i = \begin{cases} f_1(\sigma) & \text{for} \quad E_{14} > B_i, \\ f_1(\sigma) E_{14} / B_i & \text{for} \quad 0 \le E_{14} \le B_i \end{cases} \tag{6.8}$$

with

$$f_1(\sigma) = \begin{cases} f_1^* & \text{for} \quad \sigma \in [\sigma_{1,i}, \sigma_{2,i}], \\ f_1^*[\varepsilon_i \exp(-\xi_i \sigma)]^q & \text{for} \quad \sigma \notin [\sigma_{1,i}, \sigma_{2,i}] \end{cases} \tag{6.9}$$

Here, $q = (\sigma - \sigma_{3,i})/m_i$; $\sigma_{1,i}$ and $\sigma_{2,i}$ are the lower and upper boundaries of the temperature adaptation zone of the ith trophic level elements; $\sigma_{3,i} = 0.5\{\sigma_{1,i} + \sigma_{2,i}\}$; ε_i, ξ_i and m_i are constant quantities to be chosen to satisfy the condition of approximation to the measured data; B_i is the oxygen concentration level which is dangerous for the ith trophic level; and f_1^* is the portion of E_i lost due to exchange in the temperature adaptation zone. According to the data of Nitu et al. (2000a), the value of T_i for a set of Peruvian Current living elements is described by Eq. (6.7) with the following coefficients (τ_i, ω_i): *Calanoida* (1.127, 0.945), *Euphausiacea* (0.416, 0.917), *Chaetognatha* (0.037, 1.329), *Siphonophara* (0.133, 0.945), *Ctenophora* (15.67, 0.345), *Coryphaena equisetis* (1.06, 0.899), *Dactylopterus valitans* (0.673, 0.753), *Naucrates ductor* (0.633, 0.723), *Balistes* sp.

(0.823, 0.954), *Mugil* sp. (0.568, 0.615), *Nomeus* sp. (0.696, 0.77), and *Myctaphidae* (0.632, 0.905). In the case of phytoplankton, the losses due to respiration are characterized by declining assimilation of carbon in photosynthesis. For the PCE these losses attain on average as much as 74.6% of the assimilated carbon (1.45 g C m^{-2} day^{-1}).

The mortality M_i will be determined on the basis of generally adopted procedures as $M_i = (\mu_i + \mu_i' E_i)E_i$, where the factors μ_i and μ_i' reflect the dependence of mortality on the biomass density and on the environment, respectively. Calculation of the quantity H_i will be made according to the formula $H_i = h_i R_i$ with h_i as part of the nonassimilated food of the *i*th trophic level.

A special place in the model among live components is occupied by anchovy for the description of which we have identified three age stages: larvae E_6, young E_9 and commercial anchovy E_5. It is assumed that the transition rate from the *i*th into the *j*th stage is constant and is determined by the fact that all three stages are passed through by the anchovy within a year's time.

The inert components in the model are described by the following equations:

$$\frac{\partial E_{10}}{\partial t} + (v \cdot \nabla)E_{10} = \sum_{s=1}^{9} (M_s + H_s) - \rho_* E_{10} - v_* \frac{\partial E_{10}}{\partial z} + (v \cdot \nabla^2)E_{10} - W_{10} \tag{6.10}$$

$$\frac{\partial E_{12}}{\partial t} + (v \cdot \nabla)E_{12} = \rho_* E_{10} + \rho_1 (T_9 + \sum_{s=1}^{7} T_s) - \delta_1 R_2 + (v\nabla^2)E_{12} \tag{6.11}$$

$$\frac{\partial E_{13}}{\partial t} + (v \cdot \nabla)E_{13} = k_{13,0} R_2 - E_{13} k_{1,13} R_1 / P_1 \tag{6.12}$$

$$\frac{\partial E_{14}}{\partial t} + (v \cdot \nabla)E_{14} = \delta_2 R_2 - \sum_{s=1}^{7} \zeta_s T_s - \zeta_9 T_9 + (v\nabla^2)E_{14} - \delta_* \rho_* E_{10} + W_{14} \tag{6.13}$$

where

$$P_i = \sum_{j \in S_i} k_{ij} \overline{E}_j \tag{6.14}$$

$$W_{10} = -\overline{E}_{10} (k_{1,10} R_1/P_1 + k_{3,10} R_3/P_3 + k_{4,10} R_4/P_4),$$

$$W_{14} = max\,\{0, (z^* - z)/|z - z^*|\} W^*;$$

ρ^* is the rate of detritus decomposition per unit of biomass; v^* is the rate of gravity settling; v is the turbulent diffusion coefficient; ρ_1 is the part of the exchange biomass losses which transform into nutrients; δ_1 is the nutrient assimilation rate by the photosynthetic process per unit of phytoplankton production; δ_2 is the oxygen production per unit of phytoplankton production; ξ_i (i = 1–7,9) is the oxygen losses at the expense of transpiration; z^* is the maximal depth of oxygen exchange of sea water with the atmosphere; and W^* is the difference between the invasion and evasion processes.

Mangum and Winkle (1973) proposed the following correlations between the oxygen assimilation and its content in the water:

$$y_s = \zeta_s T_s = B_{os} + B\,1s\,lgE_{14} = c_{1s}\,[1 - exp\,(-\,c_{2s}\,E_{14})] = E_{14}\,[\,a_{1s} + a_{2s} E_{14}]^{-1}$$
$$= q_{0s} + E_{14}\,[q_{1s} + q_{2s} E_{14}]^{-1},$$

where b_{js}, c_{js}, a_{js} and q_{js} are given constants. The factor $x = \rho^* \delta^*$ from formula (6.13) is estimated by a value of $x = 0.13$ for $z \le 75$ m and $\sigma = 20\,°C$.

A unit of climatic and anthropogenic impacts on the PCE is described according to the procedures of the GSM. It was assumed that the scope of pollution of the PCE area is determined by the mean pollution level of the Pacific and is quantitatively characterized by $0.13 \le \theta \le 0.27$ mg m^{-2} of conventional pollutant.

Temperature variations were incorporated into the model by means of using a database in the form of charts of its distribution in space allowing for seasonal variations. In this case, the discretization of data in time and space agrees with the discretization steps adopted in the computer realization of the model. Temperature anomalies that take place in the area of the PCE during an El Niño were allowed for by setting in the model σ_ε, the temperature moments at the advent of these events, and $\Delta\sigma_\varepsilon$, the values of temperature jumps in these years. Thus, the data formed on the above scheme are corrected for the El Niño years by the magnitude of $4 \le \Delta\sigma_\varepsilon \le 6$ K. We would like to note that a more accurate simulation of El Niño is an independent task.

The temperature regime of the PCE is affected by strong variations in the solar radiation which are due to varying cloudiness near the coast and in the open ocean. In the narrow off-shore zone, where cloudiness is practically nonexistent, radiation is 1.5 times as high as in the adjoining ocean areas and attains magnitudes of the order of $E(t,\varphi,\lambda,0) = 264$ cal cm^{-2} year^{-1}. The mean energy of the solar radiation is $\bar{E} = 185$ cal cm^{-2} year^{-1}. In subsequent computations, the following dependence was used in a scenario with free parameter R: $E(t,\varphi,\lambda,0) = R\,(230 - 7\lambda/3)(1 + 0.1\,\sin\,[2t/365])$, where $R = 1$ under normal conditions of the system.

The temperature regime of the PCE along the z-coordinate is determined by an intricate complex of hydrological factors. The depth of the lower boundary of the isothermal layer varies, on the average, within a wide range from 10 m at the coast of South America to 130 m in the western part of the region. The position of the thermocline is determined by the separation boundary between surface and subsurface waters. An analysis of the measured data obtained in this region permits one to select the following model for the position of thermocline:

$$H(t,\varphi,\lambda) = H_o(t) + [\,H_1(t) - H_o(t)]\;\delta^{-1}(t)[\lambda - G(\varphi)] \qquad (6.15)$$

where H_o and H_1 are the depths of occurrence of the thermocline near the shore and in the open ocean, respectively, and δ is the distance from the shore of the strip of maximum thermocline depth described by the function:

$$\lambda = G(\varphi) = 0.145 \times 10^{-3}\,\varphi^4 - 0.0059\,\varphi^3 + 0.048\,\varphi^2 - 0.235\,\varphi + 80.1$$

The dynamics of the biogeocenotic processes in the PCE are largely dependent on the variable structure of the currents. Here, we assume $v_\varphi = |v| \cos \alpha$, $v_\lambda = |v| \sin \alpha$ with the angle α showing the direction of the current at each point of the plane (φ,λ) at depth z.

The most complex element in the construction of the model proves to be the mathematical description of the fish migration. Usually the migration process should be identified with the intense turbulent diffusion coefficient $\nu^* > \nu$. Krapivin (1996) realized this algorithm and has shown the estimation of ν^* to be a very incorrect task. The investigations by many authors have revealed that the process of fish migration is accompanied by a an external appearance of purposeful behavior. Therefore, let us formulate the law of migration which follows from the general biological principle of adaptation and long-term adaptability: migration of anchovy, predatory fishes and birds are subordinated to the principle of complex maximization of the effective nutritive ration P_i (i = 5,7,8,9), given preservation of favorable temperature conditions. In other words, traveling of migrating species takes place at velocities characteristic for them in the direction of the maximum gradient of effective food, given adherence to temperature restrictions. The PCE takes into account the following restrictions in the preferable temperatures for E_9 ($\sigma \leq \sigma_{9,max}$), E_5 ($\sigma \leq \sigma_{5,max}$) and E_7 ($\sigma \geq \sigma_{7,min}$). The birds migrate without temperature restrictions.

The elements E_3, E_4 and E_6 migrate only in the vertical direction. The diurnal migration regime of the specimen is described by the relation $c = k_a + k_c \sin (2\pi\tau)/96$, where k_a is the average diurnal velocity of active movement of the specimen (0–60 m h^{-1}), k_c is the half amplitude of diurnal variation in the velocity and τ is the time. As a result the specimen's position is defined by the following expression: $z_2 = z_1-(w + \Delta w)\Delta z/c$, where z_2 is the layer in which the specimen is situated after the ascent, z_1 is the depth where it is after the descent, w is the velocity of active movement of specimen and Δw is the variation in the velocity.

6.3 Experiments That Use the Model of the PCE

The model makes it possible to carry out computer experiments for an estimation of the PCE dynamics and its response to anthropogenic impacts on the exterior conditions. The initial data and model coefficients are set by proceeding mainly from the study of the results of other authors (Table 6.1). For the young and commercial anchovy the temperatures $\sigma \leq \sigma_{\max} = 289$ K and for predatory fishes $\sigma \geq \sigma_{\min} = 288$ K were deemed to be acceptable. The rate of fish migration was determined at 0.03 m s^{-1}, and for birds 0.14 m s^{-1}. Real distributions of the water temperature in Ω were determined according to the World Resources Institute by means of averaging the data of measurements over 10 years.

The thermocline position was described by a binary function (6.15) with $\delta = \delta_s$ in spring, summer and autumn and $\delta = \delta_w$ in winter. In further results, it was found that $\delta_s = 1°$ and $\delta_w = 30'$. Other parameters of the model were determined from literature sources. Specifically, $\alpha = 0.013$, $\beta = 5 \times 10^{-5}$, $k_2 = 0.07$, $k_5 = 0.2$, $k_1 = 10.7$, $k_4 = 0.02$, $\omega_i = 0.7$–0.95, $\mu_i' = 0.1\mu_i$, $\mu_i = 0.01$–0.02, $\rho^* = 0.1$, $\nu^* = 50$,

Table 6.1. The average values of the concentrations of the ecosystem elements at $t_o = 1900$ considering as the initial values for $E_i\ (t_o, \varphi, \lambda, z)$, $(\varphi, \lambda) \in \Omega$, $0 \leq z \leq 200$ m

Element of the PCE	Identifier	Average estimation
Bacterioplankton	E_1	10 cal m^{-3}
Phytoplankton	E_2	20 cal m^{-3}
Phytophages	E_3	5 cal m^{-3}
Predatory zooplankton	E_4	5 cal m^{-3}
Commercial anchovy	E_5	1 cal m^{-3}
Anchovy larvae	E_6	1 cal m^{-3}
Predatory fish	E_7	0.5 cal m^{-3}
Birds	E_8	5,000 individuals km^{-2}
Young anchovy	E_9	1 cal m^{-3}
Detritus	E_{10}	100 cal m^{-3}
Biomass of fished anchovy	E_{11}	250 mg m^{-3}
Nutrients	E_{12}	100 cal m^{-3}
Dissolved organic matter	E_{13}	5 ml l^{-1}
Dissolved oxygen	E_{14}	1250 t km^{-2}

$k_{1,10} = k_{1,13} = 0.3$, $k_{3,10} = 0.005$, $\rho_1 = 10^{-4}$, $k_{4,10} = 0.5 \times 10^{-3}$, $\delta_1 = 0.012$, $k_{13,0} = 0.9$, $\delta_2 = \delta^* = 0.01$, $\zeta_s = 0.05$, $z^* = 5$ m, $W^* = 0$, $\Delta\varphi = \Delta\lambda = 1°$, $\Delta z = 1$ m, $\Delta t = 1$ day and $|\mathrm{v}| = 0.2$ m s^{-1}. The producers' effort is homogeneous on Ω and is described by the function $F(t,\varphi,\lambda) = F_o F_1(t)$, where F_o is the factor such that $F_1(1970) = 1$.

The function F is represented in Fig. 6.2 where there are some results for a validation of the model.

The PCE gives stable results on a broad scale of variations in the model parameters. Fig. 6.3 gives the PCE dynamics in the plane of $E_5 \times E_7$ under different initial states. These dynamics are kept in other sections of the PCE trajectory. Stable results predicted by the PCE model under GSM control generally agree very well with the field measurements for this case. Here, we consider a hypothetical situation and obtain quantitative estimates for the PCE survivability. Accordingly, some arbitrarily chosen scenarios will be analyzed. The results obtained give the events that were observed by expeditions. For example, Figs. 6.4–6.6 present the phenomena described by many authors.

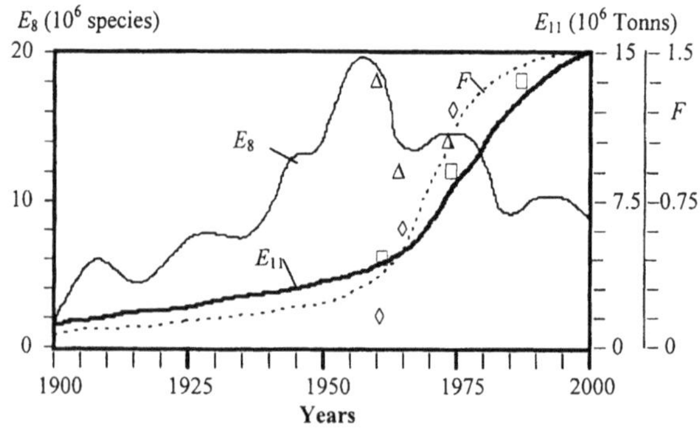

Fig. 6.2. Dynamics of elements E_8 (birds), E_{11} (anchovies) and F (relative fishing production). The *curves* are model calculations and the *symbols* are expedition measurements. *Open diamond* Fishing efforts, *open square* the anchovy production, *open triangle* quantity of birds

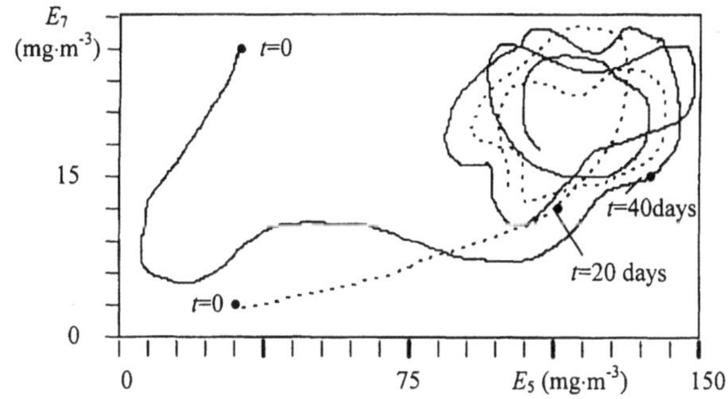

Fig. 6.3. Trajectories of the PCE on the plane of E_5 (anchovies) × E_7 (predatory fish) under different initial conditions

6.3.1 Temperature Variations

Seasonal and latitudinal temperature variations are known to occur in the PCE area. The natural cycles of such variations have been incorporated in the model by means of consulting the available real data and it appears interesting to consider herein the departures of the temperature from these model values. Figure 6.4 shows the variation of the biomass of the anchovy and predatory fishes versus temperature. As seen from Fig. 6.4, if the system is functioning under normal temperature conditions, a clear-cut division between the habitation areas of the anchovy and predatory fishes is observed. The maximum concentration of the anchovy biomass is observed in this case to occur in the offshore zone about 1.5–2° wide in longitude, whereas the biomass of the predatory fishes attains its maximum concentration in the open ocean. This pattern is observed throughout the area and, for the model distributions of temperature, appears to be stable 30–40 days later. This pattern falls apart as soon as the overall temperature background is subject to change. A reduction in the general temperature level by 1 K accounts for the fact that the habitation areas of anchovy and pre-

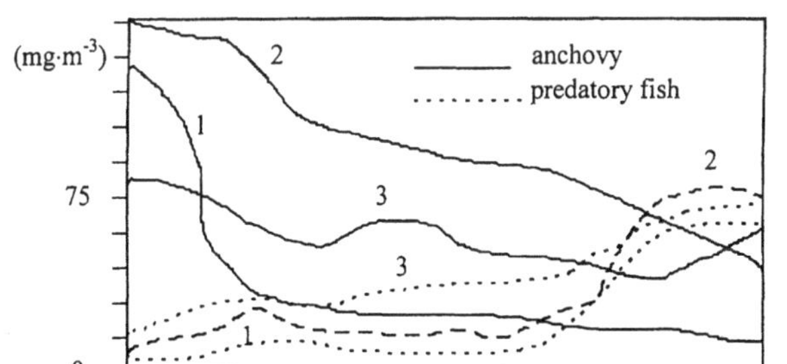

Fig. 6.4. Effect of the water temperature variations on predatory fish and anchovy biomass in the PCE at 20°S referring to the moment in time $t = 100$ days. *1* Normal temperature regime, *2* the temperature is decreased by 1 K *3* El Niño regime. The initial data for $t = 0$ were assumed to be uniform throughout the area: $E_1 = 200$, $E_2 = 400$, $E_3 = 80$, $E_6 = E_7 = E_9 = 8$, $E_4 = 35$, $E_{10} = 700$, $E_{13} = 3.5 \times 10^4$, $E_5 = 16$, $E_{12} = 200$, $E_{14} = 65$ (mg m^{-3}), $E_8 = 2$ specimens km^{-2}

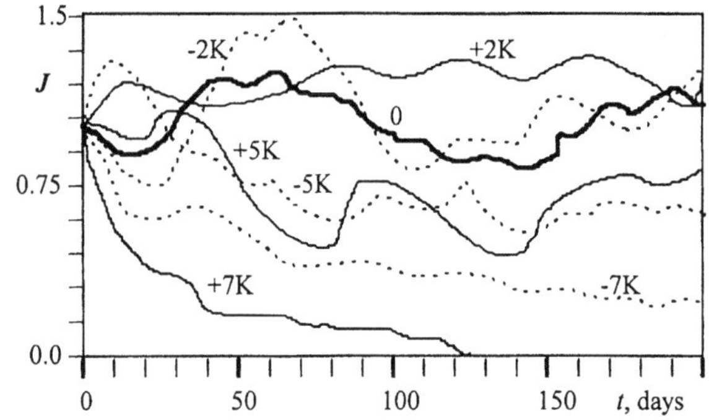

Fig. 6.5. Estimation of the PCE survivability $J(t)$ under temperature variations. The figures on the curves show the water temperature variations from the normal conditions (*thick curve*). J is measured in relative units

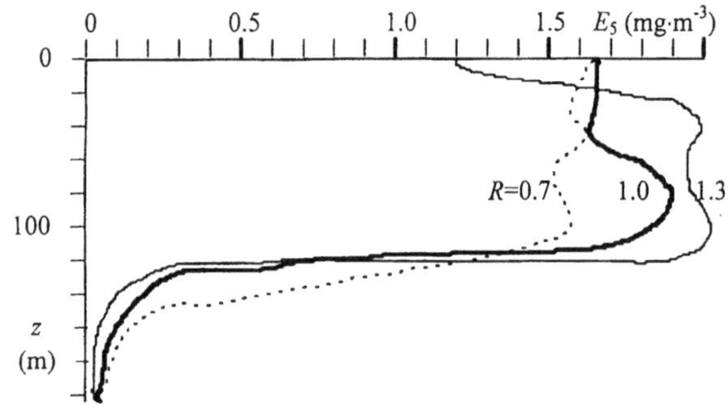

Fig. 6.6. The PCE vertical structure as a function of variations in solar radiation $E(t,\varphi,\lambda,0)$ when $t = 50$ days, $\varphi = 12$ °S, $\lambda = 82$ °W. Initial conditions are the same as in Fig. 6.5. The parameter R is explained in Fig. 6.7.

datory fishes increasingly intersect. Moreover, their total biomass rises in this case by about 10%. In the El Niño periods the effect of division of the ecosystem into the offshore and open-ocean parts disappears completely, being accompanied by a decline in the total biomass of anchovy and predatory fishes.

To estimate the survivability of the PCE, Fig. 6.5 shows the survivability function $J(t \geq t_o)$:

$$J(t) = \frac{\sum_{i=1}^{9} \iint_{(\varphi,\lambda)\in\Omega} \int_{z} E_i(t,\varphi,\lambda,u)\,d\varphi\, d\lambda\, du}{\sum_{i=1}^{9} \iint_{(\varphi,\lambda)\in\Omega} \int_{z} E_i(t_o,\varphi,\lambda,u)\,d\varphi\, d\lambda\, du} \tag{6.16}$$

for various departures of temperature from its model value. We will consider the system to be in a living state if the condition $J(t) > \kappa J(0)$ holds for $t \geq t^*$, where $\kappa < 1$ is the level of survivability and t^* is the solution of the equation $J(t) = J(0)$. It is seen that temperature fluctuations by ± 7 K bring the system to a "dead state" after 70 days for + 7 K and after 190 days for − 7 K. Water temperature fluctuations within ± 5 K turn out not to be dangerous to the system, but they may be the cause of its conversion into a different quasi-stationary state. This follows from the comparison of the phase pattern of the behavior of the system trajectories similar to that given in Fig. 6.3. The establishment of a stable cycle is observed in this case to take a longer time.

6.3.2 Variations of Illumination

It is supposed that fluctuations of solar radiation energy cause proportional variations of the water temperature. The climate unit of the GSM is used for the calculation of atmospheric temperature σ_A in Ω. The temperature regime of the PCE is described by the model of Nitu et al. (2000a). Additional conditions taken into account are as follows. Suppose that the difference $\Delta\sigma(t,\varphi,\lambda,z)$ between σ_A and the water temperature above the thermocline ($z < H$) linearly increases with higher latitudes. This difference is $\Delta\sigma(t,0,\lambda,z) = 0$ and $\Delta\sigma(t,30,\lambda,z) \le 0.2\ \sigma_A$. Further, $\Delta\sigma$ is independent of longitude. The difference between the water temperature under the thermocline ($z > H$) and the atmospheric temperature is independent of geographical coordinates. Changes in the depth of the thermocline due to light fluctuations should be neglected. Then we shall use the parameter R as a free parameter assigned to light fluctuations.

Figures 6.6 and 6.7 give instances of calculations of the vertical structure of the PCE and $J(t)$ for variations of light at the surface. These results point to the presence of a regularity observable throughout the area and are indicative of the fact that with increasing $E(t,\varphi,\lambda,0)$ the vertical distributions of all the system components do not change in shape, while only the maximum of photosynthesis acquires a slightly deeper position. This is occasionally accompanied by an increase in the total biomass of the components which takes place at the sections where the concentration gradient of biogenic elements in the layer above the thermocline is observed to be large. A decline in illumination accounts for the movement of the maximum photosynthesis layer towards the surface and for a significant reduction of the total biomass in the water column.

The computations reveal that the PCE under investigation demonstrates, on the one hand, high sensitivity and, on the other hand, high stability to variations in $E(t, \varphi,\lambda,0)$. If, generally speaking, in the case of the biosphere a 2% reduction in $E(t,\varphi,\lambda,0)$ is likely to result in glaciation in the case of the given system, fluctuations within – 60 to + 700% (Fig. 6.7) prove to be safe. This effect has a local character when the safe level is defined by the condition $J(t) \ge 0.5$. This condition

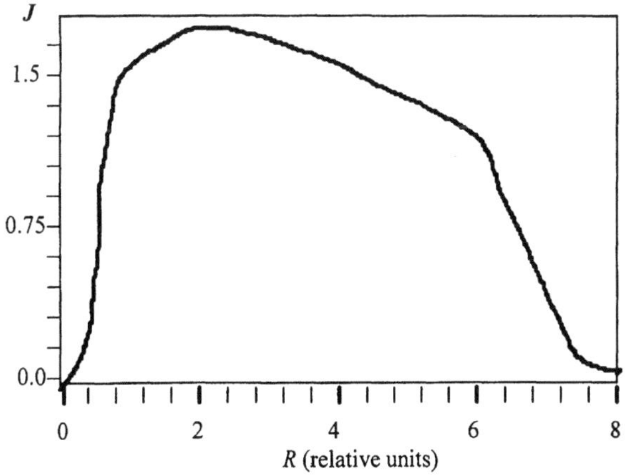

Fig. 6.7. Relationship between the survivability function $J(t)$ and the variations in solar radiation $E(t,\varphi,\lambda,0)$ in the framework of the scenario: $E(t,\varphi,\lambda,0) = R \cdot (3.04–0.03\varphi)(1 + 0.1\sin[2\pi t/365])$, where $R = 1$ under normal conditions of functioning of the system

corresponds to the situation when the system loses more than 70% of the biomass. The integral result from such effects can be understand only in the framework of the GSM.

The function $E(t,\varphi,\lambda,0)$ in Eq. (5.1) depends on the state of the atmosphere. The simulation experiments show that for the PCE survivability permissible reductions of the atmosphere's transparency must be no more than 2.5 times. Above this threshold the system is bound to perish.

6.3.3 The Effect of Water Saturation with Oxygen

The role of oxygen is taken into account by the model in terms of production, mortality rates, and expenditures of energy exchange within the environment. Simulation experiments point to great inhomogeneities in the saturation of the water with oxygen both in horizontal sections and in depth. At the places of oxygen deficit (< 6 mg m^{-3}) a drastic reduction in the biomass of the higher trophic levels is observed, but no stable pattern is noted, either with respect to waters undersaturated, or those oversaturated. Accordingly, for the region as a whole these zones are observed to be traveling, apparently due to hydrodynamic processes, on a nonstable pattern without causing any perceptible damage to the entire community.

Fluctuations in the saturation of the waters with oxygen may have an anthropogenic origin; for instance, they may be due to oil pollution or to large discharges of sewage waters. The model permits us to set some hypothetical experiments, some instances of which are given in Figs. 6.8 and 6.9. It is seen that the model displays a high degree of stability with respect to the initial saturation of the water with oxygen. From the behavior of $J(t)$ we see how rapidly the PCE proceeds to the quasi-stationary regime of functioning at $E_{14}(t_o,\varphi,\lambda,z) \geq 60$ mg m^{-3} and how long it takes to overcome the initial shortage of oxygen in the case of $E_{14}(t_o,\varphi,\lambda,z) = 16$ mg m^{-3}. For $E_{14}(t_o,\varphi,\lambda,z) \leq 6$ mg m^{-3} the PCE is unable to proceed to the stationary regime of functioning.

The model is observed to be even more sensitive to the dynamic effect on the process of saturation of water with oxygen. This clearly follows from the behavior of $J(t)$ in Fig. 6.9. A reduction of the oxygen production by 10% does

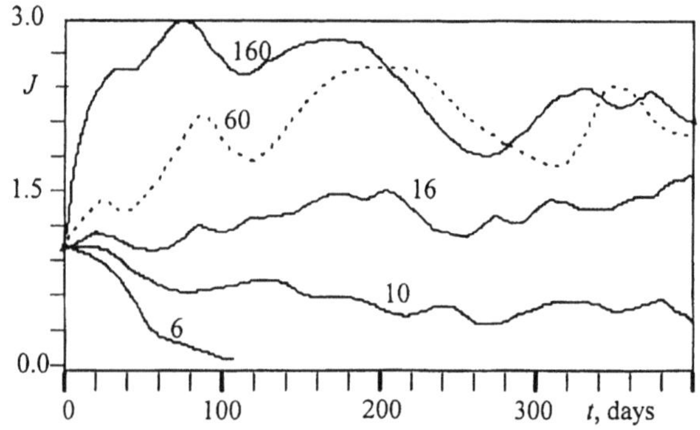

Fig. 6.8. Dynamics of the survivability function $J(t)$ versus time under different initial conditions of water saturation with oxygen (shown on the curves in mg m^{-3})

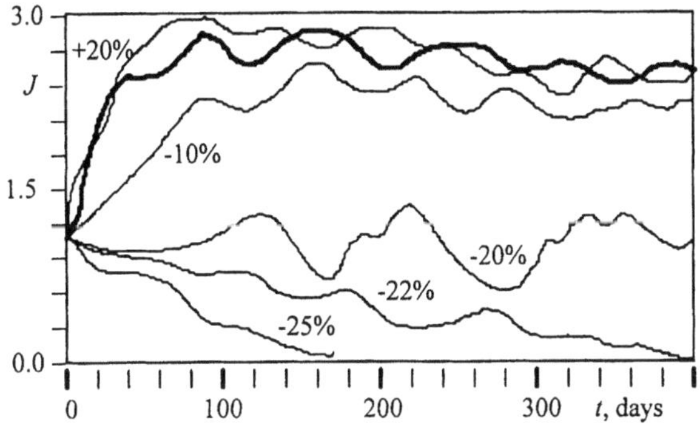

Fig. 6.9.
Dynamics of the survivability function $J(t)$ versus time under variations of the rate of photosynthetic oxygen production in percent deviation from the model value ($\delta_2 = 0.01$). The *thick curve* corresponds to normal conditions

Table 6.2. Simulation results in which various nutrient concentrations are used ($\varphi = 12°S$, $\lambda = 82°W$)

	Biomass of phytoplankton (kg m^{-2})					
	$E_{12}(t_o,\varphi,\lambda,z) = 0.05$ mg m^{-3}			$E_{12}(t_o,\varphi,\lambda,z) = 2$ mg m^{-3}		
$t-t_o$ (days): Depth (m)	10	50	100	10	50	100
Initial nutrient concentrations are held in $z \le 150$ m						
0–10	0.08	3.49	5.73	3.92	2.14	28.06
10–20	8.43	19.48	12.05	22.99	5.19	56.31
20–30	9.67	25.28	33.54	25.45	38.89	75.49
30–40	9.64	27.98	33.27	25.59	66.41	86.36
40–50	9.26	28.27	33.26	30.88	40.52	27.93
50–60	9.39	28.23	29.18	28.31	24.39	21.04
60–70	11.97	27.77	26.59	27.81	8.63	84.69
70–80	6.16	15.23	13.98	21.84	2.29	63.49
80–90	2.79	7.52	6.56	10.74	0.29	22.75
90–100	0.78	3.09	3.52	5.91	0.08	6.16
100–150	1.45	3.93	4.23	3.99	0.43	8.63
150–200	1.15	3.81	2.63	4.09	0.46	11.63
Initial nutrient concentrations are held in $z > 150$ m						
0–10	0.08	0	0	3.07	4.19	19.61
10–20	0.09	0	0	4.13	4.35	35.59
20–30	0.15	0	0	31.05	19.65	40.57
30–40	0.21	0	0	28.02	42.65	72.46
40–50	1.52	0	0	35.64	66.36	69.56
50–60	3.53	0.03	0	19.64	55.78	44.79
60–70	3.22	0.17	0	13.32	21.79	12.84
70–80	2.49	0.13	0	11.69	13.22	6.09
80–90	1.41	0.13	0	4.51	5.96	2.92
90–100	1.28	0.05	0.01	2.09	2.67	3.07
100–150	0.93	0.18	0.01	2.19	2.96	2.54
150–200	0.75	0.12	0.01	2.91	1.86	3.53

not influence the PCE dynamics. A subsequent decrease of δ_2 in Eq. (6.13) leads to a displacement in the stable state of the PCE. The PCE does not survive when the parameter δ_2 is decreased by 22%. Figure 6.9 illustrates the important qualitative result when Eq. 6.16 gives the possibility for a search of the dangerous zones in Ω.

6.3.4
The Effect of Varying the Concentration of Nutrients

Table 6.2 gives results obtained in computing the PCE dynamics in the case of a variation of the concentration of nutrient salts at the initial moment of time t_o and at the depth $z > 150$ m. From the comparison of the results contained herein it follows that a variation of the concentration of nutrients within a wide range at the moment $t = t_o$ does not practically affect the behavior of the system at the moments $t \gg t_o$. The system, so to speak, "heals" with time from the "blows" it has suffered and proceeds to the same functioning level.

It is different when at depth $z > 150$ m, E_{12} becomes close or inferior to a certain critical value $E_{12,\min}$. The system in this case is unable to make up for the fluctuations introduced, and at $E_{12} \gg 0.1$ mg m^{-3} the system begins to experience the effects of a strong limitation throughout the area. Such a reduction in the nutrient concentration is, for instance, possible in the case of contamination of bottom sediments with oil products.

6.3.5
The Effect of Variations in the Velocity of Vertical Advection

To estimate the turbulent escape of nutrients into layers overlying the thermocline it is assumed in the model that the velocity of water uprising is equal to 10^{-3} cm s^{-1}. The data obtained point to the fact that, on the average, the integrated pattern of the distribution of community elements is not subject to any significant variation within the velocity range from 3×10^{-4} to 10^{-2} and even 10^{-1} cm s^{-1}, but is observed to be drastically distorted under a higher and, what is most important, under a lower ($< 10^{-4}$ cm s^{-1}) vertical advection of water. The effect of variations in the velocity of vertical advection is estimated on the base of the average response in the total phytoplankton biomass:

$$\Delta E_2 = \frac{1}{200S}\left\{\int_0^{200}\int_{70}^{90}\int_0^{30}\int_0^{100}\left[E_5(t,\varphi,\lambda,u) - \hat{E}_5(t,\varphi,\lambda,u)\right]^2 dt\,d\lambda\,d\varphi\,du\right\}^{1/2}$$

where $S = 4712963$ km^2 is the aquatory square area for the estimation ($0 \le \varphi \le$ 30°S; 70°W $\le \lambda \le$ 90°W); $\hat{E}_5$ is the value of E_5 corresponding to the advection velocity which differs from 10^{-3} cm s^{-1}.

The dependence of ΔE_2 on the vertical advection velocity shows that the variations in the horizontal current velocity of 0.5–3.5 km h^{-1} are safe for the PCE. These variations modify the structure of the aquatory spottiness. When $v < 0.5$ km h^{-1} the plankton spots are concentrated near the local upwelling zones and upper photic layer is impoverished with respect to the system biomass. When $v > 3.5$ km h^{-1} the system begins to transform to a homogeneous distribution of biomass and for v = 5 km h^{-1} the biomass spottiness has practically disappeared.

The simulation experiments enable us to answer the question about the possibility of securing an increase in the productivity of the system through the creation of artificial upwellings. For the PCE this can be achieved up to 40%. In the open ocean the higher velocity of vertical advection improves, on the one hand, the supply of nutrients for the surface-adjoining waters and, on the other hand, it brings down the water temperature, thereby drastically restricting the productivity of the system. Moreover, on account of the lengthy trophic chain in the open ocean, the ultimate effect of increased saturation of nutrients proves to be insignificant. In the shelf zone, the limitation of the nutrients is absent.

6.3.6
The PCE Sensitivity with Respect to Variation in the Model Parameters

Tables 6.3 and 6.4 give examples of the influence of the variation in the model parameters on the PCE dynamics. Such estimations are important for planning the experimental investigations. From Table 6.3 it follows that the PCE sensitivity with respect to the variation of the model parameters fluctuates over a wide interval. It is seen that the PCE dynamics are most sensitive to variations in such parameters as h_i, μ_i, τ_i, δ_1, σ_{min}, σ_{max} and $k_{i,0}$. When there are large variations in these parameters, the system responds with considerable fluctuations in the total biomass.

Table 6.3. Simulation experiments with the variations of the PCE parameters

Parameter variation	Fluctuations in the biomass concentration caused by the parameter variation (%)			
	ΔE_2	ΔE_5	ΔE_7	ΔE_8
h_i – 20%	32	27	21	12
h_i – 10%	25	12	4	10
h_i + 10%	43	38	29	16
h_i + 20%	56	59	43	26
τ_i – 20%	27	28	25	18
τ_i – 10%	12	11	13	9
τ_i + 10%	37	44	36	18
τ_i + 20%	62	54	41	43
μ_i – 20%	28	20	22	21
μ_i – 10%	13	13	19	11
μ_i + 10%	41	38	42	25
μ_i + 20%	75	46	52	43
ζ_i – 20%	10	12	9	5
ζ_i – 10%	9	10	7	4
ζ_i + 10%	15	12	11	14
ζ_i + 20%	43	65	54	20
$k_{i,o}$ – 20%	56	59	44	25
$k_{i,o}$ – 10%	43	39	30	17
$k_{i,o}$ + 10%	24	11	5	11
$k_{i,o}$ + 20%	32	27	21	12

Table 6.4. Results of simulation experiments on the estimation of the model parameter precision

Parameter of the PCE model	Allowed variation (%)	Parameter of the PCE model	Allowed variation (%)
$k_{1,0}$	−30 to +50	$k_{3,0}$ and $k_{4,0}$	−25 to +35
$k_{6,0}$	−20 to +20	$k_{7,0}$ and $k_{9,0}$	−25 to +15
$k_{5,0}$	−17 to +12	$k_{8,0}$	±12
h_1	±107	h_3 and h_4	−28 to +36
h_6 and h_9	±27	h_5 and h_7	±15
h_8	±11	τ_1	±28
τ_3 and τ_4	±16	τ_i (i = 5–9)	±10
μ_1 and μ_3	−27 to +18	μ_i (i = 4–9)	−23 to +14
δ_2	−15 to +23	δ_1	−22 to +17
V_H	±54	Δ_z	−19 to +31

Table 6.4 gives recommendations concerning the accuracy of parameters to be measured during on-site experiments. These recommendations are a function of the following criterion: *the variation in the parameter "a" is permitted if this variation causes a deviation in B by no more than 20% and $J(t) > J_{min}$ for $t < t^*$*, where t^* is the time period when the system is considered and

$$B = \frac{1}{200t^* S}\int_0^{200}\int_0^{t^*}\int_{70}^{90}\int_0^{30} E_5(t,\varphi,\lambda,u)\,du\,dt\,d\lambda\,d\varphi$$

It follows from other computational experiments that the functioning of the ecosystem does not depend upon the initial values of E_i over a wide range of its values (5–400 cal m^{-3}) if $E_{12}(t_o,\varphi,\lambda,z) > 0.1$ µg − at · l^{-1} and $E_{14}(t_o,\varphi,\lambda,z) > 0.5$ ml l^{-1}. When $E_i(t_o,\varphi,\lambda,z) < 5$ cal m^{-3} (i = 1–9) for all $(\varphi,\lambda) \in \Omega$ the ecosystem does not survive. There are critical situations when the PCE is functioning as "living". This situation corresponds to the following initial data: $E_i(t_o,\varphi,\lambda,z) = 5$ cal m^{-3} for $(\varphi,\lambda) \in$ the upwelling zones or for $z > H$ on Ω and $E_i(t_o,\varphi,\lambda,z) = 100$ cal m^{-3} on the remaining area.

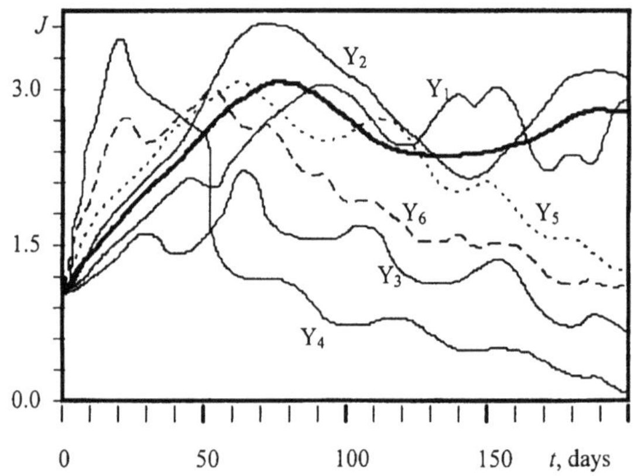

Fig. 6.10. Dynamics of the survivability function $J(t)$ versus time for the experiments in variation in the food spectra Y_s (s = 1–6) as given in the text. The *thick curve* corresponds to the model result

6.3.7
Investigation of PCE Survivability Under Variations in the Trophic Graph

The block diagram of Fig. 6.1 fixes the food chains in the PCE. There are uncertainties in this knowledge. The food spectrum of the jth element is a vector $S_j = (\eta_{ji})$, where $\eta_{ji} = 1$ when the ith element is consumed by the jth the element, and $\eta_{ji} = 0$ otherwise. We consider a set of simulation experiments the results of which are given in Fig. 6.10. These experiments correspond to the following variations of the food spectra:

$$Y_1 = \left\| \begin{matrix} 0000000001011 \\ 0000000000101 \\ 1100000001011 \\ 1111000001011 \\ 0111000000001 \\ 0111000001001 \\ 0000111011001 \\ 0000101010000 \\ 0111000001001 \end{matrix} \right\| \qquad Y_2 = \left\| \begin{matrix} 0000000001011 \\ 0000000000101 \\ 1100000000011 \\ 0110000000001 \\ 0100000000001 \\ 0100000000001 \\ 0000101010001 \\ 0000101010000 \\ 0100000000001 \end{matrix} \right\|$$

$$Y_3 = \left\| \begin{matrix} 0000000000011 \\ 0000000000101 \\ 1100000000001 \\ 0011000000001 \\ 0100000000001 \\ 0100000000001 \\ 0000100000001 \\ 0000101010000 \\ 0100000000001 \end{matrix} \right\| \qquad Y_4 = \left\| \begin{matrix} 0000000000011 \\ 0000000000101 \\ 1000000000001 \\ 0010000000001 \\ 0010000000001 \\ 0110000000001 \\ 0000111000001 \\ 0000101010000 \\ 0100000000001 \end{matrix} \right\|$$

$$Y_5 = \left\| \begin{matrix} 0000000001011 \\ 0000000000101 \\ 1100000000001 \\ 0011000000001 \\ 0100000000001 \\ 0111000000001 \\ 0000101011001 \\ 0000000000000 \\ 0100000000001 \end{matrix} \right\| \qquad Y_6 = \left\| \begin{matrix} 0000000001001 \\ 0000000000101 \\ 1100000000001 \\ 0011000000001 \\ 0100000000001 \\ 0111000000001 \\ 0000000000000 \\ 0000000000000 \\ 0100000000001 \end{matrix} \right\|$$

Here, $Y_k = \| \eta_{ji} \|$ is the matrix of the ecosystem food spectrum. Figure 6.10 demonstrates some calculations which characterize the dependence of $J(t)$ on variations in the trophic graph. We see that the extension of the food spectra in scenarios Y_1 and Y_2 practically does not change the ecosystem dynamics. Another situation arises when the correlations in the trophic graph are disturbed. For instance, the 'cutting' of high trophic levels (birds and/or predatory fish) leads the PCE to an unstable state as seen in scenarios Y_5 and Y_6. Therefore, the more detailed investigation of correlations in the trophic graph is an important task. In particular, it is important to study the trophic connections for living elements which show predatory relationships. As we see from Fig. 6.10, the consumption of the detritus by the bacterioplankton is one such important relationship. In scenarios Y_3 and Y_4, this consumption does not take place, leading to a decline in survivability of the system.

CHAPTER 7

A New Technology for Monitoring Environment in the Okhotsk Sea

7

7.1 Introduction

There are many examples of effective marine models making it possible to study the sea ecosystem dynamics. However, the sea ecosystem modeling runs into difficulties which require the development of new modeling methods. Namely, the functioning of the sea ecosystem is complex under unstable conditions of the environment. These conditions are defined for both measured and unmeasured parameters. Therefore, the sea ecosystems are studied with the use of observation data, mathematical models, and satellite observations. A combination of these approaches gives the most useful results (Krapivin 1996). The main difficulties of this method arise because of incompleteness of information.

Recently, several investigators (Krapivin and Shutko 1989; Kelley et al. 1992a; Sellers et al. 1995; Lapko and Radchenko 2000) have reported a variety of problems in monitoring complex systems for the meaningful collection and synthesis of environmental information concerning the Okhotsk Sea. In response to these difficulties, a method was developed to integrate a geographical information system (GIS) with models and field measurements. The newly developed geoinformation monitoring systems (GIMS = GIS + Model) is focused on the systematic observation and evaluation of the environment related to changes attributable to human impact on the marine system. One of the important functional aspects of the integrated system is the possibility of a forecasting capability to warn of undesirable changes in the environment. The application of mathematical modeling to the monitoring effort greatly improves the simulation of the natural processes in the complex environment.

The development of models based on biogeochemical, biocenotic, demographic, and socioeconomic information, including consideration of environmental dynamics of biospheric and climatic processes on the overall system, necessitates the formulation of requirements imposed on the GIMS structure and its database. According to guidelines proposed in the paper by Kelley et al. (1992b), the simulation of the biosphere dynamics is one of the important functions of the GIMS. Accordingly, growing importance has been attached to the value of this new monitoring approach which integrates assessment of the status of the biosphere. The basic objective of all investigations in the development of the GIMS technology is to describe the regional biogeosystem as a subsystem of the biosphere. The Okhotsk Sea represents such a subsystem. Moreover, GIMS is

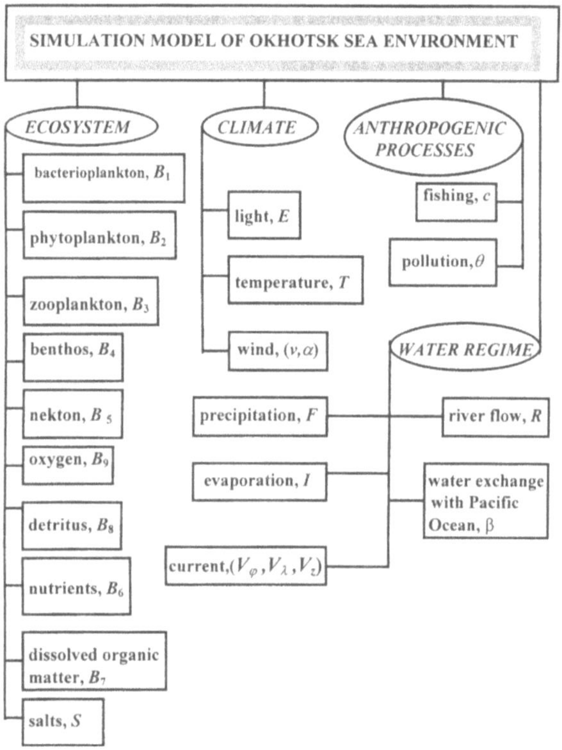

Fig. 7.1.
Structure of the SMOSE. The set of SMOSE components is divided into three types of information sources: mathematical models of the ecological and hydrophysical processes, service software, and the scenarios generator

a geographical monitoring system focused on the systematic observation and evaluation of the environment related to changes attributable to human impact on the marine system. Therefore, the application of GIMS technology to the study of the Okhotsk Sea environment (OSE) entails synthesis of a simulation model of the Okhotsk Sea environment (SMOSE) to describe the associated ecosystem dynamics. Preliminary results in this direction have been published by Aota et al. (1991a, b, 1992, 1993). This is just what helps to describe the spatial dynamics of OSE and to optimize the monitoring regime.

The OSE holds a significant position in the global natural system. At the present time, it has a low level of pollution, but fishing is the main anthropogenic influence. A correlation between the state of the OSE and global changes is one of the problems which is discussed both in the framework of regional investigations and of global studies of the environment. The OSE interacts with global biosphere processes in particular via its influence on the global climate and on the Pacific Ocean. This influence is reciprocal. This chapter suggests an approach to the estimation of such an influence. The common concept of complex system survivability is interpreted for the ecological system of the Okhotsk Sea and the criteria of survivability is defined.

The survivability of a complex system is its ability to resist the influence of external impacts and to reserve the structure and effectiveness under the realization of its "*aim*". The OSE is more than simulation of energy and nutrient flows, trophic webs, and competition of communities. It is the full simulation of interrelations among coexisting living organisms and its nonliving elements. In this study, the definition of the problem of the ecosystem "*aim*" and "*behavior*" is not discussed. These categories are determined by Nitu et al. (2000a). It is postulated that the OSE has its aim to maximize the biomass of living elements on high trophic levels and its behavior in the adaptive change of the food chains. The model developed by Aota et al. (1992) is considered as the OSE prototype. The final realization of all dynamic levels describing the interaction of ecological factors in the sea leads to a set of model units shown in Fig. 7.1. This set is considered below.

7.2 Block Diagram and Principal Structure of the Simulation Model of the Okhotsk Sea Ecosystem

Let us designate the Okhotsk Sea oceanic environment as $\Omega = \{(\varphi,\lambda)\}$, where φ and λ are latitude and longitude, respectively. The spatial inhomogeneity of the Okhotsk Sea basin model is provided for by representing Ω as spatially discrete on the set of cells Ω_{ij} with latitude and longitude steps of $\Delta\varphi_i$ and $\Delta\lambda_j$, respectively. Each of the Ω_{ij} has its square area $\sigma_{ij} = \Delta\varphi_i \cdot \Delta\lambda_j$. Accordingly, the Okhotsk Sea is considered to consist of $N = i_{\max} j_{\max}$ water bodies Ξ_m $(m = j + (i-1)j_{\max})$. These cells are the basic spatial structure of Ω for the purpose of developing computer algorithms. The water in the sea flows between Ξ_m (Kawasaki and Kono 1993). Each of Ξ_m has a vertical structure with original discretization in the depth z by steps Δz_m. Every now and then with step Δt a vertical structure is fixed based on remote sensing information about the sea surface state and temperature. A scheme of vertical structure developed by Legendre and Krapivin (1992) is taken into consideration.

The cells Ω_{ij} are heterogeneous as to their parameters and function. There is a set of cells adjacent to the river mouths (Ω_R) and to the ports (Ω_P), bordering on the land (Ω_Γ), in the Kuril-Kamchatka Straits (Ω_B) and on the boundary with the Sea of Japan (Tartar and Soya Straits, Ω_N). The distribution of depths is given as the matrix $\|h_{ij}\|$ where $h_{ij} = h(\varphi_i, \lambda_j) \in \Omega_{ij}$). As a result, the full water volume of Ω is divided into volumetric compartments $\Xi_{ijk} = \{(\varphi,\lambda,z)/\varphi_i \le \varphi \le \varphi_{i+1}; \lambda_j \le \lambda \le \lambda_{j+1}; z_k \le z \le z_{k+1}\}$ with volume equal to $\sigma_{ijk} = \Delta\varphi_i \Delta\lambda_j \Delta z_k$. Within the compartment Ξ_{ijk}, the water body is considered as a homogeneous structure. The water temperature, salinity, density, and biomass of Ξ_{ijk} are described by a point in the models at a given location. The four seasons are used to represent the effect of anthropogenic processes on the oceanic environment of Ω: τ_w – winter, τ_s – spring, τ_u – summer, τ_a – fall.

The realization of the simulation procedure is defined by the discretization structure of the OSE oceanic environment. When $\Delta z_{ij} = h_{ij}$ $(i = 1,\ldots,i_{max};$ $j = 1,\ldots, j_{max})$, the vertical mixing processes interior to Ξ_m are identified with uniform vertical distribution of all model parameters. In the case of $\Delta z_{ij} < h_{ij}$ compartment Ξ_m is considered as a vertical structure with basic characteristics.

A conceptual diagram identifying the components of the SMOSE is shown in Fig. 7.1. The dynamics of the SMOSE are supported in turn by a global biosphere

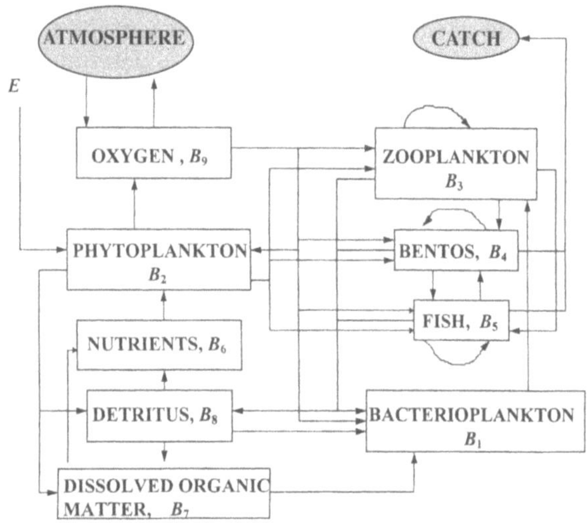

Fig. 7.2. Trophic chains and energy flows in the OSE

Table 7.1. Vertical structure of the OSE aquatic system. T_w is calculated by averaging of temperatures for mixed-water volumes

Layer and its identifier	Parameters					
	Layer thickness	Temperature	Irradiance	Turbulence coefficient	Relaxation coefficient	Reflection coefficient
Surface, Ω		T_0	E_0			
Snow, S	g	T_g	E_g	0	α_g	β_g
Floating ice, I_1	r	T_r	E_r	0	α_r	β_r
Submerged ice, I_2	f	T_f	E_f	0	α_f	β_f
Water, W	$z-f$	T_w	E_w	Δ	α_w	β_w

model. The SMOSE input data are from pollutant sources indicated in the nearshore Okhotsk Sea and sea-ice areas, as referenced on current maps. The set of SMOSE components is divided into three types of information sources: mathematical models of the natural ecological and hydrophysical processes, service software, and the scenarios generator.

The OSE functions under such climatic conditions when the greater part of the sea surface is covered with ice during several months. Therefore, the vertical structure is represented by a three-layer block diagram (Table 7.1). The trophic chains and energy fluxes of the OSE, as shown in Fig. 7.2, are completed by the specific channels given in Fig. 7.3.

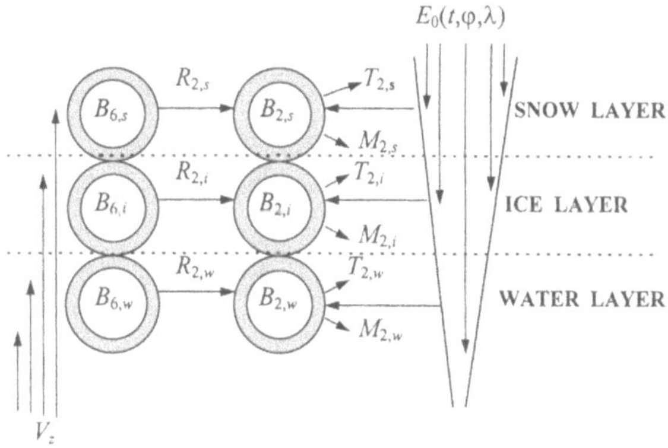

Fig. 7.3. Conceptual structure for the model of the kinetics of the phytoplankton biomass under OSE conditions. The source energy for the entire system is solar radiation energy $E(t,\phi,\lambda,z)$ the intensity of which depends on time t, latitude ϕ, longitude λ, and depth z

7.3 The Marine Biota Unit

The energy source for the entire system is solar radiation energy $E(t,\varphi,\lambda,z)$, the intensity of which depends on time t, latitude φ, longitude λ, and depth z. The equations that describe the biomass dynamics of the living elements are:

$$\frac{\partial B_i}{\partial t} + \xi_i[V_\varphi \frac{\partial B_i}{\partial \varphi} + V_\lambda \frac{\partial B_i}{\partial \lambda} + V_z \frac{\partial B_i}{\partial z}] = R_i - T_i - M_i - H_i -$$

$$\sum_{j\in\Gamma_i} C_{ij}R_j + \xi_i[\frac{\partial}{\partial\varphi}\left(\Delta_\varphi \frac{\partial B_i}{\partial \varphi}\right) + \frac{\partial}{\partial\lambda}\left(\Delta_\lambda \frac{\partial B_i}{\partial \lambda}\right) + \frac{\partial}{\partial z}\left(\Delta_z \frac{\partial B_i}{\partial z}\right) +$$

$$\beta_V \frac{\partial B_i}{\partial z}],(i = 1,...,5); \tag{7.1}$$

where $V(V_\varphi, V_\lambda, V_z)$ are the components of the current velocity; R_i is the production; T_i is the rate of exchange with the environment; M_i is the mortality; H_i is the nonassimilated food; Γ_i is the set of the trophic subordination of the ith component: $C_{ji} = k_{ji}F_i / \sum_{m\in Si} k_{jm}F_m$; S_i is the food spectrum of the jth component; k_{jm} is the index of the satisfaction of the nutritive requirements of the jth component at the expense of the mth component biomass; $F_i = max\{0, B_i - B_{i,min}\}$ where $B_{i,min}$ is the minimal biomass of the ith component consumed by the other trophic levels; $\Delta(\Delta_\varphi, \Delta_\lambda, \Delta_z)$ are components of the turbulent mixing coefficient (on the assumption of isotrophism of vertical mixing in the horizontal plane $\Delta_\varphi = \Delta_\lambda = \nu_H$); and β_V is the upwelling velocity. The functions R_i, M_i, H_i and T_i are parameterized according to the models by Krapivin (1996) and Legendre and Krapivin (1992). The equations describing the dynamics of the abiotic elements are represented in accordance with Berdnikov et al. (1989). Functions M_4 and M_5 include the biomass losses at the expense of fishing. Parameter ξ_i characterizes the subjection of the ith component relative to the current. It is supposed that $\xi_i = 1$ for $i = 1,2,3$ and $\xi_i = 0$ for $i = 4,5$.

The inert components are described by the following equations (Krapivin 1996):

$$\frac{\partial B_6}{\partial t}+V_\varphi\frac{\partial B_6}{\partial \varphi}+V_\lambda\frac{\partial B_6}{\partial \lambda}+V_z\frac{\partial B_6}{\partial z}=U_6+G_6 \tag{7.2}$$

$$\frac{\partial B_7}{\partial t}+V_\varphi\frac{\partial B_7}{\partial \varphi}+V_\lambda\frac{\partial B_7}{\partial \lambda}+V_z\frac{\partial B_7}{\partial z}=U_7+G_7 \tag{7.3}$$

$$\frac{\partial B_8}{\partial t}+V_\varphi\frac{\partial B_8}{\partial \varphi}+V_\lambda\frac{\partial B_8}{\partial \lambda}+V_z\frac{\partial B_8}{\partial z}=U_8+G_8 \tag{7.4}$$

$$\frac{\partial B_9}{\partial t}+V_\varphi\frac{\partial B_9}{\partial \varphi}+V_\lambda\frac{\partial B_9}{\partial \lambda}+V_z\frac{\partial B_9}{\partial z}=U_9+G_9+F_9 \tag{7.5}$$

where

$$G_6=\beta_V\frac{\partial B_6}{\partial z}+\rho_1\sum_{i=1}^{5}T_i,\ P_i=\sum_{j\in S_i}k_{ij}B_{j,\min},$$

$$U_6=\delta_n\rho^*B_8-\delta_1R_2+\frac{\partial}{\partial\varphi}\left(\Delta_\varphi\frac{\partial B_6}{\partial\varphi}\right)+\frac{\partial}{\partial\lambda}\left(\Delta_\lambda\frac{\partial B_6}{\partial\lambda}\right)+\frac{\partial}{\partial z}\left(\Delta_z\frac{\partial B_6}{\partial z}\right),$$

$$U_7=k_{7,0}R_2+\beta_V\frac{\partial B_7}{\partial z}-k_{1,7}R_1B_7/P_1,$$

$$G_7=\frac{\partial}{\partial\varphi}\left(\Delta_\varphi\frac{\partial B_7}{\partial\varphi}\right)+\frac{\partial}{\partial\lambda}\left(\Delta_\lambda\frac{\partial B_7}{\partial\lambda}\right)+\frac{\partial}{\partial z}\left(\Delta_z\frac{\partial B_7}{\partial z}\right),$$

$$G_8=\sum_{i=1}^{5}\left(M_i+H_i\right)$$

$$\begin{aligned}U_8=&-\rho^*B_8-\left(v^*-\beta_V\right)\partial B_8/\partial z\\&-\left(k_{1,8}R_1/P_1+k_{3,8}R_3/P_3+k_{4,8}R_4/P_4+k_{5,8}R_5/P_5\right)B_{8,\min}\\&+\frac{\partial}{\partial\varphi}\left(\Delta_\varphi\frac{\partial B_8}{\partial\varphi}\right)+\frac{\partial}{\partial\lambda}\left(\Delta_\lambda\frac{\partial B_8}{\partial\lambda}\right)+\frac{\partial}{\partial z}\left(\Delta_z\frac{\partial B_8}{\partial z}\right),\end{aligned}$$

$$G_9=\frac{\partial}{\partial\varphi}\left(\Delta_\varphi\frac{\partial B_9}{\partial\varphi}\right)+\frac{\partial}{\partial\lambda}\left(\Delta_\lambda\frac{\partial B_9}{\partial\lambda}\right)+\frac{\partial}{\partial z}\left(\Delta_z\frac{\partial B_9}{\partial z}\right),$$

$$U_9=-\sum_{i=1}^{5}\zeta_iT_i+\beta_V\frac{\partial B_9}{\partial z}$$

$$F_9=\delta_2R_2+max\{0,(z*-z)/|z-z*|\}W*-\delta*\rho*B_8$$

ρ^* is the velocity of detritus decomposition per unit of biomass; v_* is the velocity of gravity settling; ρ_1 is the part of the exchange biomass losses that transform into nutrients (Legendre and Legendre 1998); δ_1 is the nutrient assimilation velocity by the photosynthetic process per unit of phytoplankton production; δ_2 is the oxygen production per unit of phytoplankton production; ζ_i ($i = 1,...,5$) and $\delta*$ are the oxygen losses at the expense of respiration and detritus decomposition, respectively; $z*$ is the maximal depth of oxygen exchange of seawater with the atmosphere; $W*$ is the difference between the invasion and evasion processes; and k_{ij} is the effectiveness coefficient of ith element with respect to the jth element of the ecosystem.

Equations (7.1)–(7.5) are used in the complete volume only when $(\varphi,\lambda,z) \in W$. In the other cases (i.e., in the layers of ice or snow), these equations are automatically reduced in accordance with the scheme represented in Fig. 7.3.

The ice-water ergocline plays an important role in the biological productivity of OSE. According to the hypothesis of Legendre and Legendre (1998), energy ergoclines are the preferential sites for biological production in the arctic climate. The primary production in the food chains of the arctic ecosystems is determined by the phytoplankton productivity. This is connected with complex variations in the meteorological, hydrodynamic and energy parameters of the sea environment. The problem of parameterization of the phytoplankton production in the north seas has been studied by many authors. Table 7.1 gives the structure for the seasonal composition of conditions affecting the primary production in Ω. This scheme is applied for each of the $\Omega_{ij} \subseteq \Omega$.

The unit CLIMATE calculates T_w by means of the averaging of temperatures for the mixed water volumes. In addition, the following correlations are taken into account (see Sect. 5.3):

$$T_g = T_r = T_f = \begin{cases} -0.24L + 0.76T_0 + 8.38 & when \quad L \le z_\alpha, \\ -0.042L + 0.391T_0 - 0.549 & when \quad L > z_\alpha, \end{cases}$$

where $L = g + r + f$, z_α is the ergocline specific parameter (0.4–0.7 m).

Figure 7.2 shows a conceptual flow chart of the energy in the ecological system. The energy input is provided by solar radiation $E_0(t,\varphi,\lambda)$ and the up-flow of nutrients from the deep-sea layers. The concentration of nutrients $B_{8,A}(t,\varphi,\lambda,z)$ at the depth z is determined by photosynthesis $R_{2,A}$, advection and destruction of suspended dead organic matter $B_{7,A}$. As a basic scheme for water circulation, the scheme considered in Chapter 4 is accepted as correct for the conditions of the OSE. The role of hydrodynamic conditions manifests itself in the maintenance of concentration of the nutrients required for photosynthesis via their transport from other layers or aquatories of the sea where the chemical element concentration is sufficiently high.

The Okhotsk Sea environment is an ice/snow-dominated system for several months of the year. The solar radiation flux E_0 is attenuated by snow, ice, and water. Taking into account the designations of Table 7.1, we have (Legendre and Krapivin 1992):

$$E(t,\varphi,\lambda,z)=\begin{cases} E_0 & when \quad z\le -(r+g), \\ E_g & when \quad z\in[-(r+g),-r], \\ E_r & when \quad z\in -r,0], \\ E_f & when \quad z\in[0,f], \\ E_w & when \quad z>f \end{cases} \tag{7.6}$$

where the values of α_A ($A = g,r,f,w$) depend on the optical properties of the Ath medium,

$$E_g = (1 - \beta_g)\, E_o exp\{- \alpha_g z\},\ E_r = (1 - \beta_r)\, E_g(t,\varphi,\lambda,- r) exp\{- \alpha_r z\},$$

$$E_f = (1 - \beta_f)\, E_r(t,\varphi,\lambda,0) exp\{- \alpha_f z\},\ E_w = (1 - \beta_w)\, E_f(t,\varphi,\lambda,f\,) exp\{- \alpha_w z\}.$$

The irradiance E_0 arrives at the surface of Ω. The estimation of E_0 is made in the framework of the monitoring regime or is calculated by means of the climatic model. The E_0 flux is attenuated by snow, ice, and water according to the scheme in Table 7.1. In each of the cells Ω_{ij} the structure of these layers is changed corresponding to the time of year. Within each of the layers, the attenuation of the irradiance with depth is described by an exponential model (7.6). The parameters α_A and β_A are functions of the salinity, turbidity, temperature, and biomass. The form of this dependence is given as the standard functions or a scenario. The phytoplankton production $R_{2,A}$ is a function of the solar radiation E_A, the concentration of nutrients $B_{6,A}$, the temperature T_A, the phytoplankton biomass $B_{2,A}$, and the concentration of pollutant η_A. There are many models for the parameterization of the photosynthesis process (see Chap. 4). In the present study, the equation of the Michaelis–Menten type is used:

$$R_{2,A} = a_A\, k_{1,A}\, B_{2,A,max}/(E_A + k_{1,A}) \tag{7.7}$$

where $k_{1,A}$ is the irradiance level at which $R_{2,A} = R_{2,A,max}$ and $B_{2,A,max}$ is the maximum quantum yield. The coefficient a_A determines the dependence of the phytoplankton production on the environmental temperature T_A and the nutrients concentration B_6. The SMOSE realizes the following equation for the calculation of a_A:

$$a_A = a_1 K_0(T_A,t)/[1 + B_{2,A}/(a_2 B_{6,A})] \tag{7.8}$$

Here, a_1 is the maximum rate of absorption of nutrients by the phytoplankton (day^{-1}), a_2 is the index of velocity of saturation of photosynthesis, and

$$K_0\,(T_A,t) = a_3 \max\{0,(T_c - T_A)/(T_c - T_{opt}) \times \exp[1-(T_c - T_A)/(T_c - T_{opt})]\} \tag{7.9}$$

where a_3 is the weight coefficient and T_c and T_{opt} are the critical and optimal temperatures for photosynthesis (°C).

The relationships (7.8) and (7.9) make the description of the phytoplankton production more accurate for critical environmental conditions when the concentration of nutrients and the temperature have high fluctuations. The coefficients of these relationships are defined on the basis of estimations given by Legendre and Legendre (1998).

7.4 The Hydrological Unit

The circulation of the waters in the OSE is a complex system of cycles and currents with different scales. The water dynamics in Ω is presented by flows between the compartments Ξ_{ijk}. The directions of the water exchanges are represented on every level $z_k = z_0 + (k - 1)\Delta z_k$ according to Aota et al. (1992). The external boundary of Ω is determined by the coast line, the sea bottom, the Kuril Straits, the boundary with the Sea of Japan and the water–atmosphere boundary.

The data of the hydrological regimes are synthesized as a four-level structure according to the seasonal time scale. The velocity of the current in the Kuril Straits is estimated by the following binary function:

$$V(t) = V_1 \text{ for } t \in \tau_u \cup \tau_a,$$
$$V(t) = V_2 \text{ for } t \in \tau_w \cup \tau_s \tag{7.10}$$

The water exchange through the boundary with the Sea of Japan is V_3. The water temperature T_{ijk}^W in Ξ_{ijk} is a function of the evaporation, precipitation, river flows and inflows of water from the Pacific Ocean (Kawasaki and Kono 1993). Its change with time in Ξ_{ijk} is described by the equation of heat balance:

$$\varsigma C \sigma_{ijk} \frac{\partial T_{ijk}^W}{\partial t} = \sum_{s,l,m} \left(W_{slm}^{ijk} + f_{slm}^{ijk} \right) - W_{ijk} \tag{7.11}$$

where ζ is the seawater density (g cm^{-3}); C is the water heat capacity, (cal g^{-1} grad^{-1}); σ_{ijk} is the volume of Ξ_{ijk}; W_{slm}^{ijk} is the heat inflow to Ξ_{ijk} from Ξ_{slm}; f_{slm}^{ijk} is the heat exchange between Ξ_{slm} and Ξ_{ijk} caused by turbulent mixing; and W_{ijk} is the total heat outflow from Ξ_{ijk}. Equation (7.11) is used for layers $k > 1$. The temperature distribution in the layer with $k = 1$ is calculated on the basis of monitoring data (Shinohara and Shikama 1988).

It is assumed that the dissipation of moving kinetic energy, geothermal flow on the ocean bed, heat effects of chemical processes in the ocean ecosystem, and freezing and melting of the ice are not global determinants of the water temperature fields. The SMOSE does not consider these effects.

The dynamics of the water salinity $s(t,\varphi,\lambda,z)$ are described by the balance equation of Berdnikov et al. (1989). It is supposed that $s(t,\varphi,\lambda,z) = s_0$ for $z > 100$ m. The river flows, ice fields, and synoptical situations are described by scenarios formed by the user of the SMOSE. There are a set of standard scenarios for use by default in the unspecified regime. The snow-layer thickness $g(t,\varphi,\lambda)$ may be described via statistical data with given dispersion characteristics: $g = g_1 + g_0$, where the value g_1 is defined as the mean characteristic for the chosen time interval and the function $g_0(t,\varphi,\lambda)$ gives the variation of g for the given time interval.

An alternative is the parameterization of the snow-layer dynamics process in the framework of an atmospheric process simulation algorithm relating the thickness of the growth and melting of the snow-layer to the temperature and precipitation:

$$g(t + \Delta t,\varphi,\lambda) = g(t,\varphi,\lambda) + S_F - S_M$$

where S_F is the part of the snow precipitated at temperatures close to freezing ($265\ K \le T_0 \le 275\ K$) and S_M is the snow ablation (i.e., evaporation + melting).

Initially, the water space occupies the semispace $z \ge f$ where f is the depth of the submerged ice. If we designate by ρ_g, ρ_r and ρ_w the density of snow, ice and water, respectively, we obtain:

$$f = (g\rho_g + rg_r)/(\rho_w - \rho_r)$$

where r is the thickness of the floating ice.

The process of ice formation on the water surface is described by the Stephan problem:

$$\partial T(t,\varphi,\lambda,z)/\partial t = a^2(z)\partial^2 T(t,\varphi,\lambda,z)/\partial z^2 \qquad (7.12)$$

$$\lambda_1 \partial T(t,\varphi,\lambda,z)/\partial z|_{z=f} = q\,\rho_r\,\partial r(t,\varphi,\lambda)/\partial t + \lambda_2\,\partial T(t,\varphi,\lambda,z)/\partial z|_z =_f$$

with initial and boundary conditions:

$$r(0,\varphi,\lambda) = \psi_1(\varphi,\lambda), T(0,\varphi,\lambda,z) = \psi_0(\varphi,\lambda,z),$$

$$T(t,\varphi,\lambda,-r) = f_1(t,\varphi,\lambda) \le 0°C, T(t,\varphi,\lambda,f) = 0°C,$$

$$T(t,\varphi,\lambda,h) = Y(t,\varphi,\lambda)$$

Here, f_1 is a function of changeability of the near-land air temperature in the absence of snow or of the snow temperature specified in a scenario or is calculated from the climatic model, Y is the temperature of the deep water layers and $a^2(z)$ is the thermal conductivity coefficient of the medium:

$$a^2(z) = \begin{cases} a_1^2(z) & when \quad -r \le z \le f, \\ a_2^2(z) & when \quad z > f \end{cases}$$

An alternative procedure of the ice thickness formation is the empirical formula proposed by Truskov et al. (1992): $r + f = 2.1 \cdot (t_\Delta)^{0.47}$ (cm), where t_Δ is the accumulated degree-days of frost ($T_0 \ge -7°C$).

There are many efficient models to parameterize the process of ice and snow thickness formation. For example, a model developed by Ivanov and Makshtas (1990) describes various situations that cover all the meteorological and thermodynamic correlations. In spite of this, the use of this model is impossible in the SMOSE structure. A more detailed parameterization of the climate processes is necessary to use this model as the SMOSE unit.

The most complex element in the construction of the model proves to be the mathematical description of fish migrations. There is the obvious proposal that the migration process should be identified with an inverse turbulent diffusion with coefficient $\Delta^* > \Delta$. It is obvious, however, that this algorithm leads to inaccurate results. The investigations of many authors have revealed that the process of fish migration is accompanied by an external appearance of purposeful behavior. Therefore, let us formulate a law of migration which follows from the general biological principle of adaptation and long-term adaptability: the migrations of fish are subordinate to the principle of complex maximization of the effective nutritive ration P_5, given preservation of favorable temperature, salinity and pollution conditions. In other words, the traveling of migrating species takes place at characteristic velocities in the direction of the maximum gradient of effective food, given adherence to environmental restrictions.

At present, the OSE environment is exposed to pollution at a growing rate. The SMOSE subunit simulating the pollution processes of the territory Ω at the expense of the atmospheric transport, rivers and surface coastal outflow, navigation, and other human activity has a set of scenarios and some parametric descriptions. The variety of pollutants $\theta = \{O,e,\psi,\varepsilon\}$ is described by three components: oil hydrocarbons O, heavy metals (e – solid particles, ψ – dissolved fraction) and radionuclides ε. It is supposed the pollutants get into the living organisms only through the food chains. It is assumed that the rivers bring a not inconsiderable contribution to the pollution level of the OSE waters. The concentration of pollutant κ in the river γ is γ_κ. The pollutant κ enters compartment $\Xi_{ijk} \subseteq \Omega_R$ with the velocity c_γ. Subsequently, the spreading of the pollutant in Ω is described by other subunits.

The influence of water exchanges between the OSE and the Pacific Ocean on the pollution level in Ω is described by the hydrological unit. The atmospheric transport of heavy metals, oil hydrocarbons, and radionuclides is described in Chapter 5. The application of these models to the reconstruction of the pollution distribution over Ω makes it possible to estimate optimal values of $\Delta\varphi$, $\Delta\lambda$, and the step in time Δt. The present level of the database for the OSE provides for using a one-level Euler model with $\Delta t = 10$ days, $\Delta\varphi = \Delta\lambda = 1°$.

7.5 The Simulation Procedure and Experiments That Use the Simulation Model of the Okhotsk Sea Ecosystem

The complete system of equations of ecosystem dynamics is formed from the set of traditional balance, hydrodynamic and biogeochemical Eqs. (7.1)–(7.12). These equations have a set of coefficients $X = X_1 \cup X_2 \cup X_3$, where the subset X_1 contains the values of coefficients which are determined with high precision, subset X_2 consists of the coefficients determined less accurately, and subset $X_3 = X \backslash (X_2 \cup X_1)$ contains the coefficients which are estimated. The simulation procedure foresees a two-tier process of model adaptation. At the first tier the model adaptation process is realized at the expense of determination of the subset X_3 coefficients. This process consists in the variation of the values of the coefficient in the ranges fixed by the SMOSE user. The quality criterion is formulated by the user, also proceeding from the existing experimental data or from other considerations. The service unit of the SMOSE has a set of such possibilities. The user can demand to minimize the divergence between his data about some parameter of the OSE and its model estimation. The spectrum of such parameters covers the main individual and integral elements of the OSE.

At the second tier, the model adaptation process is continued by making subset X_2 more precise. The quality criterion of this process is based on the improvement of model quality reached during the first stage of adaptation process. Thus, the simulation procedure of the SMOSE is a continuous process of model adaptation via coefficient changes or trophic graph modifications. Really, there is a set of models $\{A_i\}$. This set is used with the sorting out of the model parameters. Every model A_i is characterized by the quality level Q_i. The complete model A^* is formed as the limit above the set of models. The completion of the adaptation procedure depends on the user's definition of the quality criterion Q^* and on the values $|Q^* - Q_i|$. Model A^* ensures the $\min_i |Q^* - Qi|$. Here, it is assumed that

$$Q_i = \frac{1}{h_1 \sigma} \int_{(\varphi,\lambda,z)\in\Omega} B_2(t,\varphi,\lambda,z) d\varphi d\lambda dz.$$

The value of $Q^*(=20 \text{ g m}^{-3})$ is estimated at the time t_* (middle of summer) as the average concentration of the phytoplankton biomass at the top layer with a depth of h_1 (= 100 m).

The simulations of the OSE offer the possibility for studying various aspects of how it functions. Below, there are some simulation results obtained under the values of OSE parameters given in Table 7.2. Specifically, the present status of the OSE is of great interest in order to understand the role of anthropogenic influences on it. According to the theory of complex systems, a dynamic system is in the "living state" at the time interval $[t_0, t_1]$ if the biomass of its elements is within limits $B_i \geq B_{i,min}$. Correlations between the trophic levels and nonliving elements give summary conditions for survivability criteria: $B_{min} \leq \Sigma B_i$. Finally,

Table 7.2. Values of some parameters in the framework of simulation experiments using the SMOSE

Parameter	Value
Spatial discretization by	
Latitude, $\Delta\varphi$	2°
Longitude, $\Delta\lambda$	2°
Depth, Δz	
$z \leq 100$ m	10 m
$z > 100$ m	h - 100 m
Ice thermal conductivity coefficient, λ_1	2.21 W m^{-1} degree^{-1}
Water thermal conductivity coefficient, λ_2	0.551 W m^{-1} degree^{-1}
Coefficient of vertical turbulence, Δ_z	10^2 m^2 s^{-1}
Ice-thawing specific heat, q	334 KJ kg^{-1}
Velocity coefficient of detritus decomposition, δ_*	
In the snow and ice layers	0
In the water layer	0.1
Velocity of the water current in the Kuril-Kamchatka straits, V_i:	
$i = 1$	0.25 m s^{-1}
$i = 2$	0.15 m s^{-1}
Velocity of water current between the Okhotsk Sea and the Sea of Japan, V_3	0.3 m s^{-1}
Water thermal capacity, C	4.18 kJ kg^{-1} K^{-1}
Water salinity for $z > 100$ m, s_0	34.95‰
Area of the Okhotsk Sea, σ	1,583,000 km^2
Volume of the Okhotsk Sea waters, W	1,365,000 km^3
Average depth, h_*	859 m
Critical temperature for photosynthesis, T_c	- 0.5°C

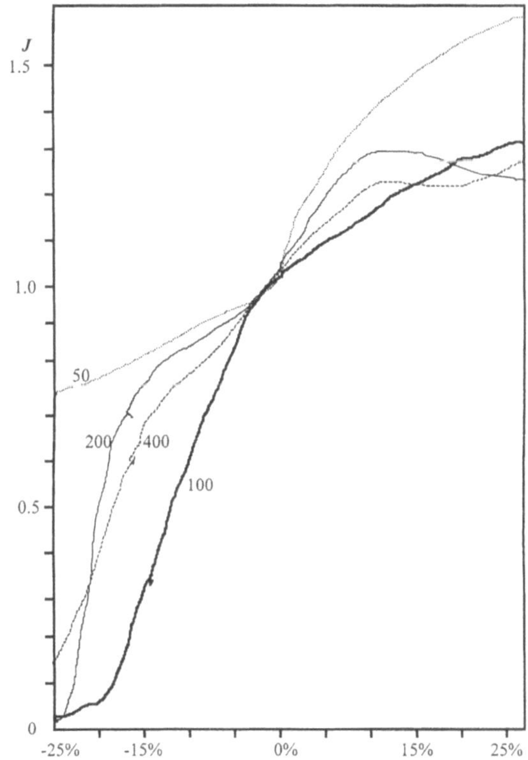

Fig. 7.4. Dynamics of the survivability function $J(t)$ under variations of the rate of photosynthetic oxygen production in percent of the model value ($\delta_2 = 0.01$). The time in days after the beginning of the simulation experiment (t_0) is marked on the curves

the survivability criteria for the OSE can be written in the form: $J(t) \geq \alpha J(t_0)$, where $J = U(t)/U(t_0)$,

$$U(t) = \sum_{k=1}^{9} \int_{(\varphi,\lambda,z)} B_k(t,\varphi,\lambda,x)dx \,,$$

t_0 is the moment of time when the value of function $A(t)$ is considered as known; and $\alpha < 1$ is the level of survivability. Actually, we consider the OSE to be in a living state if the condition $J(t) > \alpha J(t_0)$ is carried out for $t > t_0$. Calculations of $J(t)$ for $t > t_0$ demonstrate how the OSE reaction depends on the variability of various environmental conditions. For example, fluctuations in the oxygen saturation of the water may be of anthropogenic origin (oil pollution, a large discharge of sewage waters, temperature increase, etc.). The function $J(t)$ shows that the OSE displays a high degree of stability with respect to the initial saturation of the water with oxygen, then how the OSE proceeds rapidly to a quasi-stationary regime of functioning at $B_9(t_0,\varphi,\lambda,z) \geq 1.8$ ml l^{-1} and how long it takes to overcome the initial shortage of oxygen in the case of $B_9(t_0,\varphi,\lambda,z) = 1.1$ ml l^{-1}. For $B_9(t_0,\varphi,\lambda,z) \leq 0.8$ ml l^{-1}, the OSE is unable to proceed to a stationary regime of functioning. The OSE is observed to be more sensitive to dynamic effects when the water is saturated with oxygen. This clearly follows from the behavior of J(t) in Fig. 7.4. A reduction of the oxygen production by 12% does not in-

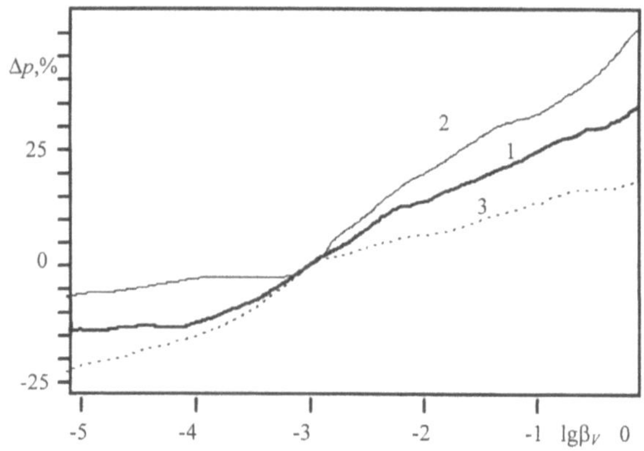

Fig. 7.5. Influence of variations in the vertical water upwelling on the biomass dynamics estimated with time. The criterion: $\Delta p = \frac{1}{W} \int_{t_0}^{t_0+100 days} \int_{(\varphi,\lambda,z)\in\Xi} [B_2(t,\varphi,\lambda,z) - B_2(t_0,\varphi,\lambda,z)] dz d\lambda d\varphi dt$ where t_0 is an arbitrary moment of time for the beginning of the simulation experiment. The values of Δ_z (m^2 s^{-1}) are: *1* 10^2; *2* 1.2×10^2; and *3* 0.8×10^2

fluence the OSE dynamics. However, the OSE does not survive when the oxygen production is decreased by as much as 20%.

Calculations of J(t) offer the possibility for detection capability in searching for zones at risk for survivability in Ω. For instance, one of the particularly dangerous anthropogenic influences is a change in nutrient concentration. Simulation experiments enable us to determine variations in the vertical uprising velocity of the water within some range.

To estimate the turbulent escape of nutrients into layers overlying the maximal depth of the photosynthetic layer ($z \leq h^*$) it is assumed in the SMOSE that the velocity of water uprising is equal to 10^{-3} cm s^{-1}. The data obtained point to the fact that, on average, the integrated pattern of the distribution of community elements is not subject to any significant variations within the velocity range from 3.5×10^{-4} to 10^{-2} and even 10^{-1} cm s^{-1}, but is observed to be drastically distorted under a higher and (what is most important) under a lower ($< 10^{-4}$ cm s^{-1}) vertical advection of water. Figure 7.5 shows the truth of this.

Table 7.3 gives results obtained in the case of a variation of the concentration of nutrients at the initial moment of time t_0. From a comparison of the results contained herein it follows that a variation of the concentration of nutrients within a wide range at the moment $t = t_0$ does not practically affect the behavior of the OSE at moments $t \gg t_0$. The system, so to speak, "*heals*" with time from the "*blows*" it has suffered and proceeds to the same functioning level. When initial nutrient conditions are limiting the OSE reaches the stable state during only 50 days. If the nutrient concentration at the time t_0 has a nonlimiting level, the OSE reaches a stable state on the 30th day after the beginning of the simulation experiment.

The SMOSE allows us to accomplish other various simulation experiments. For example, Table 7.4 considers the case when the vertical distribution of the OSE elements is given in its dynamics as functions of initial values. The concen-

Table 7.3. Simulation results in which various nutrient concentrations are used. Space-time-mean results are given. Initial data for the phytoplankton biomass are $B_2(t_0,\varphi,\lambda,z) = 2$ mg m^{-3}

Depth (m)	Phytoplankton biomass (g m^{-3})					
	$B_6(t_0,\varphi,\lambda,z) = 0.05$ mg m^{-3}			$B_6(t_0,\varphi,\lambda,z) = 3$ mg m^{-3}		
	$t - t_0$ (days)					
	10	30	50	10	30	50
0	1.1	5.4	10.7	6.3	11.3	10.9
10	1.9	5.3	12.3	5.4	11.9	15.8
20	2.4	10.2	21.9	7.1	19.2	21.1
30	1.3	6.6	18.4	7.2	14.6	14.9
40	0.6	3.1	9.2	5.5	10.1	9.7
50	0.2	1.2	8.7	4.2	7.7	8.4
70	0.1	0.9	5.3	1.7	4.8	5.1
100	0	0.4	2.1	1.1	1.9	2.2
150	0	0.1	0.9	0.8	1.1	1.3
200	0	0.1	0.1	0.2	0.3	0.2

Table 7.4. Vertical distribution of some OSE elements as the result of a simulation experiment with the SMOSE. Initial data are given for t_0 = May 15; the OSE dynamics are calculated for t_1 = September 15. The symbol "B" means "bottom"

Depth (m)	Biomass									
	B_1 (mg m^{-3})		B_2 (mg m^{-3})		B_3 (mg m^{-3})		B_4 (g m^{-2})		B_9 (mg l^{-1})	
	t_0	t_1	t_0	t_1	t_0	t_1	t_0	t_1	t_0	t_1
0	15	21	500	672	20	19	200	181	8	9.3
10	15	19	500	935	20	21	200	181	8	10.1
30	15	12	500	59	20	31	200	181	8	11.4
50	0	3	0	10	0	15	200	184	8	9.9
100	0	2	0	1	0	2	200	181	8	8.6
500	0	0	0	0	0	0	200	181	8	8.5
B	0	0	0	0	0	0	200	181	8	7.5

trations of all OSE elements have their maxima in the area with the highest density gradient, the magnitudes of these maxima have different values changing with time. Such results can be obtained for each point $(\varphi,\lambda) \in \Omega$. A comparison of these results with the data of Terziev et al. (1993) shows that the SMOSE can be a useful instrument for investigation of the OSE. Other examples demonstrating the possibilities of the model are shown in Figs. 7.6 and 7.7. A comparison of the benthos curves and biomass estimations by many authors in Fig. 7.6 indicates that the unit of the SMOSE describing the dynamics of the benthic community of the OSE does not simulate the real distribution of it in space. There are differences in the spatial variations of the biomass of the benthic

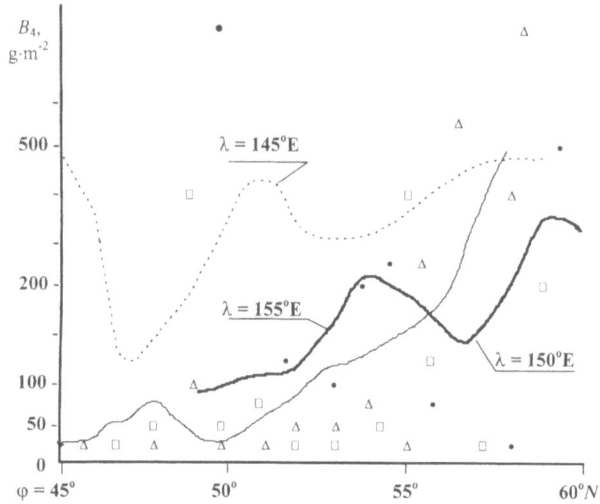

Fig. 7.6. Characteristics of the spatial distribution of the benthos biomass corresponding to benthos biomass estimation by the field experiments: *open squares* 145 °E, *open triangles* 150 °E, *closed circles* 155 °E

animals estimated by the model as compared with field observations (Anikeev and Obgirov 1993; Andreev and Gabin 2000). These variations are explained by an inaccuracy in the initial data and in the values of the SMOSE coefficients. Certainly, significant seasonal variations in the mean estimations of the SMOSE parameters are not approximated with necessary accuracy by the temporal scale adopted here based on the four seasonal periods. Shown in Fig. 7.7 is a numerical result demonstrating the model functioning for the simulation procedure of ice cover reconstruction. This map allows us to conclude that the SMOSE unit describing the ice formation process is rather accurate for simulating the general behavior of the Okhotsk Sea ice fields observed with satellite measurements.

To demonstrate the possibility of the SMOSE in studying the pollution processes an example is given in Fig. 7.8 where the dynamics of the oil hydrocarbons as a function of different processes are given. The simulation experiment shows that the intensity of the anthropogenic sources of the oil hydrocarbons (Q_o) is transformed to other forms by 47 % in the environments *g,r,f* and by 68 % in the water. The stabilization of the distribution of the oil hydrocarbons is reached 3.2 years after t_0. The hierarchy of flows H_O^i (i = 1,...,7) is estimated by the set of $H_O^2 > H_O^4 > H_O^1 > H_O^5 > H_O^7$. This set is changed for each of the OSE aquatories. The preponderance of either of the oil hydrocarbons destruction processes is defined by the seasonal conditions. The refining process at the expense of evaporation of the oil hydrocarbons (H_O^2) prevails over the other processes in the summer. In reality the oil hydrocarbons evaporated from the OSE surface with the atmosphere precipitation partly return to the Ω. The maximum destruction of oil hydrocarbons is 0.0026 g m^{-2} day^{-1}. This study indicates that the vegetative period affects a contribution to the self-clearing of the ecosystem. In the case considered, the OSE neutralizes about 25 % of oil hydrocarbons during the vegetative period. At other times, this value oscillates near 3 %.

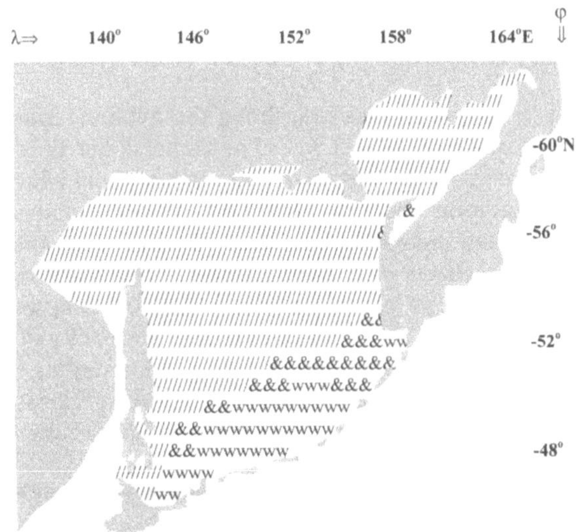

Fig. 7.7. Application of the SMOSE to the ice cover reconstruction in Ω. The input data of surface temperature are given by Shinohara and Shikama (1988); t_0 is November; the prognosis is realized to February. / Stable ice; & unstable ice; *w* water surface

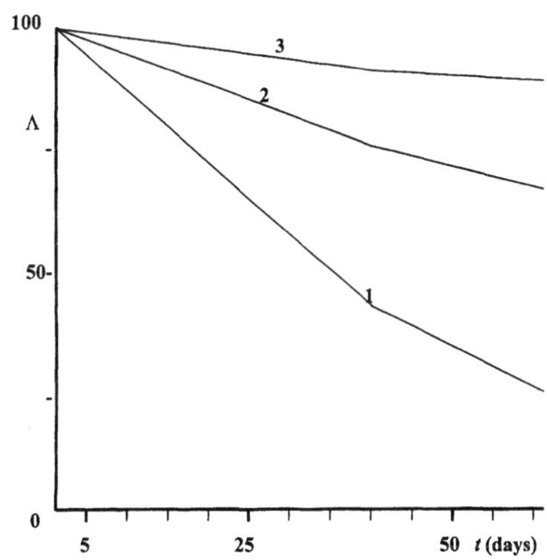

Fig. 7.8. Self-cleaning process in the OSE for the oil hydrocarbons. It is supposed that $O(t_0, \varphi,\lambda,z) = 0.05$ mg/m² when $z = 0$ and $O(t_0,\varphi,\lambda,z) = 0$ for $z > 0$; $H^1_O = 0.1$ mg m^{-3} day^{-1}; $H^2_O = 0$; $H^3_O = 0.01$ mg m^{-3} day^{-1}; $H^4_O = 0.02$ mg m^{-3} day^{-1}; $H^5_O = k_D (B_8)^{1/3}$ (k_D is the absorption coefficient equal to zero for $(\varphi,\lambda,z) \in \cup I_1 \cup I_2$ and to 0.005 day^{-1} when $(\varphi,\lambda,z) \in W$); $H^6_O = k_Z (B_3)^{1/4}$ (k_Z is the bio-sedimentation coefficient equal to zero for $(\varphi,\lambda,z) \in S \cup I_1 \cup I_2$ and to 0.004 day^{-1} when $(\varphi,\lambda,z) \in W$); $H^7_O = k_B B_1$ (k_B is the bacterial decomposition coefficient equal to 0.001 mg m^{-3} day^{-1} for $(\varphi,\lambda,z) \in \cup I_1 \cup I_2$ and to 0.005 mg m^{-3} day^{-1} when $(\varphi,\lambda,z) \in W$). Three causes are considered: *1* full speed of the OSE self-cleaning (t_0 = 5 May); *2* accumulation by ice only with the porosity equal to 80 cm³ kg^{-1} (t_0 = 15 November); and *3* decomposition by the bacterioplankton only (t_0 = 15 May). The ordinate Λ gives the oil hydrocarbon concentration relative to time zero in percent

7.6 The Biocomplexity Criteria and the Evaluation of the Okhotsk Sea Ecosystem

Biocomplexity is a complex index describing the dynamic interactions within the environmental system (see Sect. 3.6). The Okhotsk Sea Ecosystem is one of the significant elements of the biosphere, the evaluation of which demands the development of common criterion. The OSE biocomplexity index helps to explain many processes regulating the interactions between the biotic components, hydrodynamic effects and energy fluxes. Traditional estimates of the contributions from different processes within the OSE deal with the study of local or special parameters. This does not give the possibility of understanding the correlations between the OSE components and forecasting their dynamics. Moreover, it is convenient to have a simple index making it possible to evaluate the OSE state by means of ordinary calculations. In the common case, such an index is described in Section 3.6.

The OSE has a trophic graph with interactions at many levels between biological, chemical and physical processes. The OSE biocomplexity is formed by means of numerous sets of biotic regulations showing the fundamental properties of living objects. It is obvious that an important property of the OSE living components lies in the fact that all biological species exist in the form of populations. All processes and phenomena observed in the OSE are characterized by a certain degree of physical and biological stability which is a function of external and internal fluxes of energy. External fluxes of energy are defined by solar irradiation, influences of the Pacific Ocean and anthropogenic interventions. In the absence of an external flux of energy, the OSE tends toward a state of thermodynamic equilibrium which is characterized by the maximum degree of chaos possible in the given system (Gorshkov et al. 2000).

In either case, the OSE dynamics is a complex function of many parameters having different chemical, physical and biological characters. The biocomplexity index has to reflect this and to characterize the biological stability of the whole aquageoecosystem

The Okhotsk Sea belongs to those seas with high productivity, the ecosystem of which functions under a severe climate. The spatial-temporal structure fields of the basic hydrological and ecological characteristics of the Okhotsk Sea is heterogeneous. The chemical, physical and biological processes, occurring in the seawater, were studied by many authors to assess their bioproductivity. According to the investigations by Terziev et al. (1993), the following structural discretization of the Okhotsk Sea can be realized. Five ecological layers exist. Layer 1 is one of maximum photosynthesis. It is situated above the thermocline and has a depth about 20–30 m. It corresponds to the wind-mixed layer. Layer 2 occupies the water space from 30 to 150 m in depth. It has a low temperature and oxygen saturation of about 80–90%. Layer 3 is characterized by low oxygen saturation (15–20%). It lies between the depths of 150–750 m. Layer 4 extends from 750 m down to a depth of 1500 m. This layer has minimum oxygen saturation (10–15%). Lastly, layer 5 is located deeper than 1500 m. It is characterized by an oxygen saturation of 25–30%.

The Okhotsk Sea aquatory is divided into zones having specific ecological features (Suzuki 1992). The spatial distribution of the fish biomass depends on seasonal conditions and to a great extent correlates with the layers mentioned above. The use of the biological sea resources is a function of this distribution. The fishing intensity essentially depends on a knowledge of the biomass distribution in zones with specific environmental conditions. Various authors (Aota et al. 1992; Plotnikov 1996; Krapivin et al. 1990) have tried to solve this task by means of models simulating the ecosystem dynamics. However, the modeling results have not always turned out to be sufficiently representative and to reflect the classification of the sea zones by their productivity scale. The biocomplexity indicator is one of such simple forms to identify these zones. It was shown by many investigators that the Okhotsk Sea zones with high productivity are characterized by a complex, many-level trophic graph (Terziev et al. 1993). This effect is not universal to other seas. For instance, the Peruvian current ecosystem has high productivity in zones where the trophic graph is short (see Sect. 6.3). These situations are distinguished by the migration processes. That is why the biocomplexity of these ecosystems is formed in various ways.

Consider the following components of the Okhotsk Sea ecosystem mentioned in Tables 7.5 and 7.6. The trophic pyramid $X = ||x_{ij}||$, where x_{ij} is a binary value equal to "1" or „0" under the existence or absence of nutritive correlations between the ith and jth components, respectively. Define the biocomplexity function as:

$$\xi(\varphi,\lambda,z,t) = \sum_{i=1}^{20}\sum_{j=1}^{19} x_{ij}C_{ij} \tag{7.13}$$

where φ and λ are the geographical latitude and longitude; t is the current time; z is the depth; $x_{ij} = 1$ if $B_m \geq B_{m,\min}$ and 0 for $B_m < B_{m,\min}$ where $B_{m,\min}$ is the minimal biomass of the mth component consumed by other trophic levels; $C_{ij} = k_{ji}B_{i,}{*}/\Sigma_{j+}$ is the nutritive pressure of the jth component upon the ith component; $\Sigma_{i+} = \sum_{m\in Si} k_{im}B_m$ is the real food storage which is available to the ith component; $B_{m,}{*} = max\{0, B_m - B_{m,\min}\}$; $k_{im} = k_{im}(t, T_W, S_W)$ ($i = 1,\ldots, 17$) is the index of satisfaction of the nutrition requirements of the ith component at the expense of

Table 7.5. OSE components taken into consideration in the formation of the biocomplexity indicator

OSE component	Symbol	OSE component	Symbol
Phytoplankton	B_1	*Theragra chalcogramma*	B_{11}
Bacterioplankton	B_2	*Salmonidae*	B_{12}
Microzoa	B_3	*Coryphaenoides*	B_{13}
Herbivores	B_4	*Reinchardti ushippoglossoi des matsuurae*	B_{14}
Carnivores	B_5	*Clupeapallasi pallasi* Val	B_{15}
Zoobenthic animals	B_6	Crabs	B_{16}
Flat-fish	B_7	*Laemonema longipes*	B_{17}
Coffidae	B_8	Biogenic salts	B_{18}
Ammodytes hexapterus	B_9	Detritus	B_{19}
Mallotus	B_{10}	People	B_{20}

Table 7.6. Trophic pyramid of the Okhotsk Sea ecosystem taken into consideration in the formation of the biocomplexity indicator

Energy and matter consumers[a]	Energy and matter sources[a]																		
	B_1	B_2	B_3	B_4	B_5	B_6	B_7	B_8	B_9	B_{10}	B_{11}	B_{12}	B_{13}	B_{14}	B_{15}	B_{16}	B_{17}	B_{18}	B_{19}
B_1	0	0	0	0	0	0	0	0	0	0	0	0	0	0	0	0	0	1	0
B_2	0	0	0	0	0	0	0	0	0	0	0	0	0	0	0	0	0	0	1
B_3	1	1	0	0	0	0	0	0	0	0	0	0	0	0	0	0	0	0	1
B_4	1	1	0	0	0	0	0	0	0	0	0	0	0	0	0	0	0	0	0
B_5	0	1	1	1	1	0	0	0	0	0	0	0	0	0	0	0	0	0	1
B_6	1	1	1	1	1	0	0	0	0	0	0	0	0	0	0	0	0	0	1
B_7	0	0	0	0	0	1	1	1	1	1	1	1	1	1	1	1	1	0	0
B_8	0	0	0	0	1	1	1	1	1	1	1	1	1	1	1	1	1	0	0
B_9	0	0	0	0	0	1	1	0	0	1	1	1	0	0	0	1	0	0	0
B_{10}	0	0	0	0	1	1	1	0	0	0	1	0	0	0	0	0	0	0	0
B_{11}	0	0	0	0	1	0	0	0	0	1	1	1	0	0	0	0	0	0	0
B_{12}	0	0	0	0	0	1	1	0	0	1	1	1	1	0	0	1	0	0	0
B_{13}	0	0	0	0	1	1	1	1	1	1	1	1	1	1	1	1	1	0	0
B_{14}	0	0	0	0	1	1	1	1	1	1	1	1	1	1	1	1	1	0	0
B_{15}	0	0	0	0	1	1	0	0	0	0	0	0	0	0	1	0	0	0	0
B_{16}	0	0	0	0	0	1	0	1	1	1	0	1	0	0	0	1	1	0	0
B_{17}	0	0	0	0	1	1	0	0	0	0	0	0	0	0	0	0	1	0	0
B_{18}	0	0	0	0	0	0	0	0	0	0	0	0	0	0	0	0	0	0	1
B_{19}	1	1	1	1	1	1	1	1	1	1	1	1	1	1	1	1	1	0	0
B_{20}	1	1	1	1	1	1	1	1	1	1	1	1	1	1	1	1	1	1	1

[a] Definitions of the B_i are given in Table 7.5.

the mth component biomass; k_{im} ($i = 18,19$) is the transformation coefficient from the mth component to the ith component; k_{i20} is the characteristic of anthropogenic influence on the ith component; $S_i = \{i{:}x_{ij} = 1\ j = 1, ..., 19\}$ is the food spectrum of the ith component T_W is the water temperature; and S_W is the water salinity.

Designate the aquatory of the Okhotsk Sea by $\Omega = \{(\varphi,\lambda)\}$. The value of the biocomplexity indicator for any area $\omega \in \Omega$ is determined by the formula:

$$\xi_\omega(z_1, z_2, t) = (1/\sigma_\omega) \int_{(\varphi,\lambda)\in\omega} \int_{z_1}^{z_2} \xi(\varphi,, \lambda, z, t) d\varphi dz$$

where $[z_1, z_2]$ is the water layer located between the depths of z_1 and z_2.

The maximum value of $\xi = \xi_{max}$ (≈ 20) is reached during the spring-summer time when nutrition relations in the Okhotsk Sea ecosystem are extended, the intensity of energy exchanges is increased, and horizontal and vertical migration processes are stimulated. In the winter time the value of ξ is changed to near ξ_{min} (≈ 8). The spatial distribution of ξ reflects the local variability of the food spectrum for the components. Figure 7.9 and Table 7.7 show examples of such a distribution. Comparison of this distribution with the distribution of zones with industrial fish accumulations (Terziev et al. 1993) shows that there is a correlation between these distributions.

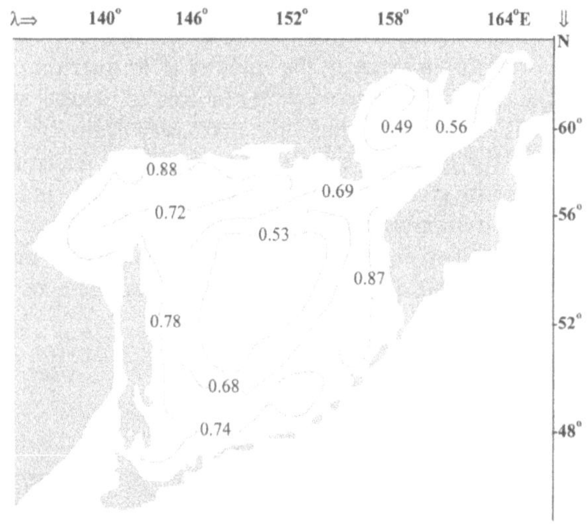

Fig. 7.9. Spatial distribution of the biocomplexity indicator $\xi^* = \xi/\xi_{max}$ for the spring–summer time

Table 7.7. Estimations of the biocomplexity indicator ξ^* for the different layers in spring-summer and in winter

Season	Layers				
	1	2	3	4	5
Spring-Summer	0.89	0.93	0.62	0.34	0.21
Winter	0.31	0.49	0.71	0.39	0.22

The indicator ξ reflects the level of complexity of the Okhotsk Sea ecosystem. A change in ξ is realized as a consequence of migration processes and the variability of the nutritive interactions. The subsystem B_{20} plays the role of an external source of change in the other components in these processes. These changes are interpreted in the terms of fishing and impacts causing variations of the components biomass.

Calculations show that the basic variability in $\xi^* = \xi/\xi_{\max}$ is caused by migration processes. Under this condition the quick redistribution of the interior structure of matrixes X and $||C_{ij}||$ occurs. For instance, according to Terziev et al. (1993), many fish migrate during spring time to the shelf zone and during winter time they move to the central aquatories of the sea. Therefore, the value of $\xi^* \to 1$ during spring and $\xi^* \to 0.6$ during winter for the shelf zone. This means that the biocomplexity of the Okhotsk sea ecosystem in the shelf decreases by 40% in winter in comparison with spring. For the central aquatories the value of ξ^* varies near 0.7 during the year. Such stability of the biocomplexity indicator is explained by the balance between nutrition correlations and productivity during spring, summer and winter seasons.

It has been established that variability in the ξ^* simulates changes of the fish congestions which are controlled by environmental conditions. Specifically, during springtime *Clupeapallasi escapes* occupies the area with $T_W < 5\,°C$. Other fish have an elective depth for their feeding and spawning (Terziev et al. 1993). All these processes influence the variability of ξ^*. A more detailed investigation of correlations between the value of ξ^* and structural and behavioral dynamics of the Okhotsk Sea ecosystem requires additional studies.

This section introduces the main idea of how to move from a verbal description of the biocomplexity to a numerical scale. In future studies it is necessary to take into consideration various factors such as bottom relief, climate trends, ice field dynamics, detailed components of trophic pyramid, bottom sediments, and current structure. In addition, it is necessary to add to formula (7.13) members describing anthropogenic impacts on the ecosystem considered in a socio-economic sense.

7.7 Concluding Remarks

Marine environmental research includes mathematical models as their main component giving the possibility to reconstruct the spatial image of the sea ecosystem on the basis of fragmentary information from observations. The findings from this study show that the SMOSE can be considered as such a unit in the case of the Okhotsk Sea environmental monitoring system. It is obvious the SMOSE is a simplified realization of a set of parametric descriptions of the OSE functions. There are many problems to be solved in the future. This study is an example of the first step in this direction.

The first problem for the future improvement of the SMOSE is the extension of the parametric description of energy and heat exchange between the OSE and the atmosphere to increase the model validity and to establish interrelationships between the model parameters and satellite measurement data (Sellers et al. 1995). The second task toward greater accuracy of the SMOSE consists in the detailed elaboration of the OSE elements in order to extend its applied significance to the fishing industry. The description in more detail of the OSE biological balance with the formation of industrial fish populations is a priority for future investigations. Finally, the third problem to improve the SMOSE touches upon the synthesis of an expert system in the structure of which the SMOSE will be the main unit.

CHAPTER 8

Pollutant Dynamics in the Angara-Yenisey River System 8

8.1 Introduction

The intensive industrial development of the northern Russian territories has led to significant environment changes in these regions (Morgan and Codispoti 1995). Oil and gas extraction on the Yamal and Taimir peninsulas in northwestern Siberia; coal and gold extraction in Yakutia and Chukotka, as well as mining industry operations on the Kola Peninsula, have led to strong anthropogenic intervention into the natural environment. Vegetation cover has been violated over large territories; the range, size, and productivity of reindeer pastures diminished; and the hydrologic regime of rivers disturbed. Major quantities of pollution substances are brought to the northern coast of Russia by rivers, thus violating the ecosystems of the northern seas. One such river system is the Angara-Yenisey river system (AYRS).

The Yenisey River flows northward to the Kara Sea along the boundary between the west Siberian flood plain and the central uplands, draining an area of about 2.6 million km^2 during its 4100-km length of travel. The flow rate of the Yenisey into the Kara Sea has large seasonal variations, averaging 19,800 $m^3 s^{-1}$, but as high as 130,000 $m^3 s^{-1}$ during the spring runoff (Gitelzon et al. 1985).

The Angara, a major tributary, accounts for about one fourth of the total flow. This river flows swiftly northward from Lake Baikal for about one third of its 1850-km length, before turning westward toward its confluence with the Yenisey. Recognizing that major sources of radionuclides, as well as other environmental pollutants found in the Kara Sea, might lie in the Siberian watersheds of the Yenisey and the Angara, a joint Russian-American expedition was undertaken in July and August of 1995. The region where the expedition was conducted contains five hydroelectric dams at Krasnoyarsk and Sayano-Shushenskoye on the Yenisey and at Irkutsk, Bratsk and Ust-Ilimsk on the Angara. The power output from these facilities has fostered rapid industrial growth in this region. Krasnoyarsk is the major industrial city located on the upper reaches of the Yenisey. Nearby is the nuclear production and processing facility, Krasnoyarsk-26, which is situated approximately 270 km upstream of the Angara-Yenisey junction. Along the Angara River, there are five cities with major industrial activities: Irkutsk, Angarsk, Usolye-Sibirskoye, Svirsk, and Bratsk. These cities have facilities producing both radionuclides and chemical pollutants that can contribute to the source terms in the two rivers.

Fig. 8.1. Map of central Siberia showing the sampling locations (indicated by *stars*): *a* Irkutsk, *b* Angarsk, *c* Bratsk, and *d* the Angara-Yenisey junction

Angara riverwater and sediment samples were taken beginning near its source at Lake Baikal and continuing at selected sites of interest downstream to its junction with the Yenisey near the lumber processing village of Strelka. Sampling was continued along the Yenisey River both upstream from just below the village of Kazachinskoe and downstream to the town of Lesosibirsk. Figure 8.1 shows the locations of the samples collected and analyzed on-site.

As shown in the paper by Krapivin and Phillips (2001a), the complex valuation of the pollution level in the Arctic basin as a whole is possible by the synthesis of a mathematical model of the pollutant transport by the rivers from adjacent territories. This chapter constructs and investigates a model of one such river system.

8.2 The AYRS Simulation Model (AYRSSM)

Following on from Chapter 5 of this volume, the block-diagram of the AYRSSM (AYRS simulation model) is represented in Fig. 8.2. The AYRSSM has three levels of units. A description of the units is given in Table 8.1. The two units of the first level act as a control for the models of different processes and ensure various regimes of the computer experiments. A unit HYD simulates the river flow with consideration of the correlation between the water regime and the ecological, topographical and synoptical parameters of the territory studied. A unit CON realizes the functions of the informational interface between the user and other AYRSSM units. The main function of CON is the formation of the database and

Table 8.1. List of the AYRSSM units in Fig. 8.2

Unit	Description
HYD	Model control of the hydrophysical and hydrological processes
CON	Interface control of the computer experiment
FLOW	Model of the river flow
STRM	Simulation procedure control of the water flows
EVAP	Model control of the evaporation processes
QUAL	Model control of the water quality
WRR	Model for description of the water regime in the reservoir (Shaw 1989)
EFM	Empirical flow model (Bras 1990)
RWS	Model of processes for spreading of the riverwater over the riverbank
INP	Model of infiltration processes (Bras 1990)
SPR	Model of surface flow with consideration of the role soil–plant formations
VMG	Model of vertical motion of the groundwaters under evaporation, drainage, exfiltration, and other water motions
FTM	Filtration model (Bras 1990)
EPM	Empirical precipitation model (Bras 1990)
THP	High precision transpiration model
KUM	Kuzmin's model (Kuzmin 1957)
TRM	Simple transpiration model (Bras 1990)
SES	Model of snow melt and evaporation from the snow
EEE	Empirical evaporation equations by Bras (1990)
KRP	Kohler's and Richard's modification of Penman's evaporation model (Bras 1990)
PEM	Penman's model (Bras 1990)
HOM	Horton's model (Bras 1990)
BIO	Simulation model of sedimentation of pollutants and biological assimilation (Shanahan et al. 2001)
POL	Simulation model of flow of pollutants from anthropogenic sources (Ward et al. 1990)
TWF	Model of water temperature formation (Bras 1990)
RAK	Model of radionuclide kinetics
CPK	Model of chemical pollutant kinetics
COD	Correction of the database
SCE	Search of the scenario for the computer experiment
VIP	Visualization procedures
AAS	Adaptation of the AYRSSM to the scenario
IAS	Identification of the AYRSSM with the spatial and temporal scales

Fig. 8.2.
Structure of the AYRSSM

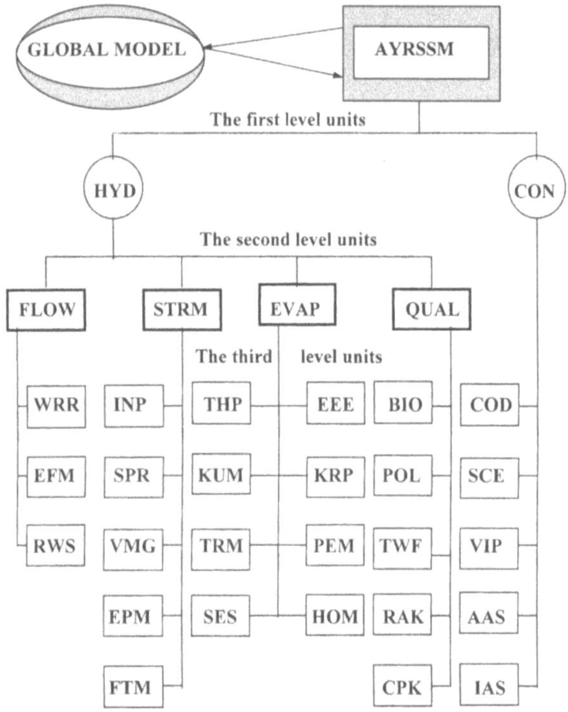

identification of the AYRS elements. The four units of the second level analyze the input information and synthesize the AYRSSM structure. The 26 units of the third level realize the concrete models and processes during the computer experiment.

It is supposed that the AYRS watershed has an area Ω. The spatial structure of Ω is determined by the spatial discretization of the AYRS surface via a uniform geographic grid with latitude φ and longitude λ, divided into steps of $\Delta\varphi$ and $\Delta\lambda$, respectively. In this study, it is supposed $\Delta\varphi = \Delta\lambda = 0.1$. As a result, the area $\Omega = \bigcup_{i=1}^{N} \Omega_k$, where Ω_k is the part of Ω with square area $\sigma_k = \Delta\varphi \cdot \Delta\lambda \left(\sum_{k=1}^{N} \sigma_k = \sigma \right)$. The cells Ω_k are situated along the AYRS beginning with Ω_1 at the Angara River source up to Ω_N at the Yenisey River mouth. The procedure of spatial discretization is provided for by the IAS unit via inclusion in the AYRSSM database the set of identifiers $A_k = \left\| a_{ij}^k \right\|$, k = 1,...,5 (Borodin et al. 1996; Krapivin et al. 1997b). The hydrology regime of the AYRS is described by the schematic diagram of Fig. 3.15. The equations for this scheme can now be written in the form of balance correlations on each of the Ω_k (k = 1,...,N):

$$\sigma_k \left(\frac{\partial W}{\partial t} + \xi_\varphi \frac{\partial W}{\partial \varphi} + \xi_\lambda \frac{\partial W}{\partial \lambda} \right) = V - B\sigma_k + D + T + L \tag{8.1}$$

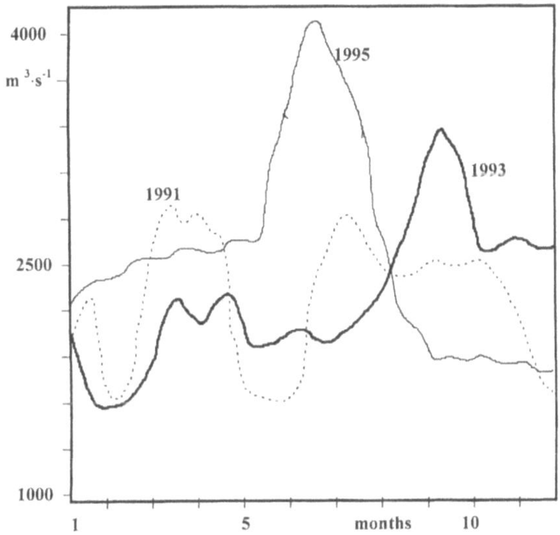

Fig. 8.3.
Annual flow rate through the Irkutsk dam for the years 1991–1995

$$\rho_k\sigma_k\left(\frac{\partial C}{\partial t}+\mu\frac{\partial C}{\partial x}\right)=\rho_k\sigma_k B+J+K-V-U-F-M-R \tag{8.2}$$

$$(1-\rho_k)\sigma_k\frac{d\Phi}{dt}$$
$$=U+F+M+N+(1-\rho_k)\sigma_k B-T-L-K-P \tag{8.3}$$

$$\sigma_k\left(\frac{\partial G}{\partial t}+v_\varphi\frac{\partial G}{\partial \varphi}+v_\lambda\frac{\partial G}{\partial \lambda}\right)=R+P-J-N-D \tag{8.4}$$

where ζ_φ and ζ_λ are the projections of the wind speed, ρ_k is the part of area Ω_k occupied by the river, μ is the river speed, v_φ and v_λ are the speed projections of the groundwater motion, x is the direction of river flow, and t is time.

The functions on the right part of Eqs. (8.1)–(8.4) are described by mathematical expressions in accordance with the papers by Bras (1990), Hong et al. (1992) and Krapivin et al. (1996b). Appropriate models are given in Table 8.1. There are many realizations for some of these functions. This gives the user of the AYRSSM the possibility to form scenarios for the computer experiments. Values of ξ, μ and v were estimated on the basis of the Irkutsk Scientific Center database. It is possible for the user to vary these parameters during the calculation process. In this study, average values of these parameters are estimated by $\xi = 3.3\ \text{m s}^{-1}$, $\mu = 1.7\ \text{m s}^{-1}$ and v = 0. Variations of the parameter μ are realized by adaptation of the left part of Eq. (8.2) to the empirical data illustrated in Fig. 8.3. Boundary conditions for Eqs. (8.1)–(8.4) are formed by the global model (see Sect. 3.2). Soil moisture transport between the cells Ω_k is neglected. Synoptical situations are described by a discrete scheme with temporal parameters $t_i (i = 1,...,4)$, where t_1 is the beginning of the summer period, t_2 is the start of winter, t_3 is the end of winter and t_4 is the time when the snow and ice are melting. Between these moments the synoptical situation does not change.

In the common case, the vertical structure of the river aquatory in Ω_k (k = 1,...,N) is described by unit SES. A snow layer of thickness g_k is formed at the expense of flow B_k:

$$\frac{dg_k}{dt}=\begin{cases}0 & when \quad t\notin[t_2,t_3],\\ B_k & when \quad t\in[t_2,t_3]\end{cases} \tag{8.5}$$

For $t\in[t_3,t_4]$ the value of $g_k(t,\varphi,\lambda)$ is decreased as a linear function from $g_k(t_3,\varphi,\lambda)$ to $g_k(t_4,\varphi,\lambda)=0$.

The functional representation of the other units from Table 8.1 is realized with the consideration of the moments $t_i(i=1,...,4)$ which are given by the user under the scenario realization.

The dynamics of pollutants in the AYRS are determined by the structure of its hydrological regime. This takes into account the transport of pollutants with water motion and the accumulation in sediments, ice, snow, and living biomass.

All of the pollutant types are divided into radionuclear elements and heavy metals. The set of radionuclear pollutants is described by the index ν. The set of heavy metals are divided into particles (index e) and the dissolved fraction (index ψ).

The ν'th radionuclide is characterized by the half-life t_ν, by the rates $f_{k\nu}$ of input to and output from the area $\Omega_k(k=1,...,N)$ and by the concentrations Q_ν, E_ν, Ξ_ν and S_ν in the water, soil, sediments, and groundwater, respectively. As a result, the main balance equations of the RAK unit are written in the form:

$$\Delta_k\left(\frac{\partial Q_{k,\nu}}{\partial t}+\mu_k\frac{\partial Q_{k,\nu}}{\partial x}\right)=H_{1,\nu}+H_{2,\nu}+H_{3,\nu}-H_{4,\nu}-H_{5,\nu}$$
$$+H_{6,\nu}-H_{7,\nu}+\rho_k H_{12,\nu}-\frac{\ln 2}{t_\nu}\Delta_k Q_{k,\nu}, \tag{8.6}$$

$$\delta_k\frac{\partial E_{k,\nu}}{\partial t}=H_{8,\nu}+H_{4,\nu}+H_{9,\nu}-H_{10,\nu}+(1-\rho_k)H_{12,\nu}-\delta_k E_{k,\nu}\frac{\ln 2}{t_\nu} \tag{8.7}$$

$$\psi_k\frac{\partial \Xi_{k,\nu}}{\partial t}=H_{5,\nu}-H_{6,\nu}-\psi_k\Xi_{k,\nu}\frac{\ln 2}{t_\nu} \tag{8.8}$$

$$y_k\left(\frac{\partial S_{k,\nu}}{\partial t}+v_\varphi\frac{\partial S_{k,\nu}}{\partial \varphi}+v_\lambda\frac{\partial S_{k,\nu}}{\partial \lambda}\right)$$
$$=H_{7,\nu}+H_{10,\nu}-H_{2,\nu}-H_{9,\nu}-H_{11,\nu}-y_k S_{k,\nu}\frac{\ln 2}{t_\nu}, \tag{8.9}$$

where $H_{1,\nu}$ is the amount of radionuclide washing away from the soil with flow K; $H_{2,\nu}$ is the radionuclide input to the river from the groundwater with flow J; $H_{3,\nu}$ is the transport of the radionuclide to the area Ω_k through boundary of Ω by the AYRS tributaries; $H_{4,\nu}$ is the radionuclide output with flows U, F and M; $H_{5,\nu}$ is the radionuclide sedimentation to the river bottom by gravitation and with dead biomass of the river ecosystem; $H_{6,\nu}$ is the washing away of the radionuclide from the sediments; $H_{7,\nu}$ is the radionuclide output with R flow; $H_{8,\nu}$ is the anthropogenic source of the νth radionuclide; $H_{9,\nu}$ is the radionuclide input to the soil from the groundwater by the N flow; $H_{10,\nu}$ is the radionuclide washing out to the groundwater by the P flow; $H_{11,\nu}$ is the radionuclide losses by sedimentation in the groundwater; $H_{12,\nu}$ is the radionuclide input by rain; $\Delta_k = \rho_k\sigma_k C_k$, $y_k = \sigma_k G_k \delta_k = l_k(1-\rho_k)\sigma_k$, $\psi_k = r_k\rho_k\sigma_k$, l_k is the thickness of the efficient soil layer on the Ω_k area and r_k is the thickness of the sediment layer.

The flows $H_{i,\nu}$ (i = 1,...,12) are parameterized by linear models according to the papers of Osterberg (1985) and Krapivin (1995). The BIO and CPK units are described by similar balance models in analogy with the models described by Nihoul and Smith (1976). The AYRS biology is given in the form of a scenario or is described by the model of Legendre and Krapivin (1992). The heavy metal flows include an assimilation of dissolved fractions by plankton $\left(H_\psi^Z\right)$ and by nekton $\left(H_\psi^F\right)$, the sedimentation of solid fractions $\left(H_e^1\right)$, the absorption from sediments by living elements $\left(H_{e.\psi}^L\right)$, sedimentation with the dead organic matter $\left(H_\psi^D\right)$ and the discharge (via erosion, diagenesis, turbulence, and anthropogenic impacts) from sediments $\left(H_{e,\psi}^a\right)$. The balance equations taking into account these flows have the same form as the Eqs. (8.6)–(8.9):

$$\Delta_k\left(\frac{\partial e_w}{\partial t} + \mu_k \frac{\partial e_w}{\partial x}\right) = \sum_{i=1}^{3} \alpha_2^i Q_{e,\psi}^i - H_e^1 + \alpha_1 H_{e,\psi}^a \tag{8.10}$$

$$\Delta_k\left(\frac{\partial \psi_w}{\partial t} + \mu_k \frac{\partial \psi_k}{\partial x}\right) = (1-\alpha_1)H_{e,\psi}^a - H_\psi^Z - H_\psi^D \tag{8.11}$$

$$\psi_k \frac{\partial e^*}{\partial t} = H_e^1 - \alpha_1\left(H_{e,\psi}^L + H_{e,\psi}^a\right) \tag{8.12}$$

$$\psi_k \frac{\partial \psi^*}{\partial t} = H_\psi^D - (1-\alpha_1)\left(H_{e,\psi}^L + H_{e,\psi}^a\right) \tag{8.13}$$

where $e_w(e*)$ and ψ_w $(\psi*)$ are the heavy metal concentrations in water (sediments) as solid and dissolved phases, respectively; $Q_{e,\psi}^i$ is the heavy metal input with AYRS tributaries ($i = 1$), by atmosphere precipitation ($i = 2$) and under industrial wastes ($i = 3$); α_2^i is the part of the solid particles in the ith flow of heavy metals; and α_1 is the part of the solid fraction in the bottom sediments. The removal of heavy metals from water by evaporation and sprays is neglected.

Approximate solutions of the initial value problem for Eqs. (8.1)–(8.13) are realized by means of the quasi-linearization method (see Sect. 3.7.4).

8.3 On-Site Measurements

For obtaining the data for the AYRSSM database, the joint US/Russian expedition to Siberia's Angara and Yenisey Rivers was realized in the summer of 1995 (Phillips et al. 1997). Some expedition results are given in Tables 8.2–8.5. The results of the radionuclide analysis are given in Table 8.2. Results of the elemental analysis of samples brought back to the laboratory are given in Tables 8.4 and 8.5. The results for each of the two categories of pollutants are discussed below.

8.3.1 Radionuclides in River Sediments

The man-made radioisotope ^{137}Cs was detected in all samples analyzed from above the Irkutsk dam, with concentrations ranging from 2 to 12 Bq kg^{-1} (dry weight). These values are consistent with background levels that can be expected from global fallout. Below the Irkutsk dam in the vicinity of the cities of Irkutsk and Angarsk, the measured ^{137}Cs concentrations ranged from <4 to 30 Bq kg^{-1}, indicating some of the samples contained ^{137}Cs concentrations somewhat higher than global fallout. Only background levels of ^{137}Cs activity were detected in samples taken near Bratsk.

The ^{137}Cs concentrations in samples taken on the Angara River upstream of the AYRS junction were determined to be at background levels, about 2 Bq kg^{-1} of dry river sediment. Downstream from the junction, the ^{137}Cs activity concentration in the Yenisey River samples ranged from 3 to 27 Bq kg^{-1}. The higher

Table 8.2. Results of on-site radionuclide measurements in river sediment (weighed average isotopic activity, Bq/kg dry weight). The ID# is given in Table 8.3. Errors given are ± 2 standard deviations. The sign "<" indicates less than the minimum detectable concentration given at the 90% confidence level

ID#	^{60}Co	^{137}Cs	^{152}Eu	^{235}U	^{238}U
Angara River					
I1	< 2.0	2.2 ± 1.0	< 3.1	< 1.9	< 280
I2	< 3.5	< 3.6	< 5.3	27.2 ± 4.6	720 ± 300
I3	< 3.7	30.6 ± 2.6	< 5.1	< 3.7	< 580
A1	< 2.0	< 1.8	< 3.1	< 1.7	< 320
A2	< 4.1	25.2 ± 3.1	< 6.0	< 4.2	< 630
B1	< 2.3	< 2.3	< 3.9	< 2.2	< 420
B2	< 2.6	3.4 ± 1.2	< 4.0	< 2.5	< 440
J1	< 1.9	2.2 ± 1.0	< 2.8	< 1.7	< 290
Yenisey River					
J2	8.6 ± 1.9	22.9 ± 2.3	6.7 ± 3.0	< 2.7	< 500
J3	241 ± 11	392 ± 12	151 ± 27	< 8.2	< 1340
J7	30.1 ± 3.0	203 ± 5	42.3 ± 9.5	< 3.8	< 680
J4	96.9 ± 5.1	211 ± 6	55.7 ± 13.5	< 4.5	< 840
J6	< 5.7	27.1 ± 3.8	< 8.2	< 6.4	< 860

Table 8.3. Location of the measurements in Tables 8.2 and 8.4

ID#	Location description
I1	Angara reservoir above the Irkutsk dam
I2	Angara River below the Irkutsk dam
I3	Angara River below Irkutsk
I4	Angara at the village Bolshaya Rechka (near the Lake Baikal outlet)
A1	Drainage ditch at Angarsk, „technical canal“
A2	Angara above Angarsk
A3	Kitoy River near Angarsk (Angara tributary)
B1	Angara at „Bratsk Sea“ (reservoir above the Bratsk dam)
B2	Angara below the Bratsk dam
J1	Angara above the Yenisey-Angara junction, near the village Kulakovo
J2	Yenisey above the junction, near the village Kazachinskoye
J3	Yenisey near Kazachinskoye
J4	At the junction, near the village Strelka
J5	Yenisey above the junction, below Kazachinskoye
J6	Yenisey below the junction, below the town of Lesosibirsk
J7	Yenisey above the junction, below Kazachinskoye

Table 8.4. Laboratory analytical results of heavy metal concentration in sediment and river-water. The concentration values are within one sigma uncertainty. The sign "<" indicates less than the minimum detectable concentration given at the 90% confidence level

ID#	As	Cd	Cr	Cu	Ni	Pb	Zn
River sediment samples (ppm)							
I4	7.2	<0.52	26	19	25	15	64
I1	5.0	<0.51	41	22	44	11	55
I2	4.5	<0.51	20	11	21	9	17
I3	8.1	<0.52	54	42	60	20	88
A1	0.9	<0.50	7.0	11	9.3	7.9	32
A2	4.3	<0.51	48	38	40	14	86
A3	2.4	<0.50	27	20	37	7.8	40
B1	<0.5	<0.50	6.4	9	12	2.3	22
B2	3.4	<0.50	31	210	39	6.1	50
J1	3.2	<0.50	14	13	18	4.5	32
J4	5.1	<0.51	47	35	37	14	100
J6	6.9	<0.51	34	31	36	17	100
J3	5.8	<0.51	54	43	390	18	150
J5	2.2	<0.50	18	8.2	20	3.9	47
Riverwater samples (ppb)							
I2	12	<0.50	<10	<20	<20	<5.7	120
I3	10	<0.50	<10	<20	<20	<5.7	92
A1	12	<0.50	<10	<20	<20	<5.7	240
B1	16	<0.50	<10	<20	<20	<5.7	87
J1	12	<0.50	<10	<20	<20	<5.7	31
J6	13	<0.50	<10	<20	<20	<5.7	92
J2	8.6	<0.50	<10	<20	<20	<5.7	140

Table 8.5. Comparison of 1995 expedition laboratory analytical results (ppm) to Russian archived Angara River water quality data at three sampling locations on the Angara River (nm, not measured; na, instrument detection limits not available). T1 = $[t_1, t_2]$, T2 = $[t_2, t_4]$

Data type	Measurements performed for Irkutsk Medical Inspection Service – Currently available archived data						US EPA certified laboratory analytical results
Year	1978		1979		1980		1995
Season	T1	T2	T1	T2	T1	T2	August
Location	Ershovsk water collector (above Irkutsk Dam)						ID# I1
Cu	0.002	0.002	0.003	0.002	0.001	0.003	< 0.02
Ni	na	na	nm	0.003	0.001	0.002	< 0.02
Location	Water collector near Sukhovskaya station above Angarsk						ID# A1
Zn	na	na	0.0053	na	na	na	0.24
Cn	0.001	0.002	0.004	0.007	0.004	0.007	< 0.02
Ni	nm	0.001	0.003	0.002	0.002	0.0005	< 0.02

value is 10–15 times the activity levels detected on the Angara River upstream of the river junction at Strelka. However, significantly higher than background levels of fission products, ^{137}Cs and ^{152}Eu, and neutron activation products, ^{60}Co, were detected in Yenisey River samples drawn from upstream of the AYRS junction: ranging from 9 to 240 Bq kg^{-1} for ^{60}Co; from 14 to 400 Bq kg^{-1} for ^{137}Cs; and from 7 to 150 Bq kg^{-1} for ^{152}Eu. These are unmistakable indications of nuclear reactor products, and are consistent with releases from the plutonium production reactors at Krasnoyarsk, approximately 270 km upstream on the Yenisey River. The large variations in activities were due to sampling locations, with the higher values coming from backwater or flood-plain areas with thick mud deposits and the lower values coming from sandy sediment where the current was swifter.

Above background levels of ^{235}U and ^{238}U were observed in several samples taken just below the dam at Irkutsk. The values observed for ^{238}U have large uncertainties due to its relatively weak gamma-ray emission. The ratios observed for ^{235}U/^{238}U are larger than the natural abundance of 0.7% and are outside the given uncertainties. However, these are counting errors only and do not include uncertainties in the interference between the nearly identical in energy 186 keV gamma rays of ^{235}U and ^{226}Ra and possible geometry errors due to nonuniformity of the samples. These samples were taken in an area undergoing active landfill. As a result, uranium-containing fills could have been brought in from elsewhere.

8.3.2 Heavy Metals in River Sediments

Using the X-Ray Fluorescence (XRF) unit as a screening instrument, a set of heavy metals were identified in all riverbank and river sediment samples collected during the expedition. After the expedition, 21 river sediments and 8 river water samples were sent for trace element analysis to the Analytical Services Center of Ecology and Environment, Inc. (an EPA-certified laboratory), located at Lancaster, New York. Part of the analytical results are shown in Tables 8.4 and 8.5. Five heavy metals (Cd, Cr, Cu, Ni, Pb) are identified in all the samples at concentrations which are within the usual ranges of background levels. Two heavy metals, As and Zn, were identified in the river water samples at concentrations much higher than the median for natural fresh water (Lal and Stewart 1994). In fact, the Zn concentration in four of the samples exceeded the range for natural fresh water. These water samples were collected near to or downstream of industrial complexes on both the Angara and Yenisey Rivers.

Table 8.5 gives a comparison of the US/Russian expedition results with archived water quality data from Irkutsk, Angarsk, and Bratsk performed for the Irkutsk Medical Inspection Service (Svender G.M., and Krechetov A.A., Chemical Department, Irkutsk State University, Russia, pers. comm.). At all locations, the expedition samples indicate an increase in heavy metal concentration levels from previous Russian measurements.

8.4 Experiments Using the AYRSSM

The AYRSSM database includes the estimates of the model coefficients, the initial information for the climatic and anthropogenic scenarios and a set of identifiers $\{A_i\}$ describing the Ω area by its boundaries, thus giving the structure of the land-water surface. The CON unit provides an interface with the database and allows the user to modify its elements.

It is assumed that the vertical distribution of pollutants in the river water is homogeneous, the pollution of the soil and plants is negligible, and the cleaning process of the atmosphere has an exponential character with half-time τ_a ($\tau_a =$ 10 days when $t \in [t_1, t_2]$, $\tau_a = 20$ days when $t \in [t_2, t_4]$). It is believed that the pollution sources located on the Ω territory support a constant level of pollutant distribution. The distribution function is uniform on the intervals shown in Table 8.6. The values of ξ, μ, and ν were determined by average estimates for the last 3 years on the basis of published data (Rovinsky et al. 1995) and data from the Irkutsk Scientific Center. The AYRS slope is taken equal to 0.21 m km^{-1}. Ecological elements of the AYRS are considered as unique levels with biomass $\beta(t)$ having a constant value at each of the intervals $[t_i, t_{i+1}]$ (i = 1,2,3). The equation for $H_{5,v}$ is written in the form: $H_{5,v} = [g_1 + m_1 m_2 \beta] Q_{k,v} \Delta$ where g_1 is the gravitation coefficient (0.05 day^{-1}), m_1 is the mortality coefficient (0.01 day^{-1}), m_2 is the pollutants capture coefficient (0.03 kg^{-1}).

Table 8.6 gives an example of using the AYRSSM for the estimation of the ^{137}Cs distribution in the Ω area where cells Ω_i are situated along the AYRS with step of discretization $\Delta x = 10$ km. The ^{137}Cs distribution along the river system has one

Table 8.6. Relative concentration of ^{137}Cs in the water (γ_w) and in the sediments (γ_d) normalized to the concentration at x = 0

Distance from Lake Baikal (km)	Time from the beginning of the computer experiment (days)							
	30		60		90		120	
x	γ_w	γ_d	γ_w	γ_d	γ_w	γ_d	γ_w	γ_d
0	1.0	1.0	1.0	1.0	1.0	1.0	1.0	1.0
250	8.2	20.4	6.7	18.7	7.4	17.8	8.3	17.1
500	9.5	14.4	9.4	12.3	9.3	15.3	8.4	14.9
1000	10.3	4.1	8.8	9.8	7.8	11.6	7.9	10.3
1500	9.7	3.0	10.2	5.2	6.3	8.5	4.8	7.6
2000	8.9	2.9	6.9	2.9	4.9	4.1	4.5	2.8
2500	6.5	2.3	5.9	2.2	3.1	2.4	3.1	1.8
3000	3.2	1.9	2.7	1.7	2.9	1.9	3.0	1.4
3500	2.5	1.6	1.8	1.1	1.8	1.2	2.4	1.1
4000	1.7	1.1	1.6	0.9	1.8	1.2	1.4	1.1

maximum, the dynamics of which are characterized by changes in the value and coordinate. A similar picture is observed for other radionuclides from Table 8.2. Such a result is not explained by the location of radionuclear pollutants sources only. It is possible the variations in the value and position of maximal ^{137}Cs concentration are caused by the high turbulence in the river system and by the existence of reservoirs and eddies. The AYRSSM gives only averaged results. The fourth level units in the AYRSSM structure are necessary for the description of hydrological processes with $\Delta \times < 100$ m.

The computer experiments show that the input of radionuclear pollution to the Kara Sea from the Yenisey River has a stable value with dispersion equal ±32%. The role of the AYRS ecosystem in the process of transformation of radionuclear pollution is neglected as small (< 3%). The percent of vertical transport by organisms varies from 0.1 to 0.7%. Such calculations can be accomplished in the framework of different scenarios.

Figure 8.4 shows the distribution of heavy metal concentration along the AYRS reconstructed by means of the computer experiment. We see that there are three maxima of heavy metal concentration located at distances from Lake Baikal equal to 200, 1200, and 2000 km. This is the result of the distribution of pollution sources along the river system. The AYRS neutralizes the pollutants over a distance of 600–1000 km from the source. The location of the maxima varies as a function of the river flow rate. For the data of Fig. 8.3 such variation is 150 km. The river system transforms the pollutant flow such that the input into the Kara Sea is estimated by a pollution level which is less than 2% of the maximal concentration of pollutant in the AYRS.

The AYRSSM allows one to estimate the dependence of the pollution level in the AYRS estuary as a function of anthropogenic activity. Suppose that the intensity of the heavy metal sources is such that its concentration in the water near Angarsk, Irkutsk, Krasnoyarsk, Bratsk, and Ust-Ilimsk is described by a stationary function supporting the heavy metal concentration at the level h at

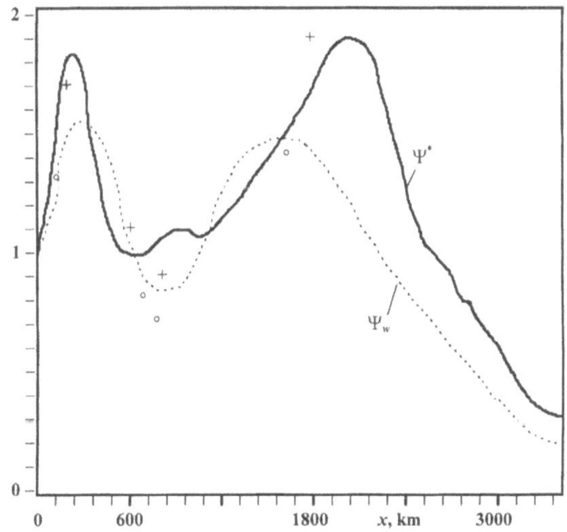

Fig. 8.4.
Distribution of the concentration of heavy metals in the water (*dashed line*) and in the sediments (*solid line*) as a function of distance x from Lake Baikal. The signs o and + correspond to the measured concentrations of metals in the water and sediments, respectively

each city mentioned above. Computer experiments show that there is a stable correlation between h, the heavy metal concentration in the AYRS estuary and the water flow rate μ. An increase in h of 10% leads to a rise of the pollution input to the Kara Sea by 2.5%. An increase in μ of 1% leads to a rise of the pollution input to the Kara Sea by 0.7%. These results are correct only when values μ and h are varied close to their average estimates. In the critical situations there are unstable estimates, and more detailed models are required.

8.5 Concluding Remarks

This chapter demonstrates the possibility of using the modeling technology in the solution of complex environmental problems demanding the combination of knowledge from different scientific fields. The AYRSSM is one example of such a combination. It gives typical elements for the synthesis of a simulation system for investigation of Arctic basin pollution (see Chap. 5).

The results given in this chapter illustrate functional features of the mesoscale simulation experiment. It is obvious that the strategy of the modeling technology is in the interplay of model calculations and on-site experiments. In the case considered, such a strategy secures economical profit as it reconstructs the distribution of pollutants along the AYRS and provides estimates of environmental consequences under the realization of scenarios.

A joint US/Russian expedition to Siberia's Angara and Yenisey Rivers detected man-made contaminants in water and sediment samples from both industrial regions and wilderness areas. On-site analysis using sensitive instrumentation revealed radionuclides, heavy metals, and volatile organic compounds. The results indicate that the nuclear production facility at Krasnoyarsk on the Yenisey River introduces radioactive contamination far downstream and is a probable source of previously detected radioactivity in the Yenisey estuary at its outlet into the Kara Sea.

The AYRSSM is a complex system having a hierarchical structure with natural and anthropogenic elements. This study of the AYRS is the first step where simulation results are based upon large-scale on-site measurements. The method used in this chapter can be applied for investigation of other northern river systems. However, to expand the experimental base, remote sensing technology must be used. Remote monitoring can give more precise information on the identifiers structure A_2 (water system) and A_3 (soil–plant formations). The authors hope for the continuation of this study. Based on the database created, it is planned to prepare a complete set of models and their corresponding software for the process of description of the transfer and transformation of pollution substances in the Arctic natural ecosystems. This set will demand the synthesis of models for the kinetics of radionuclides and chemical compounds in the food chains of the water and land ecosystems for the boreal zones and for modeling of the hydrological regime and estimation of the pollutant flows into the Arctic basin.

A final aim of the investigations is the development of an environmental technology which will be the main result of the cooperative work of scientists directed to the evaluation of the Arctic ecosystem state. The system is to be provided with a sufficiently full database to have at its disposal a ramified informational measuring network and a complete set of computer models for the main biogeochemical, climatic, and biogeocenotic processes (Nihoul and Smith 1976; Lal and Stewart 1994; Wielgolaski 1997; Rosenberg et al. 1999).

CHAPTER 9

Realization of GIMS Technology for the Study of the Aral-Caspian Aquageosystem 9

9.1 The Nature of the Problem

One of the dramatic aspects of anthropogenic activity is its influence on the biosphere water cycle (see Sect. 3.4). Presently, this influence occurs on a global level and is composed of a hierarchy of regional changes, especially in the arid districts. The control of the biosphere water systems is one element of climate system monitoring. The Aral-Caspian system (ACS) attracts a great deal of attention of scientists as an economically and ecologically significant subsystem of the biosphere which has been subject to rapid changes due to human activities. The catastrophic state of this water system is well known (Precoda 1991; Klotzi 1994; Hublaryan 1995; Pearce 1995).

In the opinion of many authors, the recent anomalous increase in the sea level of the Caspian Sea and the decrease in the sea level of the Aral Sea have global implications. According to the data of Bortnik and Chistyakova (1990), the sea level of the Aral Sea remained fairly stable, fluctuating between 50 and 53 m with inter-year deviations up to 1 m during the last 200 years prior to 1960. At that time, the Aral Sea area was $(51–61) \times 10^3$ km^2. About 55 km^3 year^{-1} evaporated from the surface, precipitation on the surface was 9–10 km^3 year^{-1}, and river inflow was between 33 and 64 km^3 year^{-1}. Some estimations of the Aral Sea water regime and other parameters are given by Bortnik and Chistyakova (1990).

A steady drop in the level of the Aral Sea began in 1961 at a rate of 21 cm year^{-1} up to 1970, then increasing to 58 cm year^{-1} until 1980 and nearly 80 cm year^{-1} after 1980. The Aral Sea level dropped from 53 m in 1960 to 41.4 m in 1985 and by 1994 had fallen 13 m from its level in 1960.

The stability of the Caspian Sea has also changed dramatically in recent years. In earlier times when anthropogenic interference in the water regime of the Middle Asia region was insignificant, the sea level of the Caspian Sea fluctuated not far from −28 m. Its area was 375,000 km^2 and evaporation from the surface was close to 380 km^3 year^{-1}. Some data of the Caspian Sea water balance are given by Terziev (1992) and data from various sources are summarized by Golitsyn (1995).

Detailed measurements of the sea level of the Caspian Sea exist from 1837 to the present. During this time, the changes in sea level can be characterized by four distinct periods:

1. During the period 1837–1928, the level fluctuated between about – 25.5 and – 26.5 m.
2. During the years 1929–1941 the level dropped rapidly from –25.96 to –27.84 m at a rate of about 16 cm $year^{-1}$.
3. During the years 1942–1977, the level dropped at a more moderate rate of 3.3 cm $year^{-1}$ to a low mark of – 29.04 m.
4. Since 1977 the level has risen at a rate of 16–18 cm $year^{-1}$.

By 1995 the sea level of the Caspian Seahad risen by 3 m to return to its pre-1929 level and the area had increased to 440,000 km^2.

The unstable hydrometeorological situation at the present time in the ACS makes the task of finding new methods for stabilizing the system urgent, including the quest for new methods of halting the above catastrophic tendencies, and for the synthesis of new modeling algorithms of the ACS which can help bring about a changeover to a controlled stable state. This chapter proposes an approach to the solution of this task. The proposed structure of the Aral-Caspian expert system (ACES) for the influence zone of the Aral and Caspian Seas is given (Fig. 9.1). This system incorporates the use of GIMS technology (see Sect. 2.2). It has remote monitoring and simulation units which function to alternately switch between measurements and prognosis procedures. Temporal parameters of the switching process are determined by means of empirical procedures (Krapivin et al. 1996a, b).

For this analysis and study, the area Ξ is chosen, limited by the geographical coordinates of (41–47)°N × (50–70)°E. Within Ξ the water regime is described by the ACS theory–information model (TIM). Outside the area Ξ it is described by the Nature–Society (NS) model (see Sect. 3.2).

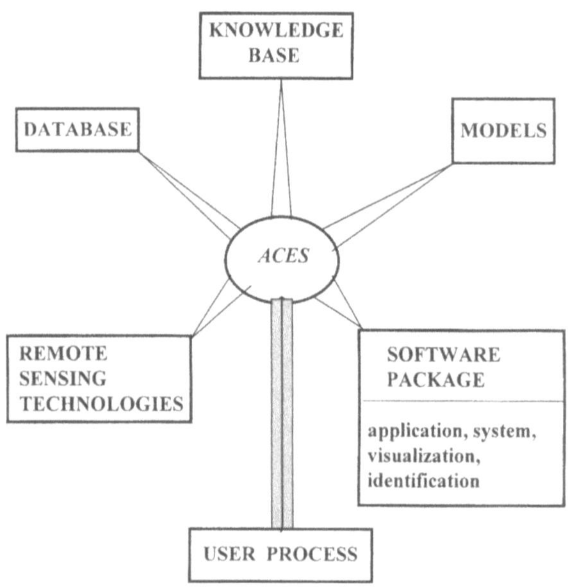

Fig. 9.1. Schematic of relationships between the elements of the Aral-Caspian expert system (ACES)

9.2
Remote Monitoring Database

The ACS territory was studied during the last 25 years by means of in-flight laboratories of the Institute of Radio Engineering and Electronics of the Russian Academy of Sciences. The measurements were obtained by microwave radiometry at wavelengths μ = 0.8, 1.35, 2.25, 3.4, 10, 18, 20, 21, 27, and 30 cm. It was shown that these microwave measurements enable the reliable classification of this area for land cover, indication of groundwater level, and the estimation of water content in the atmosphere over the ACS area (Krapivin and Mkrtchyan 1991; Shutko 1987; Shutko et al. 1994; Kutuza et al. 1998; Krapivin 2000). On the basis of these measurements carried out in central Asia, a remote database was formed with a three-level structure:

1. position measurements,
2. data processed by means of various algorithms (spectra, statistical characteristics, classification of subjects, correlation models),
3. data maps obtained using spatiotemporal interpolation methods.

All data are organized as shown in Fig. 3.9 after its processing with algorithms described in Section 3.7. The database levels were created by data processing algorithms from the results of the radiophysical monitoring (Krapivin et al. 1991, 1994). Brightness temperature contrast measurements were performed from aircraft over regions of central Asia at cm and dm wavelengths. Essential negative brightness temperature contrasts were observed in flights over moist areas. The measurements obtained at wavelengths of 0.8, 3.4, 10 and 20 cm indicated a clear correlation between soil moisture content and brightness

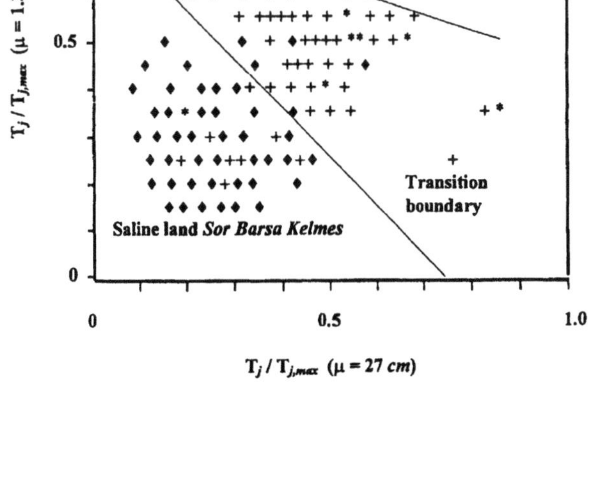

Fig. 9.2. Profile of the state system image for the *Sor Barsa Kelmes* saline land – *Ustyurt* plateau – formed on the basis of the IL-18 inflight laboratory measurements. *Closed diamonds* Saline land, *asterisk* plateau, *cross* the boundary between them

Table 9.1. An example of the reconstruction of the radiobrightness temperatures T_j with different algorithms (see Sect. 12.8) using the radiophysical measurements made along the specific route of the in-flight laboratory IL-18 in the area of the Sarykamysh reservoir (42°N, 57°E)

Measured value T_j (K) (μ = 1.35 cm)	Reconstructed value and error			
	Method of differential approximation	Error (%)	Method of data reconstruction with harmonic functions	Error (%)
247.72	324.29	31	210.67	15
249.35	316.58	27	212.06	15
150.00	172.50	15	174.01	16
229.00	190.07	17	256.48	12
243.92	217.19	11	209.92	14
139.25	164.27	18	157.76	9
234.14	203.72	13	229.46	2
248.28	196.20	21	230.92	7
152.50	181.38	19	172.26	13
223.59	268.19	20	245.89	10
234.64	194.86	8	260.38	11
223.59	279.34	25	234.74	5
235.80	188.86	20	274.65	9
244.82	198.46	19	257.02	5
258.69	181.29	30	240.63	7
141.88	164.44	16	157.39	11
252.00	264.69	5	221.76	12
262.08	288.28	10	222.78	15
252.63	272.79	8	229.95	9
146.60	175.82	19	162.66	11
249.27	199.47	20	256.74	3
257.34	226.50	12	236.78	8
258.00	221.88	14	283.81	10

temperatures (Engman and Chauhan 1995). Conducting measurements of the same regions over a prolonged time period allowed the observation of brightness temperature variations for different fields indicating an increase and decrease in moisture content.

An example of the discrimination of land cover by remote sensing is given in Fig. 9.2. This figure plots the ratio of brightness temperatures $T_j/T_{j,\max}$ at $\mu = 1.35$ cm versus the ratio at $\mu = 27$ cm and shows a clear separation of measurements made over the Ustyurt desert plateau between the Caspian and Aral Seas from those made over the saline land Sor Barsa Kelmes bordering the Aral Sea.

The data convincingly demonstrate the possibility of spatiotemporal moisture content measurements by means of microwave radiometry (Engman and Chauhan 1995). The database of the Institute of Radio Engineering and Electronics includes data obtained by radar, radiometry, opto-electronics and optical measurements. The database has sets of parameters for salt-marshes, water surfaces, watersheds, waterways of waste and drainage waters and irrigation nets as well as for processes of filtration, waterlogging, salting, bogging up, and desertification. All of the data were formed using a set of specific algorithms. Some of these are described in Section 3.7. An example of the spatiotemporal reconstruction of these data is given in Table 9.1.

9.3 Theory-Information Model of the Aral-Caspian Aquageosystem

The basic elements of the ACS water balance are given in Fig. 9.3. The functional representations of the water flows H_i ($i = 1,...,38$) are given by Bras (1990). The water flows having anthropogenic origin are described with simple models: $H_{15} = f_{15}G$; $H_{11} = f_{11}G$, where G is the population density; f_{11} and f_{15} are proportionality coefficients ($f_{11} \approx 0.25$ m^3 man^{-1} day^{-1}; $f_{15} \approx 0.85$ m^3 man^{-1} day^{-1}). Flows H_{22} and H_{23} are formed as functions of the soil-plant formation type (see Sect. 3.4). The water flows not included in this scheme are considered negligible.

A computer realization of this scheme is based on the division of the study area Ξ into discrete elements Ξ_{ij} with area $\sigma_{ij} = \Delta\varphi_i \cdot \Delta\lambda_j$. The $\Xi_{ij} \subseteq \Xi$ are characterized by specific soil-plant formation, geophysical structure and socio-economic parameters. Each element Ξ_{ij} has a subset of the water flows $\{H_k\}$ and

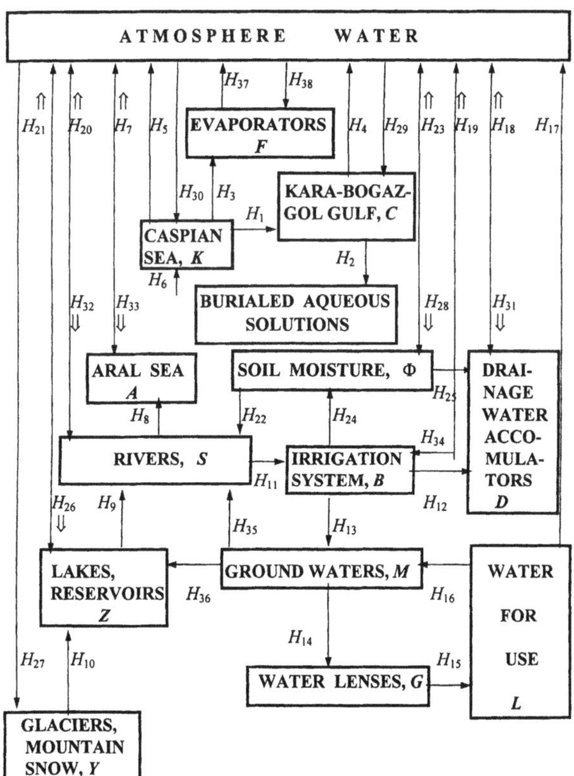

Fig. 9.3. Block diagram of the ACS water balance as represented in the framework of the TIM

Table 9.2. Set of the TIM database identifiers

Identifier	Identifier description
A_1	Categorization of Ξ; $A_1 = \|\| a^1_{ij} \|\|$, $a^1_{ij} = 0$ when $(\varphi,\lambda) \notin \Xi$, $a^1_{ij} = a_1$ when $(\varphi,\lambda) \in \Xi$; $a_1 = 1$ for the Aral Sea, 2 for the Caspian Sea, 3 for the Kara-Bogaz-Gol gulf, 4 for saline land, 5 for rivers, 6 for canals, 7 for irrigation systems, etc.
A_k	Average statistical data of wind speed ($k = 2$), atmosphere temperature ($k = 3$), precipitation ($k = 4$) and wind direction ($k = 5$)
A_6	Initial data for the all input parameters
A_7	Elevation of Ξ (absolute level)
A_8	Classification of Ξ_{ij} as an artificial evaporator ($a^8_{ij} = 1$ for "yes", $a^8_{ij} = 0$ for "no") and as an area for forced precipitation ($a^8_{ij} = 2$ for "yes", $a^8_{ij} = 0$ for "no")

associated data given by identifiers $\{A_s\}$ which are described in Table 9.2. The set of A_s exhibits an information database structure and forms the elements of the water balance equations for the water volumes W_{ij}:

$$\sigma_{ij}\left(\frac{\partial W_{ij}}{\partial t} + v_\varphi \frac{\partial W_{ij}}{\partial \varphi} + v_\lambda \frac{\partial W_{ij}}{\partial \lambda}\right) = \sum_{s \in I_{ij}} H^{ij}_s - \sum_{s \in J_{ij}} H^{ij}_s \tag{9.1}$$

$$\sigma_{ij} \frac{dE_{ij}}{dt} = \sum_{s=1}^{38} (\omega_s - \gamma_s) H^{ij}_s \tag{9.2}$$

$$\frac{dL_{ij}}{dt} = H^{ij}_{15} - H^{ij}_{16} - H^{ij}_{17} \tag{9.3}$$

where $v = (v_\varphi, v_\lambda)$ is the water velocity; I_{ij} and J_{ij} are identifiers of the evaporation and precipitation flows, respectively, on the area Ξ_{ij}; $E = (K,C,A,\Phi,D,S,B,M,Z,Y,G)$ is the specific water repository (see Fig. 9.3); ω_s and γ_s are binary identifiers reflecting the presence ($\omega_s = 1, \gamma_s = 1$) or absence ($\omega_s = 0, \gamma_s = 0$) of inflowing and outflowing processes for each element E, respectively. For example (from Fig. 9.3), if $E = A$ (Aral Sea), $\omega_s = 1$ only for $s = 8.33$ and $\gamma_s = 1$ only when $s = 7$. The repository L (water for human use) is a special case that is not associated with an area σ.

To solve Eqs. (9.1)–(9.3) on the whole territory $\Xi = \cup \Xi_{ij}$ initial and boundary conditions are needed. Initial data are either included in the database or formed by the user of the TIM in conformity with his scenario. Boundary conditions are either formed by the GSM (see Sect. 3.2) or given by the user. The water flows H_k ($k = 1,\ldots,38$) are described by the analytical, table, and graphical functions. The evaporation from land and transpiration are described by empirical correlations from Bras (1990). The evaporation from the Aral Sea surface (H_7) and shallow reservoirs ($H_4, H_{18}, H_{20}, H_{37}$) is calculated using the formula given by Bortnik and Chistyakova (1990), which is the equation of Kohler adopted to the empirical data: $H_i = 3.27 a_0 (e_s - e_2) v$ ($i = 4,7,18,19,20,37$), where e_s is the maximal saturation

Table 9.3. List of the TIM software units

Unit	Unit description
STANDARD	Calibration and scaling of the input information
FILTER	Input data filtration
APPROX	Construction of two-dimensional maps on the basis of trace measurements and fragmentary data by means of spatial-temporal interpolation methods
EULER	Solution of Eqs. (9.1)–(9.3) by the Euler method as modified in Section 12.8
CHOICE	Interface unit providing user accessibility to the database
FLOWS	Modeling of the flows H_k (k = 1,...,38)
SCENAR	Creation of the scenario
MAP	Cartographic representation of the simulation results
IBMMEN	Interface unit providing the capability of modifying the set of identifiers $\{A_i\}$
PROC	Interface unit for the control of the computer experiment and for the choice of schemes for representation of the results
ARAL	The Aral Sea water regime model
KBG	The Kara-Bogaz-Gol Gulf water regime model

vapor pressure at the temperature of the water surface (mb); e_2 is the vapor pressure in the atmosphere at the height of 2 m (mb); $a_0 = (597.3-0.57T)^{-1}/\rho_w$ where T is the atmosphere temperature (°C) and ρ_w is the water density (g m^{-3}). The evaporation from the Caspian Sea surface (H_5) is described by the Penman equation. The flow H_6 describing the river water influx to the Caspian Sea has a normal distribution with average value 290 km^3 year^{-1} and standard deviation 30 km^3 year^{-1}. The model does not include all possible inflows and outflows between the water reservoirs shown in Fig. 9.3. Those omitted are considered negligible and are lacking reliable estimations of their volume and spatial distributions.

The motion of the atmospheric moisture over the territory Ξ is described by the standard system of finite-difference balance equations including only the wind component. The parameterization of H_k and control of the TIM are realized by means of the software described in Table 9.3.

For the TIM validation, the ACS dynamics were calculated over a long period of time from 1960 to 1985. The flows $H_5, H_6, H_7, H_8, H_{30}$ and H_{33} were taken from the literature data (Bortnik and Chistyakova 1990; Golitsyn 1995) as table functions with Δt = 1 year. The other flows H_i were calculated with the models listed in Table 9.3. An adaptive criterion for the TIM was realized in the framework of the simulation procedure shown in Fig. 9.4. This is based on evolutionary modeling procedures that are described in detail elsewhere (Bukatova et al. 1991; Krapivin et al. 1991) and explained shortly in Section 2.3 of this volume. During this adaptation procedure, the parameters of the TIM are corrected to minimize the value of

$$\Delta h = \sum_{i=1960}^{1985} \left[\left| h_A(i) - h_{A,0}(i) \right| + \left| h_C(i) - h_{C,0}(i) \right| \right] / 50$$

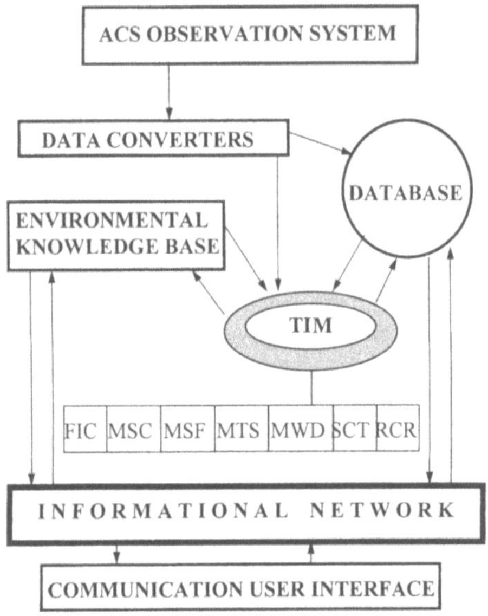

Fig. 9.4. Structure of the adaptive environmental monitoring system, based on the set of models and implementing the evolutionary computer algorithms

where h_A and h_C are the water levels of the Aral and Caspian Seas, respectively, calculated by means of the TIM; $h_{A,0}$ and $h_{C,0}$ are existing estimations of h_A and h_C, respectively (Bortnik and Chistyakova 1990; Golitsyn 1995; Hublaryan 1995). As a result of this procedure, a precision level of 23% was achieved in the approximation of the dynamics of the ACS water levels during the years 1960–1985.

9.4 Simulation System for the Study of Hydrological Fields of the Aral Sea

The Aral Sea aquatory plays an important role in the ACS water regime. There are many models describing the Aral Sea water regime (Bortnik et al. 1994; Borodin et al. 1996; Kondratyev et al. 2001a; Krapivin and Phillips 2001b). The ARAL unit of the TIM has seven subunits listed in Table 9.4 which describe the hydrophysical processes of the Aral Sea during the ice-free period of the year. Initial data were defined for the time t_w which is identified with the end of winter. The salinity field $S(t, \varphi,\lambda,z)$ is defined by taking into account the division of the Aral Sea aquatory $\Omega \subset \Xi$ into two parts with specific characteristics:

$$S(t_w,\varphi,\lambda,z) = \begin{cases} S_1 & for \quad (\varphi,\lambda) \in \Omega_D, \\ S_2 & for \quad (\varphi,\lambda) \in \Omega \setminus \Omega_D, \end{cases}$$

where Ω_D is the western deep-water aquatory of the Aral Sea and $\Omega \setminus \Omega_D$ represents the shallow aquatories (the Big Sea). $S_1 = (S_D - S_A)/V_D$ and $S_2 = (S_P - S_L)/V_P$, where V_D and V_P are the water volumes of Ω_D and $\Omega \setminus \Omega_D$ respectively, S_D and S_P are the salt storage in Ω_D and $\Omega \setminus \Omega_D$ respectively, and S_A and S_L are the salt storage of the territories which have gone dry near Ω_D and $\Omega \setminus \Omega_D$, respectively.

Table 9.4. Set of subunits of the ARAL unit of the TIM

Subunit	Subunit characteristics and functions
FIC	Unit for constructing the initial conditions for the ARAL unit taking into account the prehistory of the Aral Sea dynamics
MSC	Model of seasonal variations in the sea level of the Aral Sea
MSF	Model of the current flows in the sea
MTS	Model of the spatial-temporal distribution of the sea temperature and water salinity
MWD	Model of the water density
SCT	Simulation of the process of convective mixing in the seawater
RCR	Control of the simulation experiment via the structural and parametrical modification of ARAL

Seasonal variations in the sea level are described by the MSC subunit. Average historical data are taken into account to describe the sea water balance, considering such processes as evaporation, precipitation, and river inflows. The evaporation and precipitation processes have their principal variations at the moment of time when the water freezes. The temperature of freezing T_θ depends on the salinity as $T_\theta = 10^{-2}(0.298 + 5.269S + 4.01 \times 10^{-3}S^2 + 3.99 \times 10^{-5}S)$.

Subunit MSF simulates the currents of the Aral Sea. The basic equations of the MSF are (Bortnik et al. 1994):

$$u_\varphi = \tau_\varphi(H - z)/v + [g\rho/v] \cdot [(H^2 - z^2)/2] \cdot \partial\kappa/\partial\varphi ,$$

$$u_\lambda = \tau_\lambda(H - z)/v + [g\rho/v] \cdot [(H^2 - z^2)/2] \cdot \partial\kappa/\partial\lambda ,$$

$$\partial\kappa/\partial\varphi = -3\tau_\varphi/(2g\rho H) - [3v/(g\rho H^3)] \cdot \partial\psi/\partial\lambda ,$$

$$\partial\kappa/\partial\lambda = -3\tau_\lambda/(2g\rho H) + [3v/(g\rho H^3)] \cdot \partial\psi/\partial\varphi ,$$

where $u = (u_\varphi, u_\lambda)$ is the sea current velocity, g is the gravity coefficient, ρ is the average water density on the Aral Sea aquatory, $\tau = (\tau_\varphi, \tau_\lambda)$ is the vector of tangential wind tension on the sea surface, κ is the deviation in the sea level from the undisturbed surface of the basin, ψ is the complete flow function and v is the turbulent stickiness coefficient ($v = 0.25\, aWH/k$, where W is the wind velocity, H is depth, k is the wind coefficient and a is the proportionality coefficient).

A unit MTS simulates the spatial-temporal structures of the marine water salinity S and temperature T distributions. The aquatory Ω is divided into sub-aquatories Ω_i ($\Omega = \cup\Omega_i$) which are homogeneous as to S and T. The transport of heat and salt is realized by means of currents resulting from the difference of their gradients. The exchange processes on the "sea–atmosphere" boundary are described by linear functions (Bortnik and Chistyakova 1990). Vertical variations in the water density $\rho(t, \varphi, \lambda, z)$ are described by a linear model MWD (Bortnik et al. 1994). Unit SCT controls the criterion of stability of the water stratification and on this basis realizes the procedure of convective mixing of the water body. The stratification is considered to be stable when $\partial\rho/\partial z \geq 0$. In this case, the water mixing does not occur. When $\partial\rho/\partial z < 0$ the stratification is considered to be unstable and the process of convective mixing is identified with the

Table 9.5. Time of day corrections to the water temperature (T °C) measured on-site for use in the framework of the simulation experiments

Depth (m)	Time of day when the measurement was made (h)			
	7–13	13–18	18–23	23–7
	Open aquatories of the Aral Sea			
0	0	– 1.5	0	1.5
5	0	– 0.5	0	0.5
≥ 10	0	0	0	0
	Coastal aquatories of the Aral Sea			
0	0	– 2.5	0	2.5
5	0	– 1.5	0	1.5
10	0	– 0.5	0	0.5

procedure of averaging S and T between the water layers in contact. A unit RCR corrects the initial data received from the on-site measurements for time of day according to the procedure given in Table 9.5.

9.5 Simulation Model of the Kara-Bogaz-Gol Gulf Water Regime

The TIM unit KBG simulates the water and salinity cycles of the main evaporators of the Caspian Sea waters. The recent history of the Kara-Bogaz-Gol gulf (KBG) is an example of the complex water balance in the region (Hublaryan 1995; Pearce 1995). The KBG is a large shallow lagoon with an average normal depth of only a few meters, connected to the Caspian Sea by a narrow strait. Due to the large surface-to-volume ratio, evaporation results in a very high salinity and the KBG is a historic source of commercial salt. The level of the KBG is normally 2–3 m below the Caspian Sea, but with the decline in the Caspian Sea level, by the late 1970s the two were nearly equal in elevation. In 1980, the Soviet Union dammed the strait with the result that the KBG nearly dried up completely causing problems in the surrounding areas due to windblown salt. With the subsequent rise in the Caspian sea, an aqueduct was built in the mid 1980s to allow some water to flow back into the KBG. In 1992 Turkmenistan, in the meantime independent, demolished the dam allowing the KBG to refill to its normal level (Vera 1992).

A unit KBG realizes the schemes represented in Figs. 9.5 and 9.6. The equations of a unit KBG have the following form for the storage of salts S in the atmosphere above the KBG, B in the KBG and H in the sediments, for the atmospheric water W_A above the KBG and for the salt flows C_i:

$$\partial S/\partial t + w_\varphi \partial S/\partial \varphi + w_\lambda \partial S/\partial \lambda = C_1 + C_2 + C_{10} - C_3 - C_4 - C_7,$$

$$\partial B/\partial t + u_\varphi \partial B/\partial \varphi + u_\lambda \partial B/\partial \lambda = C_3 + C_4 + C_8 - C_1 - C_2 + C_6 - C_5,$$

$$\partial H/\partial t = C_5 - C_6 - C_9,$$

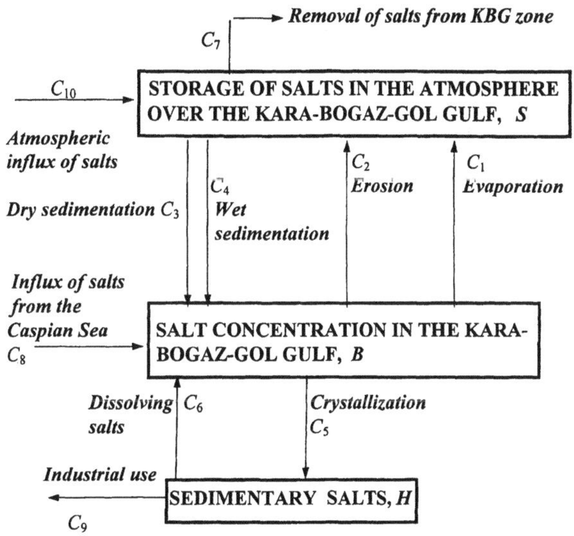

Fig. 9.5.
Principal block diagram of the salt flows in the Kara-Bogaz-Gol Gulf zone

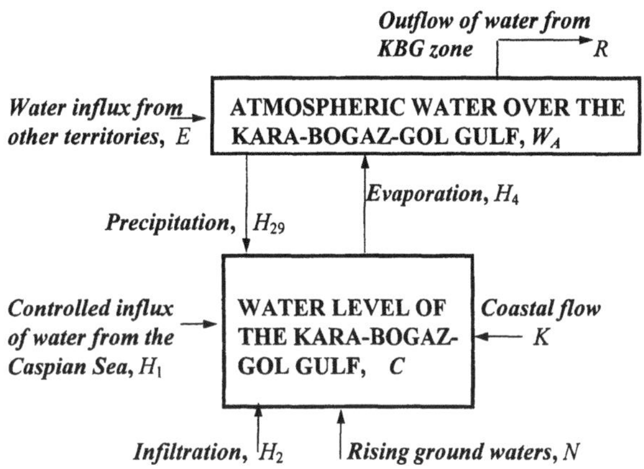

Fig. 9.6.
Implementation of the regional water balance scheme for the Kara-Bogaz-Gol Gulf zone

$$\sigma \partial W_A / \partial t = E - R + \sum_{j=1}^{n} \left(H_4^j - H_{29}^j \sigma_j \right)$$

$$\sigma_i \partial C_i / \partial t = H_1^i - H_4^i + \sigma_i H_{29}^i + K_i + N_i - H_2^i$$

where $u = (u_\varphi, u_\lambda)$ is the water velocity; $w = (w_\varphi, w_\lambda)$ is the wind velocity;

$$C_1 = \min\left\{ C_1^{\max}, \lambda_1 \theta_1^{\Delta T + \Delta W} \right\},$$

$$C_3 = aS,\ C_4 = bSH_{29},\ C_{10} = C_{10}(t),\ C_9 = f(t),\ C_8 = F(t),$$

$$C_1^{\max} = 20\ H_4,\ C_7 = \lambda_3 S(\theta_3)^{\Delta W},$$

$$C_2 = \begin{cases} 0 & for \quad a \quad wet \quad salt \quad surfce, \\ \lambda_2(\theta_2)^{\Delta W} & for \quad a \quad dry \quad salt \quad surface, \end{cases}$$

$$C_5 = \begin{cases} \beta(B - B_0) & when \quad B > B_0, \\ 0 & when \quad B \leq B_0, \end{cases}$$

$$C_6 = \begin{cases} 0 & for \quad H = 0, \\ \lambda_4 \exp[\lambda_6 \Delta T - \lambda_5 B] & for \quad H > 0, \end{cases}$$

B_o is the concentration of saturated solution; $\Delta T = T_o - T$ where T_o is the control temperature; $\Delta W = W_o - W$ where W_o is the control wind velocity; functions f and F are given as empirical estimations; $C_{10}(t)$ is scenario defined; a, b, β, θ_i and λ_i are constants; and n is the quantity of cells $\Delta\varphi x \Delta\lambda$ on the Kara-Bogaz-Gol aquatory.

9.6 Simulation Experiments

The main purpose of the computer experiments is the search of a scenario for control of the water regime in the ACS under the realization of which the Aral Sea level will increase and the Caspian Sea level will decrease. Let $\Delta\varphi = \Delta\lambda = 0.5°$, $\Delta t = 10$ days. We consider the scenario EP (evaporation + precipitation) when the Caspian Sea level is lowered by increasing the flow of its waters to other reservoirs/evaporators. Such reservoirs are the area of saline lands and depressions situated in the East Caspian Sea coast. Their absolute levels are below the recent Caspian Sea level (- 25.7 m). The following depressions correspond to this condition: "Lifeless Kultuk" (- 27 m), "Kaidok" (- 31 m), "Karagie" (- 132 m), "Kaundy" (- 57 m), "Karin Arik" (- 70 m), "Chagala Sor" (- 30 m) and "Kara-Bogaz-Gol Gulf" (- 32 m). Let us assume that the Caspian Sea water volume which is diverted to the above evaporators is ς. This water volume is evaporated at a rate which is a function of the wind and temperature. Thus, the water content in the atmosphere is increased. As a result, the atmospheric moisture over the Aral Sea can also increase. The procedure of forced precipitation over the Aral Sea watershed is realized. The flows H_i ($i = 6,8,11,15,24$) are fixed at their state in the year 1985.

The historic data indicate that northwest (NW), west (W) and southwest (SW) winds prevail over the ACS territory during 130–160 days of the year with a high degree of regularity. Hence, the atmospheric waterway in the direction of the Aral Sea zone is realized by a statistically stable system. It is considered that the evaporators (see Fig. 9.3) are filled from May to September. During the remainder of the year, the winds are variable.

The results of the simulation experiments are given in Figs. 9.7 and 9.8 and in Table 9.6. Here, the model makes use of the wind and temperature distributions recorded during 1981–1996 by the Russian Meteorological Agency. The evaporators are filled during the summer months when the atmospheric temperature is no lower than 15°C. After 10 years, the levels of the Aral and Caspian

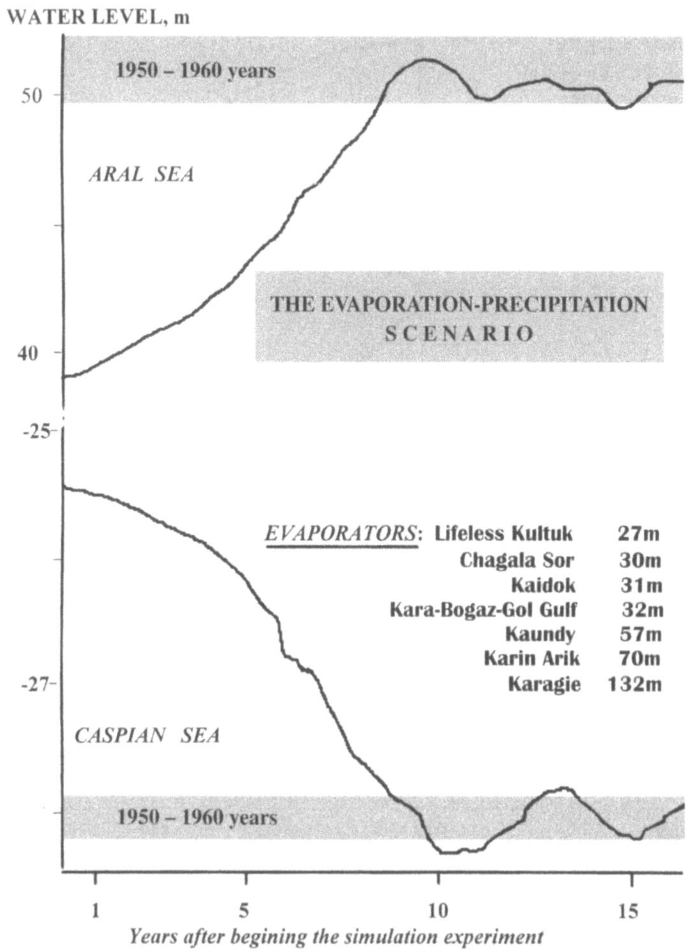

Fig. 9.7. Results of the simulation experiment for the evaporation–precipitation (EP) scenario showing the changes in the water levels of the Aral and Caspian Seas

Seas stabilize within a range comparable to that observed during the years 1950–1960. We can see that the Caspian Sea level decreases by 1.2 cm $year^{-1}$ under the following conditions: the value of ς is uniformly distributed in the interval (50–60) $km^3\ year^{-1}$, the atmospheric temperature is no lower than 15 °C and the NW, W and SW wind directions prevail 50% of the time during $\tau \geq 80$ days. Under these conditions, which occur during the summer months, 0.6–1.3 $km^3\ day^{-1}$ of atmospheric moisture comes to the Aral Sea depression directly from the Caspian Sea aquatory (flow H_5) and 0.1–0.2 $km^3\ day^{-1}$ is added at the expense of the evaporators (flow H_{38}). As a result it is possible to have precipitation of 0.5–1.5 $km^3\ day^{-1}$ in the Aral Sea depression. This allows the Aral Sea level to reach stability within the absolute range of 50–53 m in 7–12 years, as can seen in Fig. 9.7. During this time period, the Caspian Sea level stabilizes at about –28 m.

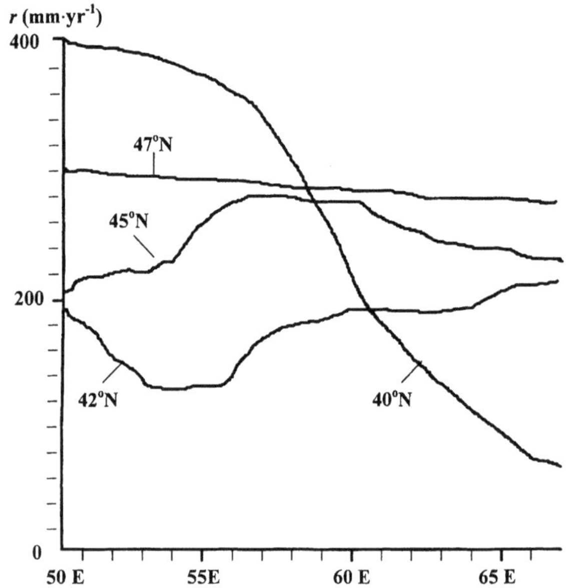

Fig. 9.8.
An example of using the ACES to estimate the spatial distribution of the precipitation intensity r on the territory Ξ

Table 9.6. Results of model estimations for some Aral Sea water balance elements under different prevailing wind directions: H_8 river flow (km^3 year^{-1}); H_{33} precipitation and H_7 evaporation (mm year^{-1})

Δt	Direction of prevailing wind								
	NW			W			SW		
	H_8	H_{33}	H_7	H_8	H_{33}	H_7	H_8	H_{33}	H_7
1	38	197	1,010	41	188	998	10	198	1,007
2	44	180	991	37	190	987	12	183	1,011
3	70	160	993	55	171	869	16	160	1,004
4	56	174	968	68	183	901	21	171	1,023
5	48	149	1,001	50	194	977	18	152	1,014
6	51	187	986	44	189	983	14	188	989
7	66	191	999	61	169	1,015	16	190	1,003
8	61	177	956	63	175	994	12	180	1,004
9	59	163	983	52	166	899	9	171	999
10	53	154	979	57	160	908	13	155	991
11	49	142	988	55	159	910	17	143	1,001
12	57	138	985	48	144	1,017	11	140	973
13	52	144	987	54	147	999	8	141	966
14	55	107	1,003	50	133	976	12	110	981
15	58	99	990	56	121	891	15	101	909

The computer experiments reveal that the potential volume of precipitation in the Aral Sea depression is quasi-invariant with respect to the NW, W, and SW wind directions. This phenomenon assures the possibility of stabilization of the ACS water regime and the conservation of it by use of remote monitoring.

As can be seen from Fig. 9.8, the precipitation in the Aral Sea depression (centered at 45°N, 60°E) increased by 12%. Moreover, the positive balance of water transport across the eastern boundary of the ACS territory increased by 4%, stimulating growth in the river flow to the Aral Sea by up to 40 km^3 $year^{-1}$ ($34 \le H_8 \le 50$ km^3 $year^{-1}$). The east wind increases the precipitation in the Aral Sea zone by 4–7% at the expense of the returning atmospheric moisture.

The dynamics of all water balance components of the ACS are stabilized in 8–9 years after the beginning of the computer experiment. An increase in the groundwater $M(t,\varphi,\lambda)$ takes place at a distance up to 100–170 km from the Aral Sea. The value of ΔM equals 1.2 cm $year^{-1}$ during this period of time. Estimations for the flows of evaporation from and precipitation to the Aral Sea are $H_7 = 51.1 \pm 9.2$, $H_{33} = 9.7 \pm 1.8$ km^3 $year^{-1}$, respectively.

The results of this study suggest the possibility to propose the following scheme for aero-orbital monitoring of the water balance in the Middle Asia arid zone under the influence of the Aral and Caspian Seas: a monitoring system registers the water balance parameters by means of episodic measurements using remote sensing systems and land observations. The TIM is used during the intervals between the measurements as shown in Fig. 9.4. The operator of the monitoring system analyzes the value of the deviation between the measured and predicted ACS parameters and makes a decision about the suitable moment for filling the evaporators so as to result in the forced precipitation over the Aral Sea zone.

The ACES allows the operator to consider different hypothetical situations for the formation of flows of atmospheric moisture and estimates the resulting precipitation on the different areas of the ACS. This problem has many important aspects for future investigations. The water balance elements of the ACS may also be obtained from a systems approach based on evolutionary modeling (see Sect. 2.3), but the set of parameters and structures, which governs the accuracy of the monitoring system, needs to be further investigated, experimentally as well as theoretically.

9.7 Summary and Recommendations

The development of remote sensing techniques and mathematical models for environmental control allows the synthesis of an expert system making possible the study of natural processes by the use of computer experiments. The ACS water regime considered in this chapter is an example of applying the combination of remote sensing and mathematical modeling for the synthesis of a regional geoinformation monitoring system. The results of this investigation show that the formation of an adaptive measurement system with the interchange of remote measurements and mathematical modeling provides a reliable evalua-

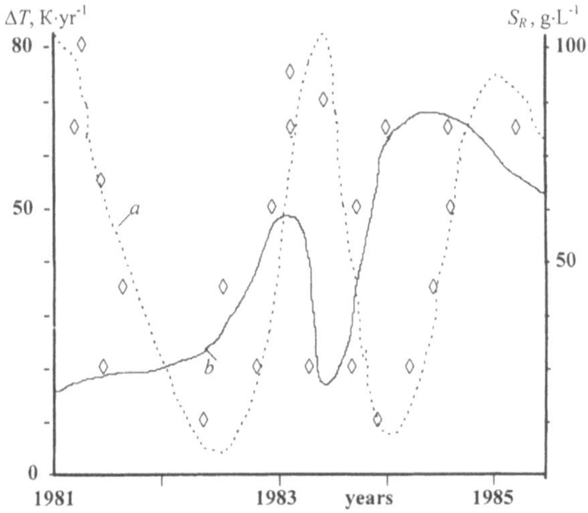

Fig. 9.9. Variations ΔT of the brightness temperatures T_j (*line a*) and mineralization level S_R (*line b*) of the Kara-Bogaz-Gol Gulf reconstructed by means of the ACES algorithms. *Open diamonds* designate the experimental data

tion of the ACS. This will lead to the creation of a system capable of predicting the dynamics of natural processes and assessing long-term consequences of large-scale global change effects on the ACS. Of course, it is necessary to further develop this approach by the expansion of the ACES structure and functions.

The ACES structure proposed in this paper makes possible the estimation of the spatial distribution of the water balance elements for the territory between the Aral and Caspian Seas. The verification problem for the ACES subunit of the theory information model (TIM) has not been solved in complete depth. Partial validation of the TIM is shown in the results of Figs. 9.7–9.9 and in Tables 9.1 and 9.6. However, these estimations are unstable with respect to variations in the climate. Therefore, the results of this study are the first step toward the future investigations of the ACS water balance problem. The next steps in the framework of this problem are:

1. the connection of remote sensing observations from satellites to the ACES database;
2. the expansion of environment elements considered in the ACES;
3. the synthesis of a climatic model for the ACS territory and the connection of it to the ACES;
4. the formation of a set of scenarios describing the various possible procedures for control of the water regime of the ACS;
5. the implementation of simulation experiments to estimate these scenarios.

In addition, it will be necessary to include many of the features of models of anthropogenic processes in future generations of the ACES.

Finally, the regular control of the ACS water balance can be realized in conformity with the scheme presented in Fig. 9.4. Microwave measurements, as we see from Fig. 9.2, make it possible to recognize the surface structures and to estimate the ACS water cycle components shown in Fig. 9.3. The main problem is in the search of the optimal procedure for alternation between the measurements and the modeling.

CHAPTER 10

Monitoring of the Seas in the Oil and Gas Extraction Zones

10

10.1 The Problem of Collection and Processing of Data in Monitoring Systems Operating in Zones of Oil and Gas Extraction

Sea oil and gas condensation deposits (GCD) are usually situated in zones where sufficiently intensive anthropogenic influences on the environment occur. Consequently, the task of designing and creating the ecological monitoring system to control the sea aquatory is to be done with consideration for the state of the atmospheric and water environments for the adjoining aquatories. The GIMS-technology helps to solve this task (Aota et al. 1992; Hong et al. 1992; Kelley et al. 1992b; Kondratyev et al. 2000).

For the sea aquatory of the oil deposit zone, the practical realization of the GIMS technology concept demands a thorough analysis of hydrophysical and synoptic characteristics and the choice of technical systems to measure the environment parameters under the local climate (low temperatures, icing, rough sea state, etc.) with consideration of the distance from land. The variations of the interannual climatic regimes that are connected with changes of the water surface state and of the insolation are also important. Climate conditions of the functioning of the ecological monitoring system in the zone of oil extraction put definite restrictions on the structure of the measurement system. These restrictions are usually connected with the absence of possibilities for the free arrangement of the measuring and transmitting devices in any arbitrary point of the aquatory; and that is why the tasks of measurement planning and the choice of effective algorithms for data processing are of great concern.

An investigation of such a dynamic environment as the water systems demands the creation of complex algorithms for the data processing of the field observations, including a set of various models giving a possibility to reconstruct the spatial image of the studied object based on the incomplete information. The water systems (oceans, seas, rivers, etc.) were investigated during the last decades by means of flying and floating laboratories. This made it possible to develop effective methods for the description of the dynamics of the water systems having different spatial scales. In most cases, the field and satellite observations are connected with the measurements of temperature, salinity, dissolved oxygen, current directions and velocities. The set of hydrophysical observations is more detailed in the anthropogenic zones.

	Micro	**Meso**	**Macro**
Centuries			Inter-century variability. Wavelengths: thousands km.
Decades			Inter-century variability. Wavelengths: thousands km.
Years			Inter-year variability: Wavelength: thousands km.
Months		Synoptical changeability, oceanic meanders and whirlwinds. Waves range: hundred km.	Fluctuations of the currents velocity and direction. Wavelengths: thousands km.
Days		Inertia currents. Waves range: hundred km.	
Hours		Rising tides. Waves range: dozens km.	
Minutes	Detail vertical kinematic structure. Waves range: m.	Interior gravitation waves. Waves range: km.	
Secundes	Turbulence, vertical kinematic microstructure. Waves range: cm and mm.	Surface waves, Waves range: m.	

Spatial scales

Fig. 10.1. Scheme for the coordination of spatial and temporal scales of the processes to be taken into consideration under the sea environment modeling in the GES zones

Let us designate a point on the water surface by φ latitude and λ longitude. A rectangular coordinate system (φ,λ,z) is selected to describe the sets of observation data where the z-axis has the value $z = 0$ on the water surface and value $z > 0$ with depth. In these coordinates any measurement is represented as function $\xi(\varphi,\lambda,z,t)$ which is a time-dependent random variable. The processing of sets of such values demands the application of special methods (Bukatova et al. 1991). As a rule, these methods include nonstationary reduction procedures using spatiotemporal discretization. The choice of spatiotemporal scales is defined by the dynamic characteristics of the water space and by the specific tasks that must be accomplished. Figure 10.1 explains a possible structure of the interconnections between spatial and temporal scales. We see that the accumulation of data sets by the monitoring system may be realized by taking into consideration discrete situations distinguished in the framework of a problem to be solved. A certain level of reliability in the assessment of the system state under study can be achieved only in this case.

The data sets containing the experimental estimations of the circulation characteristics for the sea environment always include periodic, unperiodic, stationary and nonstationary fluctuations. That is why the sets of measurements should be corrected, taking into account the scalability of the processes being studied. Hence, the planning of the special features of the measurements follows.

The study of the oil pollution of the World Ocean is one of the important problems of environmental monitoring. The global scale of this process demands the application of monitoring systems giving the possibility of controlling the water environment over such an enormous size. Satellite-based systems are one such approach. An analysis of measurement data shows that remote sensing of the water surface by means of devices using various wavelengths makes it possible to detect oil spills on the water surface, to determine the type of oil, and to estimate the oil spill parameters (area, thickness, volume). Remote sensing techniques based on microwave radiometry allow us to determine the oil pollution of the water surface under arbitrary weather conditions (Krapivin et al. 2001a,b).

An extensive application of remote sensing techniques depends on knowledge about the processes of interaction of the oil with the seawater, its optical and electric characteristics, the impact of the atmosphere, and other factors affecting the propagation of electromagnetic waves. The combination of microwave and IR (infrared) ranges with mathematical modeling techniques is an effective method to discover oil spills on the water surface. The distinction between the emittance and the temperature of polluted and fresh water surface areas is the physical base for the remote sensing of oil spills by means of microwave and IR radiometers. The emittance, κ, of the three-layer system *atmosphere–oil–water* having smooth boundaries of the sections can be calculated with the formulas suggested by Kondratyev et al. (1996a). Figure 10.2 gives an example of the dependence of the emittance variations, $\Delta\kappa$, for the *atmosphere–oil–water* system as a function of the oil spill thickness, α.

The oil spill dielectric properties occupy an intermediate position between free space (atmosphere) and water. As a result, the formation of a film leads to the effect of medium coordination (to the decrease in the reflection coefficient) and to the growth of the surface radio-brightness temperature. As the oil film thickness increases, the value $T_Y = \kappa T_0$ (T_0 is the surface temperature) at first increases, after which the interchange of maxima and minima is observed. From Fig. 10.2, it can be seen that to remove the uncertainty in the film thickness determination it is necessary to simultaneously measure the radio-brightness temperature by means of radiometers with different wavelengths.

Thin films, forming under small volumes of spilled oil or in the near-boundary area of the oil spills, do not change the emittance ability of a smooth water surface in the microwave range. However, disturbed surface areas covered by thin oil films are characterized by lowered values of T_Y, which are caused by the suppression of high-frequency components in the rough sea spectrum. The value and sign of the radiation contrast of the spills on the clean water back-

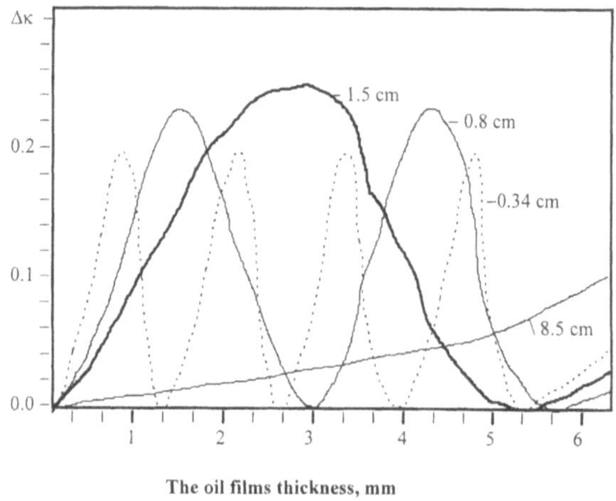

Fig. 10.2. The emittance ability of oil spills having different thicknesses (Mitnik 1977). Wavelengths (in cm) are given on the curves. The water temperature is 10 °C and the oil dielectric permeability ε is 2.2

ground depend on the thickness and optical properties of the oil films, the hydrometeorogical conditions, the time of the day, etc.

Fundamental experimental investigations of water oil pollution by means of microwave and IR radiometers have been described by many authors (Bogorodsky et al. 1976; Mitnik 1977). Field experiments have shown that satisfactory results were attained when radiometers with wavelengths of 8–12 μm, and 0.34, 0.8, 1.5 and 8.5 cm were used. The sensitivity of the radiometer relative to its antenna inputs equaled 0.1–0.3 K under the time constant of 1 s. Experiments were realized with flying laboratories at heights of 100–200 m. Calibration of the radiometers were made by using blackbody radio-brightness temperatures or by means of calculations for a calm water surface under a cloudless sky. Thin films are recognized with a high accuracy by means of IR radiometers. Most thick films are registered with a high reliability by means of microwave radiometers.

The oil film thickness can be estimated by the dependence of the radio-brightness temperature variation, ΔT_Y, on the emittance ability, $\Delta\kappa$:

$$\Delta T_Y = \Delta\kappa\,[T_0 - (1 - T_{Y,\mathrm{atm}}/T_0)],$$

where T_0 is the surface temperature, $T_{Y,\mathrm{atm}}$ is the atmosphere radio-brightness temperature calculated by radiosensing data and the value $(1 - T_{Y,\mathrm{atm}}/T_0)$ characterizes the influence of atmosphere.

The geometric parameters of the oil films are defined by means of photogrammetric methods, the base of which is the spectrozonal photo-picture on the various wavelengths. The ranges 0.4–0.5 and 0.7–0.8 μm are the most informative for solving this task. The oil products registered with wavelengths 0.4–0.5 μm are a light spot upon the dark background of the image water. The image registered by wavelengths 0.7–0.8 μm helps to decipher the water surface.

The registration of the oil spills can be realized by means of active sensing methods. Therefore, for example, the oil spill exposed by the near-ultraviolet radiation begins fluorescence in the visible range (0.6–0.7 μm). This fluorescence can be registered by the adaptive identifier in the real-time regime (Krapivin et al. 2001a, b).

The above methods allow us to consider two versions of the monitoring system for control of the gas extraction zone (GEZ). The first version corresponds to the oil extraction system situated completely below the water surface where the stationary position of remote sensing systems is impossible. In this case, the monitoring system structure has submerged measuring subsystems fixed by anchors and emerged subsystems placed on flying or floating laboratories. The estimation of the concentration of pollutants emitted to the atmosphere is realized by modeling calculations. To this aim the gradient of the gas components and the advection speed are measured in the surface layer. Also, it is possible to use the emerged measuring subsystems.

10.2 Concept of a System of Ecological Monitoring of the Sea Surface and the Atmosphere in Zones of Oil and Gas Extraction

The GEZ according to GIMS technology is to have a structure with subsystems as enumerated in Table 10.1. The design of the GEZ is connected with the interpretation and detailing of the structural and functional properties of these subsystems. The experience of such investigations shows that the first stage demands the solving of the following tasks for the design of the system of environmental monitoring (SEM):

- an elaboration of a science-based and economically covered criterion to assess the atmosphere and water environment state in the area of the GEZ influence;
- an analysis of the hydrophysical structures and aquatory formation to be taken into account by the GEZ monitoring system;
- the preparation of propositions on the structure of the software and hardware for the GEZ monitoring system with consideration of the real conditions of the initial informational field, the presence of developed technologies for monitoring the data processing, and environment quality standards;
- an elaboration of a version of the project describing in detail the hierarchy of the subsystems and their elements with instructions for the algorithmic and technical infrastructure of the SEM;

Table 10.1. List and characteristics of the SEM subsystems

Subsystem	Functions of the subsystem
Measuring subsystem	This subsystem includes contact and remote devices to measure synoptic, hydrophysical, atmospheric, ecological, and hydrochemical characteristics of the environment and to register the oil leakage from the means of its transportation
Subsystem of accumulation, analysis and interpretation of measurement data	This subsystem includes technical means for the data transmission to a single informational center which is equipped with the means for data processing, including software for realizing algorithms of environmental state assessment, identification of pollution sources, and visualization of measurement results

- preparation of the SEM technical project with the indication of the final variant chosen, and with the substantiation of the set of devices, the software, the informational network components, the control structure, and recommendations for the operation of the system;
- preparation of the technical documentation for the full set of SEM components with detailed calculation of the reliability, functional effectivity and stability, and with recommendations of possible modernization stages reflecting consideration of the system operational experience.

The SEM design depends on the conditions of its functioning. These conditions are connected basically with the hydrophysical properties of the environment. The synoptic situation influences the speed of the degradation processes and of the oil product transport through the food chains. All aquatories can be placed by the hydrophysical parameters into two classes: freezing and nonfreezing. It is evident from the viewpoint of modeling that the second class is a special case of the first class when the snow and ice cover thickness equals zero. Therefore, the first class is considered in detail. The Shtockman's GEZ situated in the Barents Sea with high variability of the synoptic situation is accompanied during the year by periods of sharp change in the hydrophysical state of the sea environment. In this zone, the air masses brought from the Atlantic Ocean and the central Arctic aquatories collide and mix. The monsoon character of the Barents Sea climate is shown by the presence of winds in the low atmospheric layers which blow from the ocean to the land in summer and in the opposite direction during the whole winter. The winter season (October–March) is characterized by strong cyclonic activity (the largest number of days with storms and maximal repetition of strong winds). The ice cover achieves its maximal size in April. During the June to August period, cyclonic activity is minimal.

The above-mentioned properties allow us to make the obvious conclusion that the technical realization of the system to measure the oil pollutants of Shtockman's GEZ demands the realization of devices protected from great physical loads and considerable temperature variations. At the same time, powerful turbulence of the lower atmosphere layer near the water surface in the winter reduces the quantity of measurement points in space which are required to give the necessary information. If the wind speed is more than 5 m s^{-1} and the atmosphere turbulence is high, the atmosphere can be considered to be well mixed with sufficient precision over a territory of 50×50 km. In this case, the measurements can be realized at two or three sites of the aquatory. In contrast, in the summer (June–August) when the intensity of the atmosphere turbulence is decreased, the measurements need to be realized in the area of influence of each oil well. The assessment of dispersion of the pollutant and the calculation of its spatial distribution are realized by means of corresponding models of atmosphere dynamics.

The most important stage of the analysis and design of the measuring system to determine the oil pollution level consists in the description of the sea environment dynamics. It is known that an oil spill spreading on the sea surface is subordinated to the superposition of two processes. The first is the spill drift due to sea current, wind and surface waves. The second is the spill spreading over a calm surface.

For the Barents Sea, the second process cannot be taken into account under the synthesis of the GEZ measurement network. This process should only be taken into account in the hydrophysical model to reflect the total range of the hydrophysical processes.

An analysis of many models describing the oil spreading over the sea surface gives the following results. The velocity of oil spill motion equals 60% of the current speed and 2–4% of the wind speed. When ice cover exists the wind component is absent.

The character of the GCD measuring subsystem securing the assessment of oil concentration into the seawater thickness essentially depends on the possibility of putting algorithms into the information processing unit and of reconstructing the forms of the oil spatial distribution based on data fragmentary in space and episodic in time. The synthesis of an adequate model to describe the oil product kinetics in the sea environment under Arctic conditions is here an important procedure.

Calculations show that oil pollution in the Arctic climate can persist for several years. Therefore, the danger of its accumulation is great. The transformation processes of oil pollution to other forms under the Arctic climate are essentially slowed down in comparison with the analogous processes in warmer waters.

It is known that contributions to the process of seawater self-cleaning equal 0–70% for evaporation, 15–30% for photo-oxidation, and 2–7% for biological utilization. Under summer conditions, the transformation of 0–60% of the raw oil mass is realized over 40 days. In winter, these estimations decrease approximately threefold. In this period the process of oil product accumulation by the ice enters into action. The intensity of this process depends on the ice porosity. A perceptible contribution to aquatory cleaning is introduced by the drifting ice, which secures oil removal in quantities estimated at 25% of its own weight. However, the participation of ice in the process of sea surface cleaning from oil pollution has had negative results. The oil captured by the ice is moved to other aquatories and practically the same volume is returned to the water environment when the ice melts. This exchange between aquatories is to be taken into consideration under the SEM design since the oil pollutants can reach a controlled aquatory by such means. Hence, in winter, control of the ice pollution level in adjoining aquatories is needed.

Shtockman's GCD zone is characterized by such generalized synoptic and hydrophysical situations which provide the conditions when this zone is covered by ice during 5 months (from February to June). During the period from February to March, the ice drifts to the north or the northwest. Under these conditions, the drift speed is estimated at 100 km $month^{-1}$. As a result, the polluted ice reaches the central Arctic Basin where it thaws and the oil is dispersed into the water.

The Shtockman's GCD zone is under the influence of the West Novaya Zemlya current branch that forms the eastern boundary of the cyclonic cycle in the region of the central depression. The configuration of the formation of the moving water masses suggests that to discover the pollutants brought to the Shtockman's GCD aquatory from other aquatories it is necessary to control its southern and eastern boundaries.

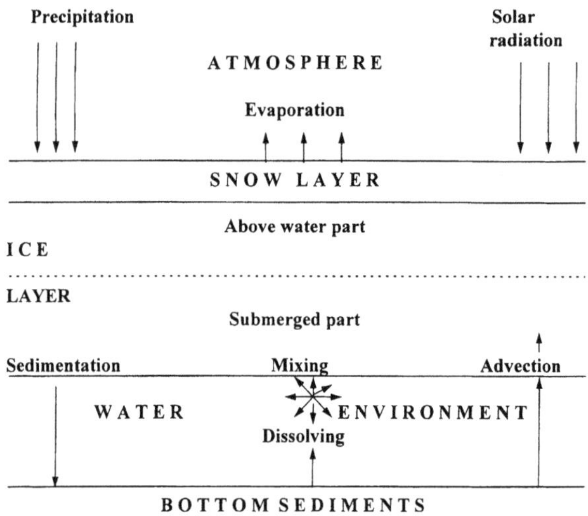

Fig. 10.3. Vertical structure of the ice-covered aquatory

The oil products are characterized by a multicomponent process that expands the situation of its behavior in the seawater. Among the most important processes of transformation of the oil products in the seawater it is necessary to indicate the following: dissolving, evaporation, spreading to the deep layers in the form of drops, oxidation, absorption by suspended organic matter, biosedimentation, and bacterial decomposition. The objective laws of formation of all these processes were studied and, therefore, it is necessary to take this into account under the SEM design. This will allow the reduction of the standards to the composition of the measuring equipment and the regime of the measurement realization (see Chaps. 5 and 8).

The biodegradation of the oil hydrocarbons is an important process of oil elimination in the sea environment. This process is connected with bacterioplankton, phytoplankton, and other sea animals. Taking this process into consideration in the SEM is possible at the expense of including the ecosystem model of Shtockman's GCD aquatory in the SEM software. The conceptual model base is represented in Fig. 10.3.

An elaboration of the Shtockman's GCD ecosystem model demands the consideration of the physical structure of the environment. This structure has a seasonal character. Parameterization of its vertical structure is possible under the consideration of results received earlier by many authors (Legendre and Krapivin 1992; Legendre and Legendre 1998; Emery and Thomson 2001; Vinogradov et al. 2001).

The scheme in Fig. 10.3 represents the typical structure of vertical division of the sea environment. This permits the realization of the scheme of Fig. 10.4.

An ecological monitoring system of practically any anthropogenic-loading object demands the consideration of all components foreseen by the GIMS technology. However, natural conditions of the SEM functioning demands the revision of the GIMS formal structure. It is quite evident that in the case considered the following standards are to be provided:

Fig. 10.4.
Principal structure of the system of ecological monitoring in reference to the gas condensation deposits zone

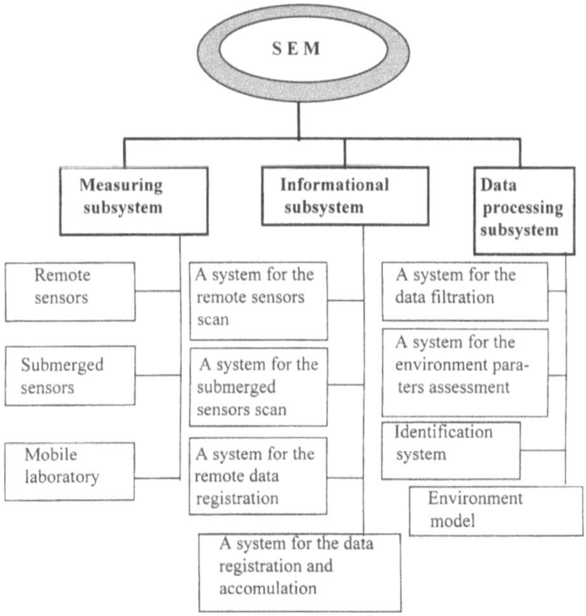

- the measuring devices are to function reliably under low temperatures;
- the measuring network infrastructure should not demand any additional construction;
- the informational network is to secure the data concept in the framework of international standards;
- the data processing subsystem is to reduce the conditions made by the demands on the measuring subsystem;
- the data presentation subsystem is to allow the GCD administration the possibility of multishape assessment of the environment state;
- the SEM is to be combined with other informational systems (regional, national, and international).

The experience of synthesis of the GIS and expert systems having the function of nature-protection calls for the necessity of observing the conception and subsequent stages under the realization of the natural environmental monitoring systems (Maguire 1991; Mackinson 2000). This experience permits us to propose the SEM infrastructure in the form represented in Fig. 10.4.

The SEM software package is to be oriented on the realization of an algorithmic set giving the possibility of realization of the following minimal set of functions:

- integrated assessment of the ecological state of the aquatory;
- assessment of the ecological state of the aquatory for each of its components;
- assessment of the integrated ecological state of the aquatory on the spot, on the square area, and for the total space;
- identification of the pollution source;
- prediction assessment of the ecological state of the aquatory.

The answers to questions arising under the realization of these functions and the decision-making demand transformation of the measurement data into an acceptable format. This can be realized by means of the software package listed below:

- measurement data calibration;
- measurement data filtration;
- technical tools for scanning and blending functions of the measuring devices with the informational unit;
- decision-making regarding the existence of an external situation;
- spatiotemporal coordination of multiple data types;
- spatial interpolation of the measurement data and formation of spatial images;
- reconstruction of the spatial distribution of environmental characteristics on the basis of measurements which are fragmentary over the aquatory and the depth;
- calculation of kinetic characteristics of the pollutants in the seawater under climatic conditions of the sea region;
- calculation of gas contaminants and solid particle content in the atmosphere over the GCD aquatory;
- assessment of the ecological situation of the controlling zone in accordance with the given criterion;
- synchronization of informational fluxes and securing of its dumping to the data processing center in a volley;
- realization of physicochemical processes in the system *atmosphere-sea-GCD*;
- formation of short- and long-term forecasts of common ecological situations in the GCD zone;
- identification of sources causing the disturbance of ecological standards on all controlled parameters taking into account the accepted criteria set;
- realization of computer cartography algorithms;
- choice of the program system for database control, the elaboration of data maintenance in the SEM informational network and its accumulation to the database;
- software for the user interface;
- software for database maintenance;
- reconstruction of omitted functions when information is incomplete or inexact;
- accumulation of knowledge about different specific and typical situations;
- formation of the operative data under emergency situations.

Algorithmic support of the SEM database has a double load. On the one hand, the accumulation of data regarding the functioning of the ecological system of the GCD zone allows us to increase the reliability of receiving assessments of the ecological situations and to decrease the demands to the measuring subsystem. On the other hand, the SEM database can be used as part of the regional or national database.

The SEM functioning is secured by the correlation between the measuring subsystem, the regenerated database and the model. According to Aota et al. (1993), continuous monitoring can be realized with sufficient stability only under adaptive algorithmic support. The dynamic correlation of the functioning

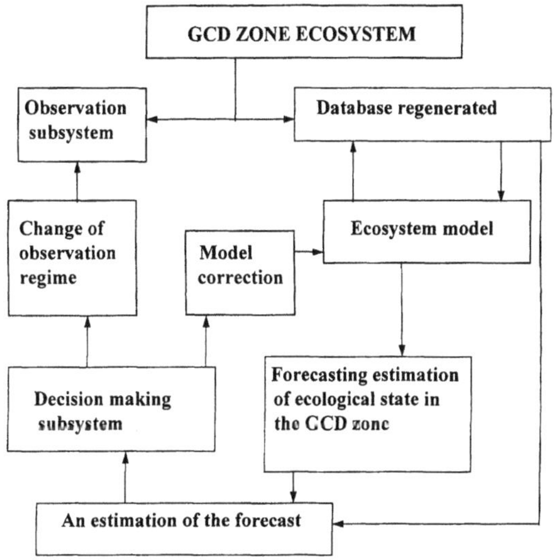

Fig. 10.5. A mutual adaptive scheme using the ecosystem model and measurements for realizing the GIMS technology to the SEM functioning regime

regimes for all SEM subsystems is represented schematically in Fig. 10.5. According to this scheme, the measuring subsystem can work in discrete regimes together with the model. Detectors of emergency situations work in the continuous regime only.

10.3 Estimation of Oil Hydrocarbon Pollution Parameters in Seawater

The problem of identification and assessment of oil pollution parameters in the water environment has been studied sufficiently. Existing methods of oil discovery in the water may be divided into two groups, surface film and suspension, corresponding to the state of the oil hydrocarbons. The oil and oil products, due to their physical-chemical properties, can form films, clots, emulsions, and solutions in seawater. The oil films have thicknesses from fractions of a millimeter to several centimeters. All this determines the methods of oil pollution monitoring.

The choice of the monitoring method is determined by the oil pollution level. It is known that oil pollution exercises influence on the physicochemical processes taking place in seawater. Specifically, the surface strength for oil and oil products is two to four times less than for nonpolluted water. The thermal conductivities and thermal capacities of water and oil are distinct and equal, respectively, 0.599 W m^{-1} K^{-1} and 4.187 KJ kg^{-1} K^{-1} for water and 0.15 W m^{-1} K^{-1} and 1.7–2.1 kJ kg^{-1} K^{-1} for oil. These distinctions influence the many processes in the *atmosphere–water* system. Oil pollution films decrease thermal conductivity and thermal capacity of the upper sea layer. They change the evaporation process decreasing it 1.5 times or more, they disturb the gas exchange between the atmosphere and seawater, and they change the water temperature. All these effects are used for the design of measuring devices to estimate the oil pollution level (Klyuev 2000; Marhaba et al. 2000). Microwave methods are described by Krapivin (2000). Here, we consider optical methods.

Fig. 10.6. Spectral reflection coefficients for the oil and water

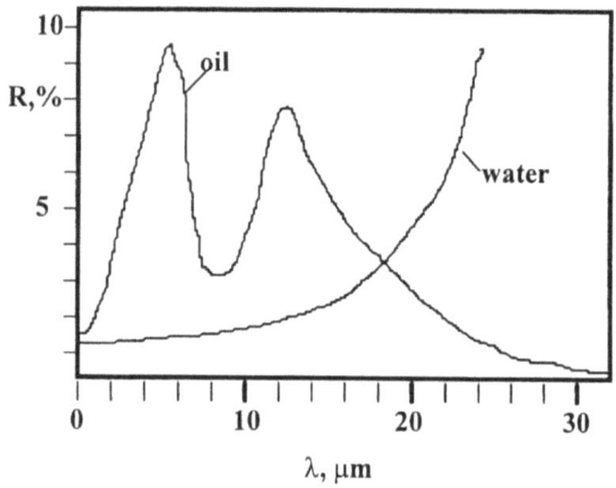

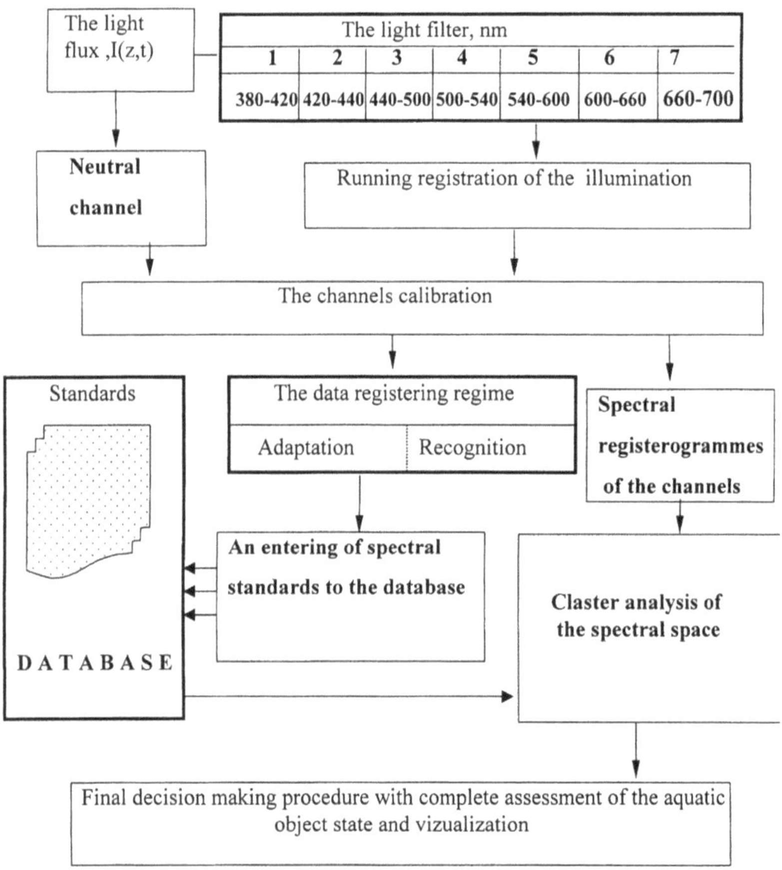

Fig. 10.7. Scheme of the functioning of the adaptive identifier. The results of the measurements are represented by the scale of the analog-to-digital converter (a range from 0 to 2^{12}). AI-1at first transforms the light signal to an electric one in the range ± 5 V and then to digital code in the interval [0,4096]

On a level with the above-mentioned effects of seawater oil pollution, an optical effect exists which is determined by the change in sea surface albedo and by the variation of the interior optical properties of seawater. Theoretical approaches to the problem of light spreading in the seawater environment or its reflection from the sea surface are connected with the consideration of various tasks. For example, calculation of the reflection coefficient under different sea environment states is given by Kabanov et al. (2000). Theoretical and experimental studies show the existence of contrasts in the reflecting properties of oil films and a nonpolluted water surface. Figure 10.6 shows these contrasts. Certainly these contrasts are functions of many parameters: wavelength, oil film thickness, vision angle, salinity, water roughness, light intensity, and content of other contaminants. As was shown by Hong et al. (1994), the use of spectral measurements at a range of 380–700 nm allows one to have a reliable technique for the detection of oil pollution on the water surface. The adaptive identifier (AI-1) described in the paper by Aleshin and Klimov (1992) is designed on the principle shown in Fig. 10.7. AI-1 consists of a multichannel device registering the light scattered in the water or reflected from the surface in the visible range, an informational interface, and software. A principle of the functioning of the system consists in the registration in numerical code $n \in [7,128]$ of spectra $\{T_n(\lambda)\text{: } \lambda \in [\lambda_i, \lambda_{i+1}], i = 0,1,\ldots,n-1\}$ on the basis of which the solution about the existence of oil pollution is taken and the assessment of its parameters is realized. Specifically, for $n = 7$, the basic spectral characteristics of the channels are given in Fig. 10.8. We see that the spectral characteristics of the AI-1 interference

Fig. 10.8.
Basic spectral characteristics of the AI-1 light filters

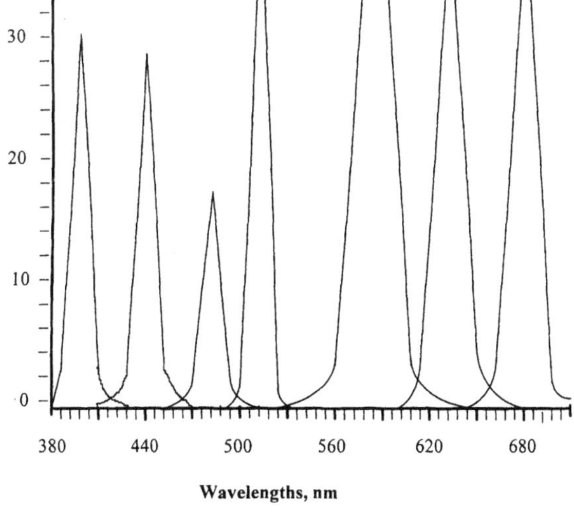

Table 10.2. Averaged correlation matrix of AI-1 reflecting the sky emission in December in Ho Chi Minh City (Hong et al. 1994). Channels, N, are characterized in Fig. 10.7.

N	Channel						
	1	2	3	4	5	6	7
1	1.00	- 0.11	- 0.24	0.03	0.29	- 0.06	0.02
2	- 0.11	1.00	- 0.01	0.13	0.09	0.10	0.06
3	- 0.24	- 0.01	1.00	0.24	- 0.27	- 0.03	0.06
4	0.03	0.13	0.24	1.00	0.05	0.09	-10^{-3}
5	0.29	0.09	- 0.27	0.05	1.00	- 0.07	0.04
6	- 0.06	0.10	- 0.03	0.09	- 0.07	1.00	0.01
7	0.02	0.06	0.06	-10^{-3}	0.04	0.01	1.00

filters have pronounced maxima and insignificant intersections. The maximal conductance of the channels is observed under $\lambda_1 = 398$ nm, $\lambda_2 = 439$ nm, $\lambda_3 = 480$ nm, $\lambda_4 = 510$ nm, $\lambda_5 = 583$ nm, $\lambda_6 = 631$ nm, and $\lambda_7 = 680$ nm. A typical correlation matrix with the addition of the solar radiation spectrum $[\lambda_0, \lambda_n]$ is represented in Table 10.2. The stability of this matrix equals 5%.

An increase in the size of n certainly expands the possibilities for oil pollution classification. The second modification of AI-1, described by Krapivin et al. (2001a, b), gives such possibilities.

The combined use of real-time spectrometry measurements and data processing methods has been implemented for the first time in an adaptation identifier AI-2 (Krapivin et al. 2001a, b). Its creation becomes possible due to a new approach in the field of polarization optics. An effective elemental basis for polarization optics and a method of discrete modulation of the polarization mode have been developed (Kovalev et al. 1999). The application of simple and highly effective polarization switches and sets of silicon photodiodes with arbitrary access to them has essentially simplified the problem of creating compact and cheap polarization-optical devices: spectro-photometers, spectro-polarimeters, spectro-ellipsometers, etc.

The technology of the combined use of spectrometry and algorithms of identification and recognition has, for the first time, allowed the creation of a standard integral complex of instrumental, algorithmic, modular, and software tools for the collection and processing of data on the aquatic environment with forecasting and decision-making functions. The adaptive identifier of a second version has a set of modifications intended for application under different conditions. A stationary version covers the complete set of technical and algorithmic tools providing for the real-time measurements. This version can be used if a power supply line with a voltage of 220 V is available. A handheld field version of the AI-2 provides two applications. If a notebook computer is available under field conditions (in the absence of a power supply line), the whole spectrum of the adaptive identifier is realized in real-time mode. Otherwise, the results of the measurements are stored in an off-line storage unit and are processed later. The same regime exists in AI-1.

The algorithmic support of AI-1 and AI-2 is based on a complex application of methods of recognition and classification of discrete images formed on the basis of 35 spectra registered during a fixed period of time. A time interval of 1 s is usually established and provides about 60 values of brightness for each of the 35 optical channels. The spectra obtained are sources of sets of statistical parameters and different characteristics united into vector spaces for their comparison with standard samples stored on the computer. The technology of this comparison depends on the diversity of identification methods.

The adaptive identifier is designed to learn from the measurement of spectral characteristics and the simultaneous independent measurement of the content of chemical elements in the aquatic environment. As a result, a standard bank is formed in the knowledge base, comparison with which provides the solution of the identification problem. The software of the adaptive identifier provides different algorithms for the solution of this problem, and cluster analysis is among them.

The adaptive identifier can be used in different fields where the quality of water should be estimated or the presence of a particular set of chemical elements should be revealed. The adaptive identifier solves these problems by real-time monitoring of the aquatic environment. In the stationary version it allows the tracking of the dynamics of the water quality in a stream, and when placed on a ship, it allows the measurement of water pollution parameters along the route.

The functionality of the adaptive identifier can be extended by increasing the volume of standards in the knowledge base. The use of a natural light source allows the examination of soils, the identification of oil products on a water surface, the determination of the degree of pollution of atmospheric air and the estimation of the conditions of other objects of the environment, whose spectral images may change. That is why the adaptive identifier is a universal measuring device the use of which in the SEM allows one to have operative information, not only on the arising of oil pollution, but about other pollutants which can appear in the GCD zone.

10.4 Expert System for the Identification of Pollutant Spills on the Water Surface

The technology of adaptive identification of environmental elements allows us to synthesize an expert system for adaptive identification of environmental parameters (ESAIEP). The ESAIEP components are described by Krapivin et al. (1997a,b, 2001a) and Nitu et al. (2000a, b). A compact multichannel spectropolarimeter (MSP), an informational interface with the computer (IIC), software (STW), and an extending database (EDB) are the main ESAIEP components. The STW realizes a set of algorithms for processing the data flows being received from the MSP and secures the service functions such as visualization and control of the measurement regime. The EDB consists of the set of standards for spectral images of pollutant spots represented by points in the multidimensional vector space of indications calculated beforehand on the strength of teaching data retrieval sets.

The ESAIEP functioning principle is based on the fixation of light flux fluctuations on the MSP output and the transformation of it to digital code.

Subsequent processing of these data depends on the STW structure including different algorithms of two-dimensional image recognition. The adaptivity of the recognition procedure is determined by the knowledge accumulation level concerning the fluctuation features of the intensity and polarization properties of the light reflected from the water surface. The STW includes the algorithms allowing, in the case when the pollution spot identification is indefinite, an expert decision to be made on the basis of a visual analysis of the spectral image. This procedure is realized in the dialogue regime with the ESAIEP. In this case, the operator can set the decision taken in the knowledge base as standard.

The principal scheme of the STW unit securing the identification procedure is some transformation F (see Fig. 11.5). The light intensity ξ_i^j registered at the time t_i by the channel λ_j comes to the algorithm F where the procedure of distinction between two hypotheses H_0 and H_1 is realized. The ESAIEP operator determines the initial data γ, α and β and also decides what parameters $u_i = (u_1,...,u_\tau)$ will be calculated on the basis of the measurements ξ_i^j. The service unit IIC gives the possibility of forming the vector u_i using the statistical characteristics of the sets ξ_i^j or using direct measurements. A priori information characterizes the type of distribution $f_\alpha(u_i)$. The function

$$L_i = \sum_{j=1}^{n} \Psi_j = \sum_{j=1}^{n} f_{\alpha 1}(u_i^j) / f_{\alpha 0}(u_i^j)$$

is compared with its limit values $L_{i,\min}$ and $L_{i,\max}$. At the first stage, these values are selected to be rather arbitrary, but then they are changed until a maximal precision of distinguishing between hypotheses H_0 and H_1 is reached. We have $L_{i,\min} \to L^*_{i,\min}$ and $L_{i,\max} \to L^*_{i,\max}$. The values $L_{i,\min}$ and $L^*_{i,\max}$ are stored in the EDB.

After teaching the expert system, functioning is restricted only by the measurement volume fixed by the operator, who proceeds with regard to achieveability of statistical reliability and real-time regime keeping. The operator has two possibilities to regulate this regime: establishing the volume of the set $\{\xi_i^j\}$ or fixing the time interval for its accumulation. Usually, the time interval is fixed. For example, AI-1 and AI-2 reliably distinguish between the competed hypotheses under a fixed time interval of observation equal to 1 s. This procedure is illustrated in Fig. 11.2. Operator contacts with the different ESAIEP units occur via the man-machine interface IIC, which secures a selectivity in the control for the functions of all units.

When an oil spill on the water surface exists, the ESAIEP analyses its thickness, age, source, and geometrical configuration. In this case, the most informative channels have wavelengths $\lambda = 398, 439$ and 480 nm. In the case when there are dissolved and suspended oil components, the system estimates their concentration. If data exist in the EDB for the hydrodynamic parameters, it calculates the spatial distribution of the concentration of the oil components using the methods described by Nitu et al. (2000a, b).

10.5
Monitoring of the Gas Extraction Zone in the South China Sea

The ESAIEP was tested more than once during several Russian/American and Vietnam/Russian hydrophysical expeditions (Hong et al. 1992, 1994; Rochon et al. 1996; Phillips et al. 1997). The results of these expeditions include many contaminants of anthropogenic origin. Here, we consider only one example of oil pollution, namely, a matter of investigation by means of AI-1 and AI-2 of some water bodies situated in the territory of South Vietnam. The most important aquatory of South Vietnam is the area Ω of the South China Sea bordering on Vung Tau City where there is exploration and extraction of oil from the sea bottom. In 1975, the Vietnam Oil and Gas General Directorate was set up. The majority of the oil companies (Vietso Petro, PSCT, Gas Company, Petroleum Assembling Enterprise Cooperation) realize national and international projects concerning the exploration and exploitation of the GEZ.

The quality of seawater in this area mainly depends on the GCD zone and the flows of the Mekong and Saigon rivers. The aquatory Ω is restricted in the north by the coastline running in a northeasterly direction, in the west and east by the meridians 106° and 109°, respectively, and in the south by the parallel 8.5°N latitude. The water current scheme of this aquatory is considered as shown. According to the existing data, it can be approximated by a seasonal scheme with the attachment to winter (τ_w), spring (τ_s), summer (τ_u), and fall (τ_a). On average, during the fall and winter, the currents are directed along the coastline in a northeasterly direction ($\alpha = 45°$). During the remaining time, the currents are in the opposite direction ($\alpha = 225°$). The speed value varies insignificantly both in

Table 10.3. Results of measurement data processing during the field study of oil products content in the water (mg l^{-1}) on the South Vietnam territory

Depth (m)	Saigon River			Dongnai River		
	Day hours					
	9	15	18	9	15	18
0.1	0.052	0.101	0.047	0.019	0.019	0.011
0.5	0.061	0.094	0.06	0.012	0.013	0.015
1.0	0.063	0.085	0.053	0.012	0.012	0.013
1.5	0.066	0.103	0.061	0.010	0.010	0.011
2.0	0.054	0.077	0.088	0.009	0.010	0.007
2.5	0.051	0.049	0.064	0.008	0.007	0.006
3.0	0.062	0.068	0.048	0.006	0.008	0.005
4.0	0.043	0.094	0.034	0.004	0.009	0.004
5.0	0.048	0.083	0.026	0.001	0.002	0.006
6.0	0.059	0.074	0.055	0	0	0
7.0	0.062	0.052	0.084	0	0	0
8.0	0.041	0.043	0.054	0	0	0
9.0	0.042	0.049	0.077	0	0	0
10.0	0.037	0.055	0.038	0	0	0

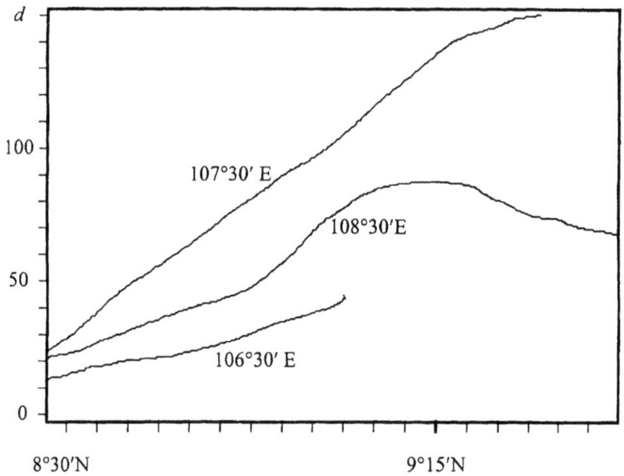

Fig. 10.9. The ratio, *d*, of oil hydrocarbon content in the upper sea layer of 1 m depth to the content in the thick water layer as a function of geographical coordinates. The modeling results are averaged by all seasons. It is supposed that the oil hydrocarbon concentration in $\Omega_R \cup \Omega_P$ and in GCD zone is constant and is equal to 0.06 mg l^{-1}

space and time. Near the coastline this speed equals 2–6 km h^{-1}. Further from coastline, the speed variations are 0.7–4.8 km h^{-1}. It is supposed that the depths h_{ij} are constant on the whole aquatory Ω. The vertical gradient of the water temperature is considered as negligibly small. The synoptic regime during the year is approximated by a binary situation: the rainy season exists from May to October, the dry season is from November to April.

Table 10.3 consists of some results of oil pollution measurements by means of AI-1. Simulation results are given in Figs. 10.9 and 10.10. As follows from Fig. 10.9 under the above suppositions, the average annual distributions of oil hydrocarbons on the aquatory Ω are essentially nonuniform. Taking an average of the seasons eliminated the influence of pollution drift by the currents and revealed

Fig. 10.10. The map-scheme of yearly averaged distribution of the oil hydrocarbons in the GCD zone of Vung Tau City in 2001. These estimations are calculated by means of the model based on the initial data of 1994. The scale step is 0.01 mg · l^{-1}

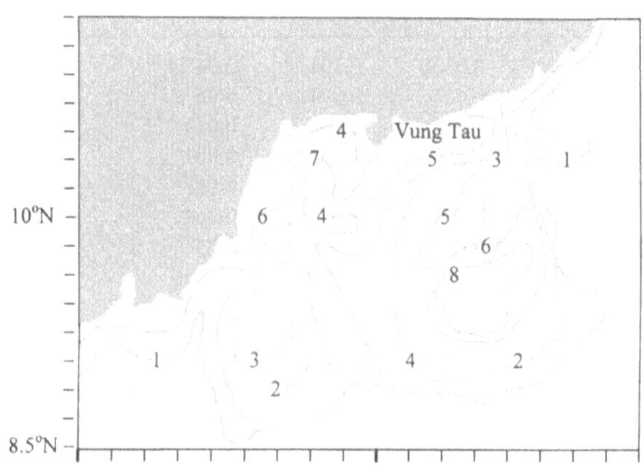

the places where oil hydrocarbons always come to the upper sea layer. The degree of oil hydrocarbon accumulation in the surface-water layer varies between the limits of 25–130 mg m^{-2}. This is several times higher than the oil hydrocarbon concentration in the open ocean and significantly lower than in the very polluted waters.

Figure 10.9 gives the forecast of oil hydrocarbon distribution on the whole aquatory Ω. Here, it was supposed that the oil hydrocarbon concentration on the boundary Ω_B is a constant equal to 0.001 mg l^{-1} and is brought into Ω in conformity with the scheme of the currents stored in the ESAIEP database.

As seen from Fig. 10.9, the stabilized distribution $\int O(\varphi,\lambda,z,t)dz$ is sufficiently well coordinated with the direction of the currents. The formation of a field with increased oil hydrocarbon concentration situated near the Vung Tau shelf zone is explained by the supposition that in the GCD site the following conditions occur: $O(\varphi,\lambda,z,t) = 0.09$ mg l^{-1} in the upper layer, and $O(\varphi,\lambda,z,t) = 0.05$ mg l^{-1} in the remaining water depth. Under other suppositions the structure of the distribution $O(\varphi,\lambda,z,t)$ is unchanged. Only the pollution scales are changed.

The creation of the SEM for the GCD situated in the South China Sea foresees the detection of oil pollution sources. That is why AI-1 was used to estimate the oil pollution of the main rivers of South Vietnam (Saigon and Dongnai). Apart

Table 10.4. Results of model-based calculations of pollution levels of the Saigon River based on the measurements with AI-1. The river velocity is equal to 3 km h^{-1}. Measurements were taken on 15 November 1994 from 10:00 to 15:00 h. Designations: x is the distance from Ho Chi Minh River port along the Saigon River downstream (km), Δ is the model error (%), TDS (total dissolved solids), TSS (total suspended solids), BOD (biological oxygen demands), symbols M and A correspond to the model and experiment, respectively

x	Water quality factors					
	NO_3 (mg l^{-1})	P_2O_5 (mg l^{-1})	TDS (mg l^{-1})	TSS (mg l^{-1})	BOD_5 (mg O_2 l^{-1})	Oil (mkg l^{-1})
0	0.80	0.10	50.1	15.5	2.2	0.40
1	0.81	0.12	50.1	15.5	2.4	0.40
2	0.83	0.15	49.9	15.4	2.5	0.41
3	0.92	0.21	50.1	15.5	2.8	0.42
4	0.96	0.24	52.3	15.9	2.9	0.44
5	1.11	0.35	54.2	16.7	3.3	0.45
6	1.26	0.42	60.9	17.2	4.0	0.46
7	1.35	0.44	65.8	19.8	5.1	0.47
8	2.09	0.91	70.3	25.1	6.2	0.53
9	3.41	1.23	77.7	30.2	6.8	0.56
10	4.18	2.42	80.1	37.3	7.7	0.58
11	6.01	3.09	90.4	41.8	8.3	0.60
12	6.12	3.17	91.3	42.7	8.4	0.61
13	5.99	3.18	91.2	41.9	8.3	0.63
14	6.03	3.21	91.1	41.8	8.5	0.65
15 M	6.07	3.19	91.1	41.9	8.6	0.65
15A	6.8	7.32	4.01	130.5	58.3	0.92
Δ	17.08	20.45	30.19	28.13	30.08	29.34

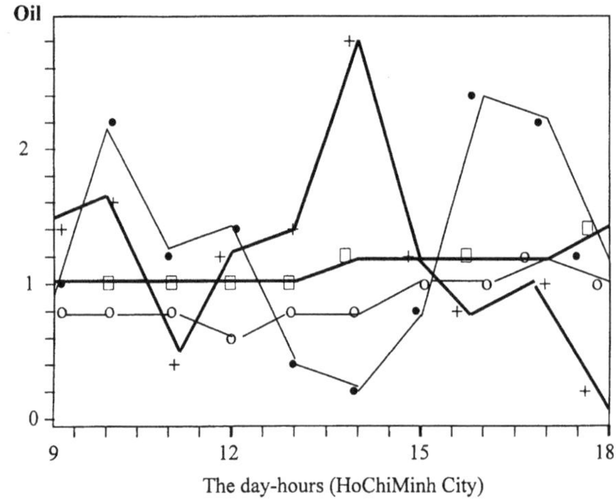

Fig. 10.11. The results of field experiments during the Vietnamese/Russian hydrophysical expedition of 1994 when the oil pollution of the Saigon and Dongnai Rivers was studied by means of AI-1. The results of the measurements are marked by closed circles and open circles for the Saigon River and by + and open squares for the Dongnai River on 14 and 15 November, respectively. The scale step is 0.05 mg l^{-1}

from this, the task of distinguishing between oil and salt spills on the water surface of reservoirs, where industrial biological production is realized, was solved. In addition, the control of water quality in the Dongnai river is important for the ecological service of Ho Chi Minh City.

Table 10.4 and Fig. 10.11 present some experimental data. It can be seen that the Saigon River is characterized by unstable formations of oil clots moving randomly in depth. The predictability of the Saigon River pollution level equals 60–70% during 1 day. It is clear that including the SEM functions in the control of Saigon River pollution demands additional information about the dynamics of the rise and fall of the tides. This information is to be parameterized and stored in the SEM knowledge base, or it is to come from the measuring subsystem.

At present, the Dongnai River has a low pollution level. Measurement data received from AI-1 show a high stability of oil pollution distribution in this river. Therefore, the realization of the SEM function to control oil pollution of the Dongnai River can be considered at a future stage.

Finally, Fig. 10.11 presents hourly dynamics of the oil pollution level for both rivers. The measurements with AI-1 are realized for the Saigon River at its midpoint in Ho Chi Minh City center and for the Dongnai River in the drinking water station at a distance of 30 m from the waterfront. A comparison of the curves from Fig. 10.11 shows that control of the oil pollution near this station is realized by means of AI-1 in the continuous regime and can control the water standard. It is more difficult to control the water standard in the Saigon River where hourly variations of the pollution level oscillate between the limits of 100–300%. Such oscillations are caused by complex hydrodynamic conditions of the spreading of the pollutants. A combination of high current velocity, rise and fall of the tides, and high turbulence causes the formation of pollutant clots, which move as closed water volumes.

Decision-Making Procedures in Environmental Monitoring Systems

11

11.1 The Problem of Statistical Decision-Making and Basic Definitions

The environmental monitoring regime can foresee decision-making situations in real time based on the data accumulated up until the moment of the decision-making or as a result of the prior data analysis without correlation to the current time. Statistical analysis of the events accompanying the functioning of the monitoring system can be realized by means of many methods, the applicability of which is determined in each case by the probable combination of parameters characterizing the process studied. However, nonstationary and parametrical uncertainty demand a search for new methods capable of making the decision using fragmentary time and space data. These kinds of methods are proposed by a sequential analysis and evolutionary technology (Bukatova et al. 1991; Krapivin et al. 1997b; Nitu et al. 2000b).

11.1.1 Correlation Between Classical and Sequential Decision-Making Procedures

The classical decision-making procedure is based on the given volume, n, of the measurements. The size of n is determined by a priori information about the probability density function $f_a(x_1,\ldots,x_n)$ where the random variables $\{x_i\}$ are the observation data. The hypotheses H_0 and H_1 are such that a 0 or 1 is prescribed, respectively. The distinction between these hypotheses is based on the synthesis of the boundary for the optimal critical area, E_1, in the hypersurface of the form:

$$L_n = L_n(x_1,\ldots,x_n) = f_{a1}(x_1,\ldots,x_n)/f_{a0}(x_1,\ldots,x_n) = C \qquad (11.1)$$

where

$$f_a(x_1,\ldots,x_n) = \prod_{i=1}^{n} f_a(x_i)$$

$f_a(x)$ is the probability density function of the variable x with the unknown parameter a and C is a constant determined by the conditions that E_1 has a given level of the error of the first kind α.

The ratio of conditional probabilities in Eq. (11.1), called the likelihood ratio, provides the final choice between the above hypotheses: (1) if $L_n \le C$ then hypothesis H_0 is accepted; (2) if $L_n > C$ then hypothesis H_1 is chosen. In general, many criteria exist: Bayes, minimum error, minimax, etc.

The content of hypotheses H_0 and H_1 depends on the specific conditions of the task. In reality, there are two steps to use for these criteria. The first step is the synthesis of the empirical function of the probability distribution for the observed variable x. The second step is its transformation to $f(x)$ using, for example, the Neyman-Pearson criterion. The sequential procedure of the decision-making process does not separate these steps.

Let us consider the case of a uniform data set where the selected values x_i ($i = 1,...,n$) are independent realizations of the same casual value ξ having the density $f_{a0}(x)$ for the hypothesis H_0 and the density $f_{a1}(x)$ when the hypothesis H_1 is true. For this case, the parameter a of the real density f_a cannot be equal to a_0 or a_1.

The errors of the first kind, α, and the second kind, β, satisfy the following formulas:

$$\alpha = \exp[-0.5\{(E_{a1}\xi - E_{a0}\xi)(D_{a0}\xi)^{-1/2}\}^2 n], \tag{11.2}$$

$$\beta = \exp[-0.5\{(E_{a1}\xi - E_{a0}\xi)(D_{a1}\xi)^{-1/2}\}^2 n], \tag{11.3}$$

where

$$E_a\xi = \int_{-\infty}^{\infty} \ln\left[f_{a1}(x)/f_{a0}(x)\right]f_a(x)dx \tag{11.4}$$

$$D_a\xi = \int_{-\infty}^{\infty} \{\ln\left[f_{a1}(x)/f_{a0}(x)\right]\}^2 f_a(x)dx - \left(E_a\xi\right)^2 \tag{11.5}$$

The sequential procedure has the following basic characteristic:

$$L(a) \approx [A^{h(a)}-1]/[A^{h(a)}-B^{h(a)}], \tag{11.6}$$

where A and B are the boundaries for the $L_n(x)$ and $h(a)$ is the solution of the equation:

$$\int_{-\infty}^{\infty} \left[f_{a1}(x)/f_{a0}(x)\right]^{h(a)} f_a(x)dx = 1 \tag{11.7}$$

The values of A and B have the estimations:

$$B \approx \beta/(1-\alpha),\ A \approx (1-\beta)/\alpha \tag{11.8}$$

Accordingly, $L(a_0) = 1-\alpha$ and $L(a_1) = \beta$. From this, it follows that the average number of observations in a sequential procedure equals:

$$E_a\,\nu = \alpha)\ln[\beta/(1-\alpha)] + \alpha\ln[(1-\beta)/\alpha]/E_{a0}\,\xi, \text{when } a = a_0,$$

$$E_a\,\nu = [(1-E_a\,\nu = [\beta\ln[\beta/(1-\alpha)] + (1-\beta)\ln[(1-\beta)/\alpha]/E_{a1}\,\xi, \text{ when } a = a_1 \tag{11.9}$$

For the value $a = a^*$ and when $E_{a*}\,\xi = 0$ and $E_{a*}\,\xi^2 > 0$ we have:

$$E_{a*}\,\nu \approx [-\ln[\beta/(1-\alpha)]\ln[(1-\beta)/\alpha]/E_{a*}\,\xi^2 \tag{11.10}$$

11.1.2
Distribution of the Sequential Analysis and Its Universality

According to (11.9) and (11.10), the number of observations of a sequential procedure is a random variable, ν, the average value of which ($E_a\nu$) can be smaller or larger than n. To judge the possible values of ν, it is necessary to have the distribution $P(\nu = n) = P_a(n)$:

$$E_a\, \nu\, P_a(n) = W_c(y) = c^{1/2} y^{-3/2} (2\pi)^{-1/2} \exp[-0.5c\,(y + y^{-1} - 2)], \tag{11.11}$$

where $0 \leq y \leq n \mid E_a\xi \mid < \infty$, $c = K \mid E_a\xi \mid / D_a\xi = (E_a\,\nu)^2 / D_a\,\nu > 0$, $D_a\,\nu = K\,D_a\xi / (E_a\,\xi)^3$, $E_a\,\nu = K / E_a\,\xi$, $K = \ln A$ for $E_a\,\xi > 0$ and $K = \ln B$ for $E_a\,\xi < 0$.

The distribution function $W_c(y)$, according to (11.11), has the form:

$$W_c(x) = \int_0^x w_c(z)\, dz \tag{11.12}$$

where $w_c(z) = (c/2\pi)^{1/2} z^{-3/2} \exp[-0.5c(z + z^{-1} - 2)]$.

The universality of the distribution (11.12) follows from its duality to the Gaussian distribution. As far back as 1960, Wald showed that if $|E_a\,\xi|$ and $D_a\,\xi$ are sufficiently small in comparison with $\ln A$ and $\ln B$ then the distribution of $\nu / E_a\,\nu$ defined by the expression (11.11) will be a close approximation to the real one, even for ξ not distributed by the Gaussian law.

Theoretical aspects as to the universality of the distribution $W_c(x)$ are important for the integrated estimation of the sequential procedure efficiency. However, these are not the principal aspects for experimental applications. That is why, as a rule, the synthesis of the decision-making system is realized without these considerations. In reality, the measurement volume, as a rule, is small and the central limit theorem does not work. The statistical difficulties arising under this can be overcome by evolutionary modeling (Bukatova et al. 1991), intelligent technology (Nitu et al. 2000b) and the use of other algorithms.

11.1.3
Scheme of the Decision-Making Procedure Using Sequential Analysis

In contradiction to the Neyman-Pearson criterion, the sequential procedure does not separate the stages of measurement and data processing, rather it alternates these stages. Schematically, this is represented in Fig. 11.1. For comparison, the classical procedure is characterized by Fig. 11.2. From these figures, it is seen that the algorithmic load of the sequential procedure is dynamically changed while at the same time the classical procedure realizes the data processing stage only on the finishing step of the experiment. In this connection, the synthesis of the efficient decision-making system poses the following problems:

1. selection of the criterion for the parameters estimation;
2. revealing of the probabilistic characteristics of the process studied;
3. a priori estimation of the possible losses concerning the precision of the decisions taken;
4. prognosis of the dynamic stability for the experiment results.

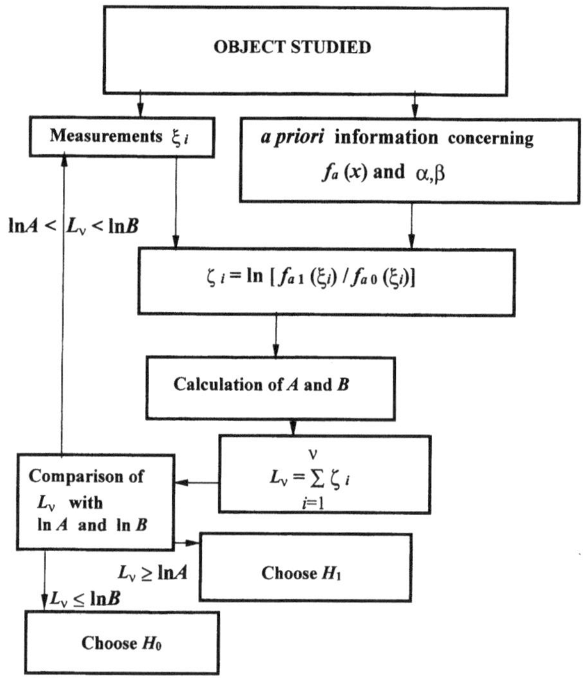

Fig. 11.1.
Scheme of the experimental realization using the sequential analysis procedure to decide between hypotheses H_0 and H_1

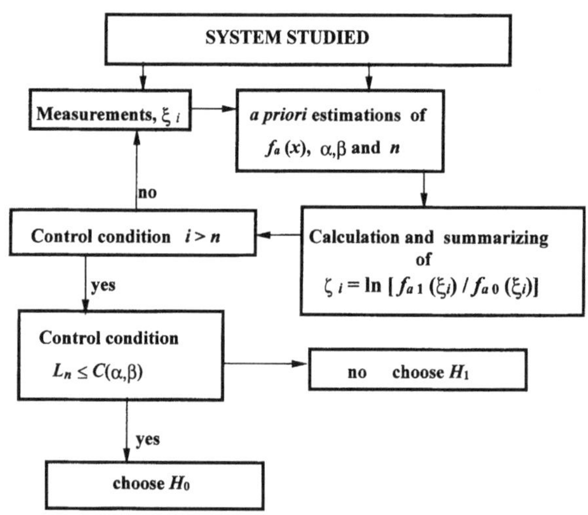

Fig. 11.2.
Classical Neyman-Pearson decision-making procedure to choose between the two hypotheses H_0 and H_1

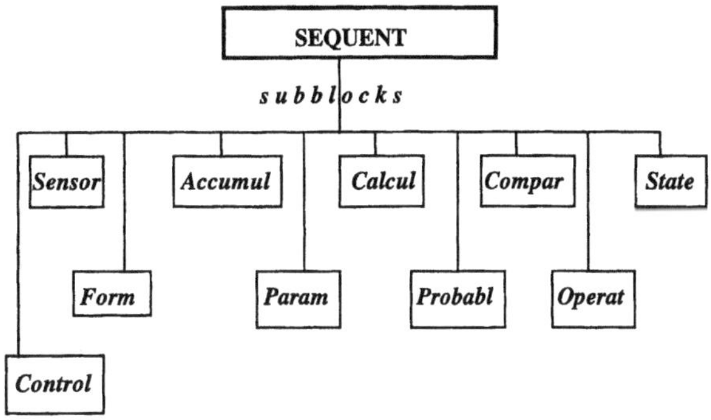

Fig. 11.3. Structure of the unit SEQUENT. A description of the subunits is given in Table 11.1

The decision-making system is to have a wide spectrum of functions:

1. visualization of the measurement data in the form of direct soft copy, discrete distribution and statistical parameters;
2. calculation of the statistical characteristics (mean, central second- and third-order moments, asymmetry and excess coefficients, expression of the entropy, etc.);
3. synthesis of the empirical and theoretical distribution functions;
4. valuation of the parameters to be used in the Neyman-Pearson and the sequential procedures of the hypothesis decision;
5. realization of the user access to all the functions of the decision-making system.

According to the scheme represented in Fig. 11.3 the decision-making system is to have an expert control level. Namely, the unit SEQUENT realizes through its inputs and outputs the control of the decision-making procedure. According to the functions of the subunits described in Table 11.1, the user can interfere promptly in any arbitrary stage of the computer experiment, correcting the parameters of the decision-making procedure or ceasing it. The subunit

Table 11.1. Description of the subunits of the SEQUENT unit

Subunit	Description
Sensor	Inputting of data for the values of α, β, a, $\{x_i\}$
Accumul	Accumulation of measurements as the likelihood ratio
Calcul	Evaluation of the thresholds A and B
Compar	Comparison of the likelihood ratio with the thresholds A and B
State	Visualization of the sequential procedure state
Form	The THEOR unit query with respect to the form of $f_a(x)$ and the activation of the appropriate knowledge base level
Param	Evaluation of E_v, D_v, c
Probabl	Computation of the probability of completion of the sequential procedure
Control	Control of the type of task for the choice between H_0 and H_1 with consideration of the errors of the first and second kind
Operat	Management of the operative intervention to the functioning of the SEQUENT unit

Fig. 11.4.
An example of the dynamics of the accumulated sum of the likelihood function logarithm visualized by the subunit *State*

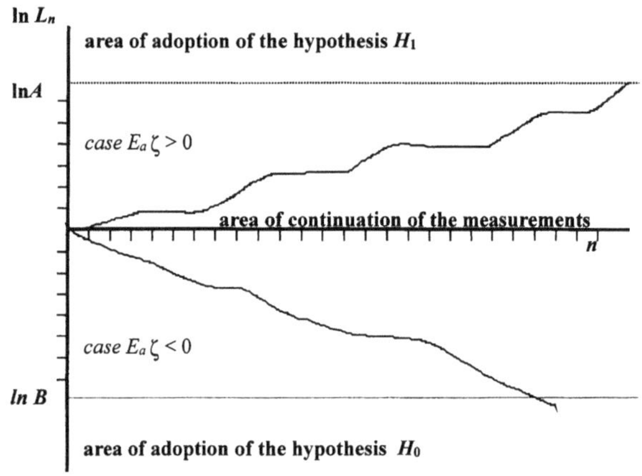

CONTROL manages the calculation process taking into account the character of the task. It forms the variants that correspond to the concrete combination of the errors of the first and second kind, α and β. Based on this combination, the subunit CONTROL produces a set of parameters to manage the other subunits. Depending on α and β, the simplified procedures are possible. For example, two variants for the asymmetric thresholds A and B are: (1) $B = \beta/(1-\alpha) \to 0$, $A = (1-\beta)/\alpha \to$ const; and (2) $B = \beta/(1-\alpha) =$ const, $A = (1-\beta)/\alpha \to \infty$. In other words, the errors α and β are unequal in value, namely: (1) $\beta \to 0$, $\alpha =$ const; (2) $\beta =$ const, $\alpha \to 0$. In these cases, the sequential procedure will be finished with the probability equal to 1 if the following conditions are in place: (1) $E_a\,\zeta > 0$, $\zeta = \ln[\,f_{a1}(\xi)/f_a(\xi)]$; (2) $E_a\,\zeta_0 < 0$. The probability of completion of the procedure is small when one of the following conditions are not realized: (1) $\beta \gg \alpha$, $E_a\,\zeta > 0$; or (2) $\alpha \gg \beta$, $E_a\,\zeta < 0$. The user can visualize the state of the procedure as shown in Fig. 11.4.

The unit MEASURE coordinates the system input with the separate channels of measured data. Each realization of a random process $\{x_i\}$ performs analyses to eliminate errors and to represent inputting signals in the form that is acceptable for the other units. The unit STATIST calculates signal characteristics:

$$M_1 = \frac{1}{n}\sum_{i=1}^{n} x_i,\; M_2 = \frac{1}{n}\sum_{i=1}^{n}\left(x_i - M_1\right)^2,\; r_1 = M_3 M_2^{-3/2},$$

$$V = \frac{\sqrt{M_2}}{M_1}\cdot 100\%,\; M_3 = \frac{n}{(n-1)(n-2)}\sum_{i=1}^{n}\left(x_i - M_1\right)^3,$$

$$r_2 = \frac{M_4}{M_2^4} - \frac{3(n-1)^2}{(n-2)(n-3)},\; M_4 = \frac{n(n+1)}{(n-1)(n-2)(n-3)}\sum_{i=1}^{n}\left(x_i - M_1\right)^4,$$

$$H = -\int_{-\infty}^{\infty} f_a(x)\ln\left[f_a(x)\right]dx,\; R_i = x_{i,\max} - x_{i,\min},\; M_6 = \frac{M_4}{M_2^2} - 3$$

These characteristics and other standard parameters are used to reconstruct $f_a(x)$. For instance, if $M_3 \approx 0$ and $r_1 \approx 0$ then f_a should be searched in symmetric

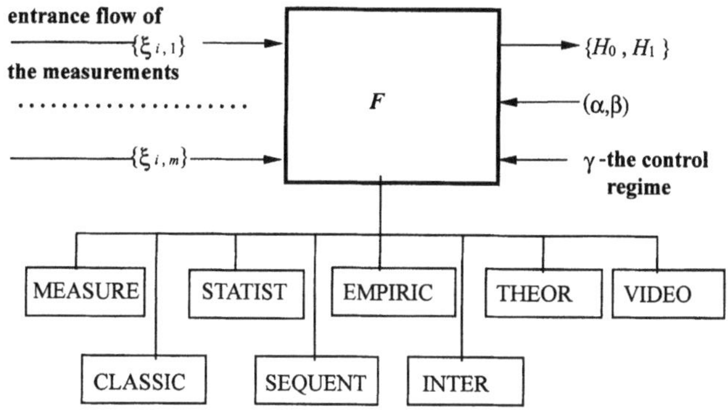

Fig. 11.5. Principal scheme of the decision-making system based on the sequential analysis procedure

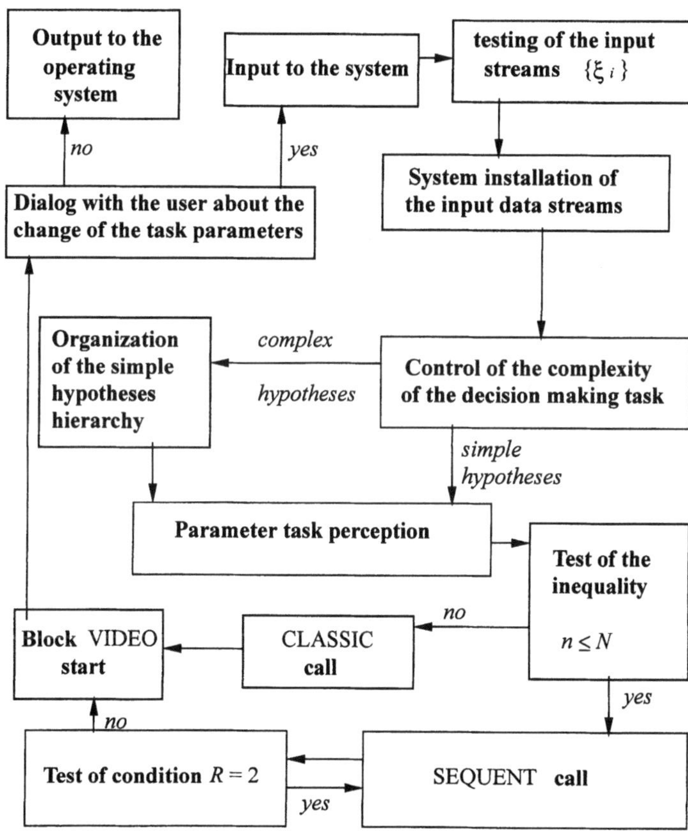

Fig. 11.6. Block scheme of the data processing in the sequential analysis procedure

distributions. The unit EMPIRIC realizes this search selecting from the knowledge base continuous distributions and estimating their closeness to the empirical distribution using the criterion:

$$X^2 = \sum_{i=1}^{L} \frac{\left[m_i - f_a(x_i)\Delta x_i\right]^2}{f_a(x_i)\Delta x_i}$$

The continuous distribution f^* is chosen from the following condition:

$$\min_{f_a} X^2(f_a) = X^2(f^*)$$

The decision-making system is synthesized according to the principal scheme of Fig. 11.5. Its functioning scheme is characterized by Fig. 11.6.

11.2 Parametrical Estimations for Sequential Analysis

An investigation of the sum x_n of independent random values having the same distribution imposes a double task on the distribution function, $P(x_n < x) = F_1(x)$, both for fixed n and for the variable case. The situation of comparison between x_n and some level C arises in both cases. However, in the second case this task is transformed into the study of the distribution function, $P(\nu < n) = F_2(n)$, of the chance number ν of components by which x_ν first exceeds the level $C = C(\alpha,\beta)$: $x_i < C$ $(i = 1,..., \nu\text{-}1)$, $x_\nu \geq C$.

In conformity with the central limit theorem, in the first case, the distribution $F_1(x)$ under $n \to \infty$ approaches the normal distribution. In the second case, we have the distribution represented by expression (11.12). The following correlation between these distributions exists:

$$W_c(x) = \Phi[(x-1)(c/x)^{1/2}] + \Phi[-(x+1)(c/x)^{1/2}] \exp\{2c\}, \qquad (11.13)$$

where Φ is the normal distribution function.

The expression (11.13) makes it possible to study the sequential analysis distribution using the characteristics of Φ. As seen from (11.11), the distribution function $W_c(x)$ is defined on the half-space $[0,\infty]$ and it has one maximum at the point $x = m_c$. In fact, we have:

$$dW_c(y)/dy = (d/dy)\{c^{1/2}/[y^{3/2}(2\pi)^{1/2}] \exp[-0.5c(y + y^{-1} - 2)]\} = 0$$

After the set of transformations, this equation is solved:

$$y = m_c = [(9 + 4c^2)^{1/2} - 3]/(2c) \qquad (11.14)$$

The position of the $W_c(y)$ maximum changes depending only on the parameter c remaining less than one ($m_c < 1$). Moreover, $m_c \to 0$ when $c \to 0$ and $m_c \to 1$ under $c \to \infty$. Comparing Eqs. (11.1) and (11.14), we find:

$$W_c(m_c) = (3/(2\pi)^{1/2})(1/[m_c(1 - m_c^2)^{1/2}]) \times \exp\{-3(1-m_c)/[2(1 + m_c)]\} \qquad (11.15)$$

Hence, the function $W_c(y)$ degenerates under $c \to 0$ or $c \to \infty$ into the delta-functions $\delta(y)$ or $\delta(y - 1)$, respectively.

Taking into account that $W_c(x) \to 1$ when $x \to \infty$, from Eq. (11.12) we have:

$$W_c(x) = \int_0^x W_c(y)dy = \int_{1/x}^{\infty} zW_c(z)dz = 1 - \int_0^{1/x} zW_c(z)dz \tag{11.16}$$

The important parameters of the distributions Φ and W_c are solutions of the following equations:

$$\Phi(u_p) = p,\ W_c(x_p) = p \tag{11.17}$$

It is easy to see that when $c > u_p^2 / 4$ we find:

$$1 + (u_p - \varepsilon)c^{-1/2} + (u_p^2 - \varepsilon)c^{-1}/2 + (u_p^3 - \varepsilon)c^{-3/2}/8 - (u_p^5 - \varepsilon)c^{-5/2}/128 < x_p(c) < 1 + u_p c^{-1/2} + u_p^2 c^{-1}/2 + u_p^3 c^{-3/2}/8 \tag{11.18}$$

where

$$\varepsilon = \Phi(-2c^{1/2})\exp\{2c\} < 0.5(2\pi c)^{-1/2} \tag{11.19}$$

For $c < u_p^2 / 16$ we have:

$$c / u_a^2 < x_p(c) < c / u_b^2 \tag{11.20}$$

where $a = 0.5\,p\exp(-2c)$ and $b = p/2$.

When $c \to \infty$, we have:

$$x_p(c) = 1 + u_p c^{-1/2} + O(1/c) \tag{11.21}$$

From Eqs. (11.13) and (11.21) we obtain:

$$W_c(x) = \Phi[(x-1)c^{1/2}] + O(1/c) \tag{11.22}$$

Thus, the random value v/E_v is asymptotically normal with average value and dispersion equal to 1 and $1/c$, respectively, when $c \to \infty$. The expressions (11.13) and (11.22) can receive various analytical approximations of W_c. For example, let us represent $\Phi(y)$ by the following:

$$\Phi(y) = \Phi(y_0) + \varphi(y_0)(y - y_0)[1 - 0.5y_0(y - y_0)], \tag{11.23}$$

where $y = (x-1)(c/x)^{1/2}$, $y_0 = (x_0 - 1)(c/x_0)^{1/2}$ and x_0 is some point where the value of Φ was estimated.

The following formula can be easily derived using Eqs. (11.12), (11.13), and (11.23):

$$W_c(x) = W_c(x_0) + \varphi(y_0)(y - y_0)[1 - 0.5y_0(y - y_0)] + \varphi(v_0)(v - v_0)\exp(2c)[1 - 0.5v_0(v - v_0)], \tag{11.24}$$

where $v = -(x+1)(c/x)^{1/2}$, $v_0 = -(x_0+1)(c/x_0)^{1/2}$ and $\varphi(z) = \varphi(z_0)\,[1 - z_0(z - z_0) + 0.5(z_0^2 - 1)(z - z_0)^2]$.

Let us designate

$$H(x) = \frac{2}{\sqrt{\pi}} \int_0^x \exp\left(-t^2\right)dt \tag{11.25}$$

We have

$$H(x)=\begin{cases} \dfrac{2}{\sqrt{\pi}}\sum\limits_{k=1}^{\infty}(-1)^{k+1}\dfrac{x^{2k-1}}{(2k-1)(k-1)!} & \text{when} \quad |x|\leq x_1, \\ \dfrac{2}{\sqrt{\pi}}\exp(-x^2)\sum\limits_{k=0}^{\infty}\dfrac{2^k x^{2k+1}}{(2k+1)!!} & \text{when} \quad x_1<|x|<x_2, \\ 1-\dfrac{1}{\pi}\exp(-x)\sum\limits_{k=0}^{\infty}(-1)^k x^{-(k+1/2)}\Gamma(k+1/2) & \text{when} \quad |x|\geq x_2. \end{cases}$$

The functions Φ and H are connected by the obvious correlation:

$$\Phi(x) = [1 + \sin(x)H(2^{-1/2}\,|\,x\,|)/2, \tag{11.26}$$

from which we have:

$$\Phi(x)=\begin{cases} g(x)\, for \quad x\leq 0, \\ 1-g(x)\, for \quad x>0, \end{cases}$$

where

$$g(x)=\begin{cases} 1/2-\dfrac{1}{\sqrt{2\pi}}\sum\limits_{k=1}^{\infty}(-1)^{k+1}\dfrac{|x|^{2k-1}}{2^{k-1}(2k-1)(k-1)!} & for \quad |x|\leq x_1, \\ 1/2-\dfrac{1}{\sqrt{2\pi}}\exp(-x^2/2)\sum\limits_{k=0}^{\infty}\dfrac{|x|^{2k+1}}{(2k+1)!!} & for \quad x_1<|x|<x_2, \\ \dfrac{1}{\sqrt{2\pi}}\exp(-x^2/2)\sum\limits_{k=0}^{\infty}(-1)^k\dfrac{(2k-1)!!}{|x|^{2k+1}} & for \quad |x|\geq x_2. \end{cases}$$

Thus, formula (11.26) allows us to calculate $W_c(x)$ for various values of c and x. The free parameters x_1 and x_2 influence the calculation error. Practically, the values $x_1 = 2.2$ and $x_2 = 7.5$ are acceptable. Also, there is some problem of convergence of the above rows when parameter c is increasing. It is easy to see that $W_c(x) \approx \Phi[(x-1)c^{1/2}]$ for $c \gg 1$ and $W_c(x) \approx 1 - \exp(-cx/2)$ when $x \to \infty$.

The formula (11.13) is the basic expression for the calculation of $W_c(x)$. The large factor, $\exp(2c)$, can be neutralized by the following expression:

$$\exp(2c)\Phi\left[-(x+1)\sqrt{c/x}\right]=\frac{1}{\sqrt{2\pi}}\exp(2c-x^2/2)\sum_{k=0}^{\infty}(-1)^k\frac{(2k-1)!!}{x^{2k}}$$

The calculation of $W_c(x)$ can be also realized using the Bessel (J) and Whittaker (Ψ) functions:

$$W_c(x)=\sqrt{\frac{c}{2\pi}}\exp\left[c(1-\frac{1}{2x})\right]\sum_{k=-\infty}^{\infty}\frac{x^l}{c^m}J_k(-c)\Psi_{-l,-s}(-c/x)$$

where $l = (2k + 3)/4$, $m = (6k + 9)/4$ and $s = (1 + 2k)/4$. Here, also the following correlations are very useful:

$$\exp\left[(1/z - z)c/2\right] = \sum_{k=-\infty}^{\infty} J_k(-c)z^k, \int_u^{\infty} e^{-x}x^{-v}dx = u^{-v/2}e^{-u/2}\Psi_{-v/2,(1-v)/2}(u)$$

Finally, let we consider the following algorithm to evaluate the distribution function $W_c(x)$:

$$w_c(x) = A_0\varphi_0(x) + A_1\varphi_1(x) + \ldots + A_m\varphi_m(x) + \ldots,$$

where

$$A_i = \frac{(-1)^i}{i!} \int_{-\infty}^{\infty} w_c(z)R_i(z)dz$$

and $R_i(z)$ is the Chebyshev-Ermite polynomial: $R_i(z) = (-1)^i \varphi_i(z)/\varphi_0(z)$. Here, it is designated by $\varphi_0(z) = (2\pi)^{-1}\exp(-t^2/2)$, $\varphi_i(z) = d^i \varphi_0(z)/dz^i$, $(i = 1,2,\ldots)$.

Also, the normalization requirement is accepted:

$$\int_{-\infty}^{\infty} \varphi_0(z)R_i(z)R_j(z)dz = \begin{cases} 0, & i \neq j, \\ 1, & i = j \end{cases}$$

Thus, the sequential analysis distribution should be written in the form:

$$W_c(x) = \varphi_0\left[\sqrt{c}(x-1)\right] + \sum_{k=1}^{\infty} (-1)^k c^{-k/2} \frac{(2k+1)!!}{(k+2)!} \varphi_{k+1}\left[\sqrt{c}(x-1)\right]$$

11.3 An Algorithm for Multichannel Data Processing in the Decision-Making Task

The schematic diagram of a monitoring system for detecting anomalies on the Earth's surface involves many levels. The search organization structure may contain a more profound hierarchy including the processing of information from satellites, airborne and floating laboratories, and stationary systems. In spite of the possibility of the definite realization of this structure and the applicable computer technique, there exists, nevertheless, one general problem whose solution determines the detection efficiency. This task involves the organization of in-line data processing at all detection system levels under real-time conditions, ensuring an uninterrupted matching of the operation of all the cybernetic devices of the system, taking into account the limitation on their functional characteristics.

Statistical analysis of the information at each level of the monitoring system is completed by making the following two subsequent decisions: (1) storage of features of the landscape elements for the accumulation of additional information about them, and (2) completion on the presence of an anomaly and transmission of its characteristics to the next data processing level.

These two stages determine the efficacy of the detection system. At the first stage, the selection and storage of parameters of the elements of landscapes suspected of being anomalous are dependent on the character of the algorithm being used for singling out the two-dimensional signal against the background noise and are associated with the gradual filling of the main storage. The second stage determines the magnitude of the detection probability depending on the criteria of registration of the landscape elements selected at the first stage for further analysis. As a result, there arises a problem, that of matching the flow of information between the processing stages under monitoring conditions and at minimal losses. The solution of this problem involves two procedures of parallel component processing, forming the intermediate information delay either as a constant computer memory buffer size, or for a fixed storage time.

11.3.1 Organizational Scheme of the Statistical Analyzer Operation

Let the on-board computer memory have M cells. The statistical analyzer realizes with time the simultaneous parallel selection of components of the "suspected" variant of the vectorial parameter $b = b^*$. This is achieved by the work of the K consecutive devices or by algorithms of selection of component values b_i ($i = 1,..., K$) corresponding to the space of anomalous features. To "stabilize" the process of selecting the anomaly variant, it is possible to create constant delays of two types between the selection devices of the anomalous components: constant delay with number of "candidates" and constant delay with time (see Fig. 11.7).

The two delay modes allow one to have a more flexible organization of the distribution of M memory elements for storing the intermediate values of the "suspected" components. Because of the random character of the whole process,

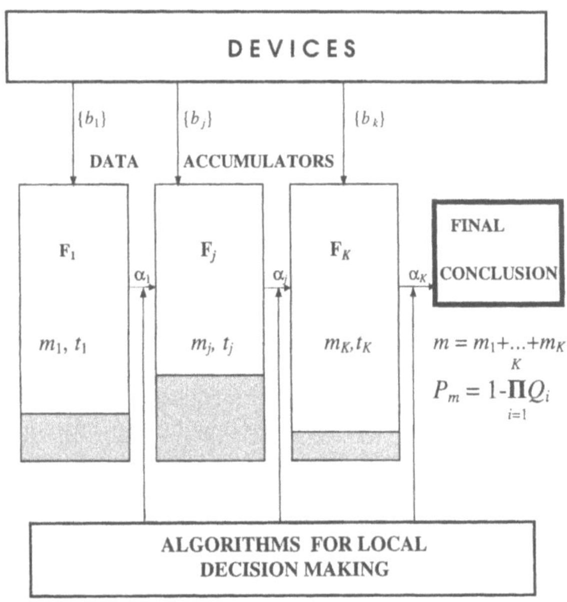

Fig. 11.7. Schematic diagram of the consecutive simultaneous exhaustive search for numerical values of components of the vectorial parameter

the statistical analyzer is apt to make transient errors due to the overflow of the computer memory capacity intended for the delays. Therefore, it is necessary to obtain an assessment for the probability of appearance of errors and to find memory capacities for delays $M_{1+}\cdots + M_k < M$ and for the distribution of time delay intervals $t_1,\ldots, t_K$.

11.3.2 Error Probability Assessment of the System and the Requisite Delay Memory Capacity with Constant Expectation Time

Let selection b_i^* be realized in the volume sample n_i, basing each selection on N_i reasonable values b_i. In this case, the nonanomalous value b_i is taken as the anomalous value with probability α_i and is rejected with probability $1 - \alpha_i$. The probability of appearance of "candidates" during time $t_i = r_i\, \alpha_i\, S_i$ equals:

$$P\left\{\mu_i = S_i\right\} = C_{r_i}^{S_i} \alpha_i^{S_i} \left(1-\alpha_i\right)^{r_i - S_i} = v_i\left(S_i\right)$$

When $\mu_i \le r_i\ \alpha_i$, the "candidates" arrive rarely and have time to be processed without delay. If $\mu_i > r_i\ \alpha_i$, the "candidates" $\{b_i^*\}$arrive often and do not have time to be analyzed before the arrival of $\{b_{i+1}^*\}$. Therefore, the variants are delayed at F_i The probability that the number of „candidates" will not exceed the mean values $r_i\ \alpha_i$ by more than ε_i equals:

$$P\left\{\mu_i \le r_i\alpha_i + \varepsilon_i\right\} = \sum_{s=0}^{m(i)} v_i(s)$$

where $m(i) = r_i\ \alpha_i + \varepsilon_i$. Considering r_i to be sufficiently large, we obtain according to Laplace's limit theorem:

$$P\{\mu_i = s_i\} = \phi[(s_i - r_i\, \alpha_i)\{r_i\, \alpha_i\, (1 - \alpha_i)\}^{-1/2}]$$

where $\phi(u) = (2\pi)^{-1/2} \exp(- u^2/2)$.

Denoting through M_i the memory capacity intended for delay of the ith component of b and fixing the condition emerging from the limitation, we obtain: $r_i\ \ \alpha_i + \varepsilon_i \le M_i\ (i = 1,\ldots, K)$. Then the probability of nonoverflow of memory M_i will equal: $P\{\mu_i \le M\} = \Phi(u_i), (i = 1, \ldots, K)$, where

$$u_i = \varepsilon_i\{r_i\, \alpha_i(1 - \alpha_i)\}^{-1/2} = (M_i - r_i\alpha_i)\, \{r_i\, \alpha_i\, (1 - \alpha_i)\}^{-1/2} \qquad (11.27)$$

Utilizing the Boulean formula, let us calculate the probability of nonoverflow of memory M_i on the ith component during delay with time t_i and the uninterrupted transfer of the "candidates" from F_i to F_{i+1} without delay. Let us denote this probability by $P(1, 2,\ldots, N_i)$:

$$P(1, 2,\ldots, N_i) \ge 1 - N_i\, [1 - \Phi(u_i)], i = 1,\ldots, K;$$

where N_i is the number of possible variants of values b_i. Let us now demand that this probability differs from unity by not more than δ_i. Then we obtain an equation for determining the delay value t_i $(i = 1,\ldots, K)$:

$$\Phi(u_i) = 1 - \delta_i/N_i \qquad (11.28)$$

From Eqs. (11.27) and (11.28) we obtain:

$$\Phi[(M_i - r_i\alpha_i)\{r_i\,\alpha_i\,(1-\alpha_i)\}^{-1/2}] = 1 - \delta_i/N_i \tag{11.29}$$

With $u \gg 1$ we may write approximately: $\Phi(u) \approx 1 - \exp(-u^2/2)$. Then Eq. (11.29) assumes the form:

$$r_i^2 - 2r_i\left[\alpha_i / M_i + (1-\alpha_i)\ln(N_i/\delta_i)\right] + (M_i/\alpha_i)^2 = 0, i = 1,\ldots,K \tag{11.30}$$

For an unambiguous determination of the numerical value of r_i let us find the probability of "nonemptying" of the delay, i.e., the probability that during the exhaustive search for N_i values of the ith component, at least one "candidate" will be under delay, and then demand that this probability be within the permissible limits. Thus,

$$P\{\mu_i > r_i\,\alpha_i - v_i\,[r_i\,\alpha_i\,(1-\alpha_i)]^{1/2} = 0\} = 1 - \Phi(-v_i) = \Phi(v_i) \tag{11.31}$$

The probability $Q(1,2,\ldots,N_i)$ of fulfilling the inequality in Eq. (11.31) during the procedure of an exhaustive search for b_i according to the Boulean formula, is estimated by the inequality:

$$Q(1,2,\ldots,N_i) \geq 1 - N_i\,[1 - \Phi(v_i)] \tag{11.32}$$

Thus, by solving Eq. (11.30) and inserting its root into Eq. (11.32), we obtain two probable values of emptying the delay: $q_i = N_i\;[1 - \Phi(v_i)]$. Hence, that root of Eq. (11.30) is selected which is reasonable with respect to the quality of the numerical delay value.

In cases when the binomial distribution $v_i\,(s)$ cannot be approximated precisely enough by the normal distribution, we can use the Poisson distribution: $v\,(s) \approx p\,(s) = (r\alpha)^s\exp(-r\alpha)/s!$ We obtain:

$$p_i\,(s_i) \approx (2\pi s_i)^{-1/2}\exp\{-r_i\,\alpha_i - s_i\{1 - \ln(r_i\,\alpha_i\,s_i)\}$$

Further, all the arguments regarding the inference from this equation are similar to Eq. (11.29); for determining t_i they remain the same.

Generally, the calculation procedure for t_i can be realized with the computer for any expression $v_i\,(s)$. However, since $N_i\,\alpha_i \gg 1$ is practically always the same, the agreements presented are extremely real and simple.

Thus, the probability of error of the whole anomaly detection system at a given time delay is specified by the expression:

$$P_t = 1 - \prod_{i=1}^{K} Q(1,2,\ldots,N_i)$$

11.3.3
Evaluation of the System Error Probability and the Requisite Memory Capacity Delay with a Constant Number of Computer Storage Registers

Let us consider another variant of constant delay, namely: delay of a constant number of "candidates". The expectation time for a complete filling of this delay is a random value with a Pascal distribution:

$$P\{\tau_i = t_i^*\} = C_{R_i-1}^{m_i-1}\alpha_i^{m_i}(1-\alpha_i)^{R_i-m_i}$$

where $R_i = t_i^* / n_i$ is the number of b_i variants surveyed for time t^*_i and n_i is the sample capacity.

In the case of $\tau_i \geq m_i n_i / \alpha_i$, the variants from F_i enter F_{i+1} more rarely than on average. According to the matching condition of the whole flow, they have time to pass by without delay. At $\tau_i < m_i n_i / \alpha_i$, the b_i^* variants do not have time to be processed at F_{i+1} and are delayed at F_i. It is necessary to determine the numerical values of m_i ($i = 1, ..., K$) so that $m_i \leq M_i$ and in the process of an exhaustive search for the ith component there should be a definite probability of no overflow and no delay emptying.

Let us take advantage of the approximate expression of Pascal's distribution via distribution (11.12). At $R_i\,\alpha_i \gg 1$, the following expression is known

$$P\{\tau_i = t_i^*\} = \alpha_i m_i^{-1} w_{c_i}(y_i)$$

To ensure the predetermined probability of nonoverflow of M_i memory, it is necessary to choose m_i in such a way that the time between the arrival of the "candidates" should be close to average. Let us calculate the probability that τ_i exceeds the value $m_i n_i / \alpha_i$ by more than some constant d_i

$$P\{\tau_i \geq m_i n_i / \alpha_i - d_i\} = 1 - W_{c_i}\left[(m_i n_i - d_i\alpha_i)m_i^{-1}n_i^{-1}\right] \tag{11.33}$$

The $Q(1,2,..., N_i)$ probability of fulfilling inequality (11.33) during the whole procedure of an exhaustive search for values of the ith component has the evaluation:

$$Q(1,2,...,N_i) \geq 1 - N_i W_{c_i}\left[(m_i n_i - d_i\alpha_i)m_i^{-1}n_i^{-1}\right].$$

Let us try to make this probability differ from unity by not less than the value ε_i. Then the equation for determining the delay value m_i will take the following form:

$$W_{c_i}\left[(m_i n_i - d_i\alpha_i)m_i^{-1}n_i^{-1}\right] = \varepsilon_i / N_i \tag{11.34}$$

At $c_i > 0$, involving the designated approximation of the normal distribution function, we have $m_i \approx d_i\alpha_i\,[2c_i \ln(N_i/\varepsilon_i)]^{-1/2}/n_i$ ($i = 1,..., K$). Further, let us demand that the probability of appearance during time $m_i n_i/\alpha_i - d_i$ being greater than M_i "candidates" be anywhere near zero. Then we have:

$$P\{\mu_i \leq M_i\} \approx \Phi\left[(M_i - R_i\alpha_i) / \sqrt{R_i\alpha_i(1-\alpha_i)}\right] \tag{11.35}$$

The probability of realizing $\mu_i \leq M_i$ during the whole procedure of an exhaustive search for values of the ith component according to the Boulean formula is estimated by the inequality:

$$\Omega(1,2,...,N_i) \geq 1 - N_i\left[1 - \Phi\left\{(M_i - R_i\alpha_i) / \sqrt{R_i\alpha_i(1-\alpha_i)}\right\}\right] = 1 - \Delta_i.$$

Given that a certain small value is expressed by Δ_i, we obtain in addition to condition (11.34) the second condition for determining m_i:

$$\Phi\left[\left(M_i - q_i\alpha_i\right) / \sqrt{q_i\alpha_i\left(1-\alpha_i\right)}\right] = 1 - \Delta_i / N_i \tag{11.36}$$

where $q_i = m_i / \alpha_i - d_i / n_i$.

Thus, if the delay capacities are calculated from Eq. (11.36), the probability of error for the whole system will equal:

$$P_m = 1 - \prod_{i=1}^{K} \Omega\left(1, 2, \ldots, N_i\right)$$

because of the overflow of the memory.

The above-considered evaluation of probability of the on-board computer memory overflowing under monitoring conditions of processing information through the K channels allows one in definite situations to calculate the parameters of the system, to estimate its efficacy, and to choose one of the two processing methods indicated above. Both delay variants discussed above are equivalent in the requisite additional memory capacity and operation time. Indeed, let $\Delta_i = \sigma_i$, then $r_i = R_i$, i.e., the delay in the first variant equals the number of variants considered in the second. Similarly, assuming $q_i = \varepsilon_i$, we obtain the equality $r_i = R_i$. Therefore, the choice of the kind of delay should be determined by technical considerations of its realization.

11.4 Applications of the Sequential Decision-Making Procedure

The sequential decision-making procedure is used to estimate the parameters of the land surface based on the remote data received by means of the flying

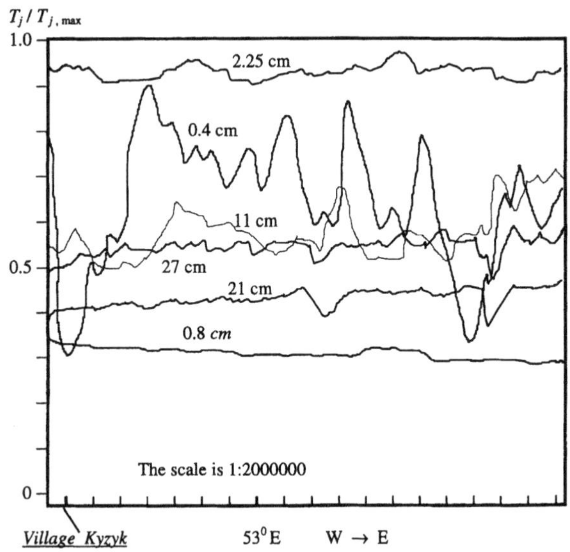

Fig. 11.8. Sample of the radio-brightness temperature registration on board the airborne laboratory IL-18 flying east from the village of Kyzyk on the eastern shore of the Caspian Sea

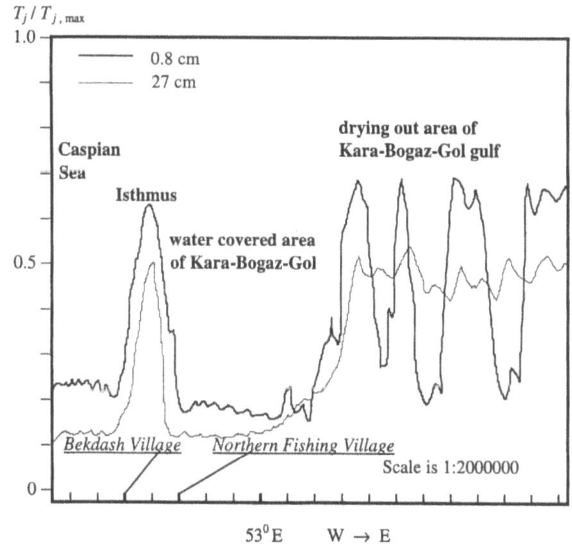

Fig. 11.9. Radio-brightness contrasts in the area of the Kara-Bogaz-Gol gulf on the eastern shore of the Caspian Sea as registered on board the flying laboratory IL-18

laboratory (Krapivin 2000). The measurements are obtained by radiometers of 0.4, 0.8, 2.25, 11, 21, and 27 cm wavelength. The sets of radio-brightness $\{T_j\}$ are registered and $T_{j,\min}$, $T_{j,\max}$ and $T_{j,\text{mean}}$ are calculated. As a result they are transformed to the normalized sets $\{T_j / T_{j,\max}\}$ or $\{\Delta T_j = T_j - T_{j,\text{mean}}\}$.

Figure 11.8 gives examples of such transformations. The registered radio-brightness temperatures are typical for the land covers of the east Caspian Sea coast. It is a region of saline lands, dry sands, saline waters, dense bushes, and populated landscapes. As can be seen from Fig. 11.9, multichannel measurements provide the data sets making it possible to distinguish between the surface cover. Moreover, an application of the sequential analysis procedure guarantees the solving of this task in real-time on-board the flying laboratory.

The radio-brightness temperature sets, made discrete by the quasi-uniform subsets, are transformed to combinations of distribution characteristics which

Table 11.2. Data processing results of multichannel radiometric measurements by means of the airborne laboratory IL-18 in the area of the saline land Sor Barsa Kelmes bordering the Aral Sea. Designations: *R* is the Rayleigh distribution, *E* is the exponential distribution, *L* is the log-normal distribution, *G* is the Gaussian distribution, *V* is the Weibull distribution, *W* is the sequential procedure, *N–P* is the Neyman-Pearson procedure

Channel (cm)	Distribution			Procedure used for the decision-making		
	Steppe	Saline land	Boundary	Steppe	Saline land	Boundary
0.8	*R*	*G*	*E*	*W*	*W*	*W*
1.35	*R*	*L*	*G*	*W*	*N–P*	*W*
2.25	*R*	*L*	*G*	*W*	*N–P*	*W*
21	*R*	*L*	*E*	*W*	*W*	*W*

Fig. 11.10. Empirical distributions of the precipitation in four regions of Russia and Kazakhstan estimated by means of the sequential procedure. The distribution parameters are given in Table 11.4

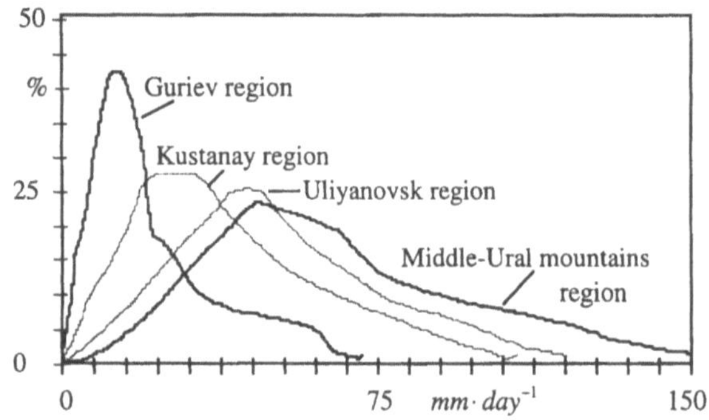

Table 11.3. Results of processing temperature anomalies for the arctic atmosphere (during the last 10 years). Designations: M_1 is the first-order moment (mean), M_2 is the central second-order moment (dispersion), M_5 is the asymmetry coefficient, and M_6 is the excess coefficient, *NW* is the number of the measurements in the framework of the sequential procedure, Δ is the advantage of the sequential procedure in comparison with the fixed level of 100 measurements in the classical procedure, *G* is the Gaussian distribution, *K* is the Cauchy distribution, *Si* is the Simpson distribution and *St* is the Student distribution

Season	M_1	M_2	M_5	M_6	*NW*	Δ (%)	Distribution
Winter	0.02	0.29	0.16	0.41	44	56	*K*
Spring	0.08	0.87	- 0.52	0.05	108	0	*St*
Summer	0.00	0.39	- 0.01	- 0.65	51	49	*Si*
Autumn	0.09	0.64	0.11	4.74	73	27	*G*
Year	0.05	0.55	0.26	- 0.75	59	41	*G*

Table 11.4. Estimations of the distribution parameters for precipitation (mm month^{-1}) in three regions of the Russian Federation and in Kustanay, Kazakhstan

Region	Variation coefficient	Distribution	Mean	Dispersion
Uliyanovsk	16	Pearson	310	51
Middle-Ural mountains	15	Pearson	435	67
Guriev	31	Pearson	109	33
Kustanay	31	Pearson	211	61

make possible the formation of the multi-dimensional image state of the land surface. An example of the cluster analysis application is given in Fig. 9.2. The statistical parameters of the distributions are given in Table 11.2. We see that the varied distributions do not correspond to the land surface structure. However, the parametrical space expansion increases the efficiency of the decision system by including the distribution characteristics.

The optimal procedure of the functioning of the decision system is in the joint use of the classical and sequential algorithms for data processing (see Fig. 11.6).

This procedure saves time and gives the operative parameters of the distributions, which can be used in the framework of the other procedures. Table 11.3 shows the advantage of such an approach. Figure 11.10 gives the results of reconstructions of the empirical distribution for precipitation in three regions of the Russian Federation and Kustanay in northern Kazakhstan. The parameters for these distributions are given in Table 11.4.

References

Abalkin LI (ed) (1999) Russia-2015: optimistic scenario. MMVB, Moscow, 414 pp (in Russian)

Adamenko VN, Kondratyev KYa (1999) Global climate changes and its empirical diagnostics. In: Izrael YuA, Borisenkov YuP, Kondratyev KYa (eds) Anthropogenic influence on the nature of the North and its ecological consequences. Kolskyi Sci Centre of Russian Acad Sci, Apatity, pp 17–34 (in Russian)

Aleshin VA, Klimov VV (1992) An analysis of optical heterogeneity. Proc Int Symposium on Ecoinformatics Problems, 14–18 Dec 1992, Zvenigorod, Moscow Reg, pp 202–204 (in Russian)

Alexeeva LB, Strachan WMJ, Shlychkova VV, Nazarova AA, Nikanorov AM, Korotova LG, Koreneva VJ (2001) Organochlorine pesticide and trace metal monitoring of Russian rivers flowing to the Arctic Ocean: 1990–1996. Mar Pollut Bull 43(1–6):71–85

Allen M, Raper S, Mitchell J (2001) Uncertainty in the IPCC's Third Assessment Report. Science 293(5529):430–433

Andreev AG, Gabin IA (2000) Distribution of the freons and dissolved oxygen in the intermediate waters of the Okhotsk Sea. Meteorol Hydrol (Moscow) 1:61–69 (in Russian)

Andreev IL (2001) Once again about Russia's rehabilitation. Herald Russian Acad Sci 71(1): 39–44 (in Russian)

Anikeev VV, Obgirov AI (1993) Influence of low-temperature hydrotherms on the gas content of the near bottom waters in the Okhotsk Sea. Oceanology (Moscow) 33(3):360–366 (in Russian)

Aota M, Shirasawa K, Krapivin VF (1991a) Simulation model for study of the Okhotsk sea. Proceedings of the Int Symp on Engineering Ecology-91, 22–24 Oct 1991, Zvenigorod, Russia, pp 110–113

Aota M, Shirasawa K, Krapivin VF, Mkrtchyan FA (1991b) The system for data processing in Okhotsk sea monitoring. Proceedings of the Sixth Int Symp on Okhotsk Sea and Sea Ice, 3 Feb 1991, Mombetsu, Japan. Okhotsk Sea and Cold Ocean Res Assoc, Mombetsu, pp 317–318

Aota M, Shirasawa K, Krapivin VF, Mkrtchyan FA (1992) Simulation model of the Okhotsk Sea geoecosystem. Proc of the 7th Int Symp on Okhotsk Sea andand Sea Ice, 2–5 Feb 1992, Mombetsu, Japan. Okhotsk Sea and Cold Ocean Res Assoc, Mombetsu, pp 311–313

Aota M, Shirasawa K, Krapivin VF, Mkrtchyan FA (1993) A project of the Okhotsk Sea GIMS. Proc 8th Int Symp on Okhotsk Sea and Sea Ice and ISY/Polar Ice Extent Workshop, 1–5 Feb 1993, Mombetsu, Japan. Okhotsk Sea and Cold Ocean Res Assoc, Mombetsu, pp 498–500

Apps MJ, Kurz WA, Luxmoore RJ et al (1993) Boreal forests and tundra. Water Air Soil Pollut 70(1–4):39–53

Arbatov AG (2000) Russia's national security in the multi-polar world. Herald Russian Acad Sci 70(11):94–993 (in Russian)

Armand NA, Krapivin VF, Mkrthyan FA (1987) Methods for data processing in the radiophysical investigation of the environment. Nauka, Moscow, 270 pp (in Russian)

Armand NA, Krapivin VF, Shutko AM (1997) The GIMS technology as a new approach to the information control in the environmental investigations. Prob Environ Nat Res (Moscow) 3:31–50 (in Russian)

Arora KL (1962) A generalized problem in air defense. Naval Res Logist Q 9(3–4):256–261

Asrar G, Kaye JA, Morel P (2001) NASA research strategy for earth system science: climate component. Bull Am Meteorol Soc 82(7):1309–1330

Balvanera P, Daily GC, Ehrlich PR, Ricketts TH, Bailey S-A, Kark S, Kremen C, Pereira H (2001) Conserving biodiversity and ecosystem services. Science 291(5511):2047

Bard S (1999) Global transport of anthropogenic contaminants and the consequences for the Arctic marine ecosystems. Mar Pollut Bull 3(5):356–379

Barlybaev HA (2001) The way of humanity: self-destruction or sustainable development. Publ House of the State Duma, Moscow, 143 pp (in Russian)

Barnett J (2001) The meaning of environmental security. Ecological politics and policy in the new security era. Zed Books, New York, 184 pp

Bazilevich NI, Rodin LE (1967) Diagrammatic map of productivity and biological cycle of basic terrestrial plant types. Proc All-Union Geogr Soc 99(3):190–194 (in Russian)

Berdnikov SV, Dombrovsky YuA, Ostrovskaya AG, Prichodko MV, Titova LI, Tjutjunov YuV (1989) Simulation model of basic components of the Okhotsk Sea ecosystem. Mar Hydrophys J (Vladivostok) 3:52–57 (in Russian)

Bjorkstrom A (1979) A model of CO_2 in interaction between atmosphere, oceans and land biota. In: Bolin B, Degens ET, Kempe S, Ketner P (eds) Global carbon cycle. SCOPE 13, Wiley, Chichester, pp 403–457

Blumenstein O, Schachtzabel H, Barsh H, Bork H-R, Küppers U (1999) Grundlagen der Geoökologie. Erscheinungen und Prozesse in unserer Umwelt. Springer, Berlin Heidelberg New York, 250 pp

Bodenbender J, Wassmann R, Papen H, Rennenberg H (1999) Temporal and spatial variation of sulfur-gas-transfer between coastal marine sediments and the atmosphere. Atmos Environ 33(21):3487–3502

Boehmer-Christiansen C (2000) Who and how determines policy on climate change? Proc Russ Geogr Soc 132(3):6–22

Bogorodsky VV, Kondratyev KY, Rabinovitch YI (1976) Microwave remote indication of the sea surface by the oil pollutants. Ann Main Geophys Observ Leningrad 371:22–36 (in Russian)

Bokov VA, Lushchik AV (1998) Basis of ecological safety. Sonat Press, Sympheropol, 222 pp (in Russian)

Bolin B, Sukumar R (2000) Global perspective. In: Watson RT, Noble IR, Bolin B, Ravindranath NH, Verardo DJ, Dokken DJ (eds) Land use, land-use change, and forestry. Cambridge Univ Press, Cambridge, pp 23–53

Borodin LF, Krapivin VF, Long BT (1996) Application of the GIMS technology to the Aral-Caspian aquageosystem monitoring. Prob Environ Nat Res 10:46–61 (in Russian)

Bortnik VN, Chistyakova SP (eds) (1990) Hydrometeorology and hydrochemistry of the USSR seas, vol 7. Aral Sea. Gidrometeoizdat, Leningrad, 196 pp (in Russian)

Bortnik VN, Lopatina SA, Krapivin VF (1994) Simulation system for the study of hydrophysical fields in the Aral Sea. Meteorol Hydrol (Moscow) 9:102–108 (in Russian)

Bourke RH, Paquette RG, Blythe RF (1992) The Jan Mayen current of the Greenland Sea. J Geophys Res C 97(5):7241–7250

Boysen M (2000) Biennial report 1998/1999. Potsdam Institute for Climate Impact Res Press, Potsdam (Germany), 129 pp

Bras RL (1990) Hydrology. Addison-Wesley, New York, 643 pp

Bratimov OV, Gorsky YuM, Delyagin MG, Kovalenko AA (2000) Globalization practice: games and a new era rules. INFRA-M Press, Moscow, 344 pp (in Russian)

Broecker VS, Takahasi T, Simpson HJ, Peng T-H (1981) Transformation and fate of CO_2 resultant from combustion of fossil fuel. Is it possible to close the global carbon balance? In: Borisenkov Yu P (ed) Problems of atmospheric carbon dioxide. Gidrometeoizdat, Leningrad, pp 46–78

Brown LR, Flavin C, French H et al (2001) State of the world 2001. Earthscan Publ, London, 275 pp

Bui TL, Krapivin VF (1997) System of survey and simulation for air pollution over large industrial regions. Proc Int Conf Pollution Control'97, Bangkok, Thailand, 12–16 Nov 1997, p 122

Bukatova IL, Michasev YI, Sharov AM (1991) Evoinformatics. Sci Publ House, Moscow, 205 pp (in Russian)
Carvalho GO (2001) Sustainable development: is it achievable within the existing international political economy context? Sustainable Dev 8(2):61–73
Champ MA, Makeyev VM, Brooks JM, De Laca TE (2000) Contaminants in the Arctic. Mar Pollut Bull 40(10):801–802
Chen W, Chen J, Cihlar J (2000) An integrated terrestrial ecosystem carbon budget model based on changes in disturbance, climate, and atmospheric chemistry. Ecol Model 135(1):55–79
Cherny IV, Raizer VYu (1998) Passive microwave remote sensing of oceans. Wiley, London, 300 pp
Chistobaev AI (2001) Sustainable development: global context and local specific character. Proc Russian Geogr Soc 133(4):22–27 (in Russian)
Clark JS, Carpenter SR, Barber M et al (2001) Ecological forecasts: an emerging initiative. Science 293(5530):657–660
Claussen M, Brovkin V, Ganopolski A, Kubatzki C, Petoukhov V, Rahmstorf S (1999) A new model for climate system analysis: outline of the model and application to palaeoclimate simulations. Environ Model Assess 4(4):209–216
Crane K, Galasso J, Brown C, Cherkashov G, Ivanov G, Vanstain B (2000) Northern ocean inventories of radionuclide contamination: GIS efforts to determine the past and present state of the environment in and adjacent to the Arctic. Mar Pollut Bull 40(10):853–868
Danilov-Danilian VI, Zalikhanov MC, Losev KS (2001) Ecological safety. General principles and Russian aspects. MNEPU Press, Moscow, 330 pp (in Russian)
Demers S, Legendre L, Therriault JC, Ingram RG (1986) Biological production at the ice-water ergocline. In: Nihoul JCJ (ed) Marine interfaces ecohydrodynamics. Elsevier, Amsterdam, pp 31–55
Demirchian KS (2001) Russia again in the darkness. What to do? Human aspects of the anti-crisis strategy. Industr Vedomosty 8/9:19–20 (in Russian)
Demirchian KS, Kondratyev KY (1998) Energetics development and the environment. Ann Russ Acad Sci Energ 1:1–39 (in Russian)
Demirchian KS, Kondratyev KY (1999) Scientific reason of the forecast of the energetics influence on climate. Proc Russ Acad Sci Energ 6:3–46 (in Russian)
De Rivero O (2001) The myth of development. Zed Books, London, 224 pp
DeWitt DP, Nutter GD (eds) (1988) Theory and practice of radiation thermometry. Wiley, New York, 1137 pp
Diatlov SA (1988) Basis of sustainable development. St Petersburg State Univ of Economy and Finance, St Petersburg, 155 pp (in Russian)
Dilao R, Domingos T (2000) A general approach to the modelling of trophic chains. Ecol Model 132(2):191–202
Dimitroulopoulou C, Marsh ARW (1997) Modelling studies of NO_3 night-time chemistry and its effects on subsequent ozone formation. Atmos Environ 31(18):3041–3057
Dobson A (1998) Justice and the environment. Conceptions of environmental sustainability and dimensions of social justice. Clarendon Press, Oxford, 292 pp
Doos R (2000) Increasing food production at the expense of tropical forests. Integrated Assess 1:189–202
Drake F (2000) Global warming: the science of climate change. Arnold Publ/Oxford Univ Press, Oxford, 273 pp
Emery WJ, Thomson RE (2001) Data analysis methods in physical oceanography. Elsevier, Amsterdam, 328 pp
Engman ET, Chauhan N (1995) Status of microwave soil moisture measurements with remote sensing. Remote Sensing Environ 51(1):189–198
Ernst WG (ed) (1999) Earth system. Processes and issues. Cambridge Univ Press, Cambridge, 704 pp
Ferm M, Hultberg H (1999) Dry deposition and internal circulation of nitrogen, sulfur and base cations to a coniferous forest. Atmos Environ 33(27):4421–4430
Fink DG (1966) Computers and the human mind. Wiley, New York, 253 pp

Fisher F, Hajer M (eds) (1999) Living with nature. Environmental politics and cultural discourse. Oxford Univ Press, Oxford, 284 pp

Fogel LJ, Owens AJ, Walsh MJ (1966) Artificial intelligence through simulated evolution. Wiley, New York, 230 pp

Forrester JW (1971) World dynamics. Wright-Allen Press, Cambridge, 189 pp

French H (2000) Vanishing borders. Protecting the planet in the age of globalization. Norton, New York, 257 pp

Friend AD (1998) Parametrization of a global daily weather generator for terrestrial ecosystem modelling. Ecol Model 109(2):121–140

Gaffin M, Harrison F, Titova G (2000) Behind the scenes of coming-to-be-economical theories. From the theory to corruption. BFK Press, St Petersburg, 310 pp (in Russian)

Gerard S, Turner RK, Bateman IJ (2001) Environmental risk planning and management. Edward Elgor Publ, Cheltenham, 656 pp

Girusov EV, Bobylev SN, Novoselov AL, Cherepnykh NV (2000) Ecology and economics of nature base. UNITI Press, Moscow, 455 pp (in Russian)

Gitelzon II, Abrosov NS, Gladyshev MI et al. (1985) Yenisey: problems of the largest Siberian river. In: Degens ET, Kempe S, Herrera R (eds) Transport of carbon and minerals in major world rivers. Wiley, Hamburg, pp 471–485

Golitsyn GS (1995) The Caspian Sea level as a problem of diagnosis and prognosis of the regional climate change. Atmos Ocean Phys 31(3):366–372 (English translation)

Gorshkov SP (1998) Conceptual basis of geoecology. Smolensk State Univ Press, Smolensk, 445 pp (in Russian)

Gorshkov VG (1990) Biosphere energetics and the environment stability. VINITY Press, Moscow, 238 pp (in Russian)

Gorshkov VG (1995) Physical and biological basis of sustainability of life. VINITY Press, Moscow, 472 pp (in Russian)

Gorshkov VG, Makarieva AM (1999) Impact of virgin and disturbed biota on the global environment. Earth Res Space (Moscow) 5:3–11 (in Russian)

Gorshkov VG, Gorshkov VV, Makarieva AM (2000) Biotic regulation of the environment. Springer / PRAXIS Publ, Chichester, 364 pp

Goudriaan J, van Keulen H, van Laar HH (eds) (1990) The greenhouse effect and primary productivity in European agro-ecosystems. Pudoc, Wageningen, 90 pp

Grigoriev AlA, Kondratyev KY (2001) Natural and anthropogenic environmental disasters. St Peterburg Science Centre of Russian Acad Sci Press, St Peterburg, 688 pp (in Russian)

Gupta J (2001) Our simmering planet. What to do about global warming? Zed Books, London, 192 pp

Harris NRP, Hudson RD (2001) Report of the SPARC/IOC Workshop on Understanding Ozone Trends. SPARC Newslett 17:10–13

Harrison WG, Cota GF (1991) Primary production in polar waters: relation to nutrient availability. Polar Res 10(1):87–104

Hidayat H (1994) Inventorying and monitoring forestry in Indonesia. Spot Mag 22:13–14

Holmberg M, Rankinen K, Johansson M, Forsius M, Kleemola S, Ahonen J, Syri S (2000) Sensitivity of soil acidification model to deposition and forest growth. Ecol Model 135(2/3): 311–325

Hong NS, Yen LB, Thao LH et al (1992) On the problem of ecological monitoring in Vietnam. IREE Press, Moscow, 46 pp (in Russian)

Hong NS, Yen LB, Son DT et al (1994) An investigation of the spatial-temporal structure of spectra for water objects of South Vietnam with adaptive identifier. In: Krapivin VF (ed) Ecoinformatics problems. IREE Press, Moscow, pp 128–135 (in Russian)

Hotuntsev YL (2001) Humanity, technology, environment. Stable World, Moscow, 224 pp (in Russian)

Houghton JT, Ding Y, Griggs DJ (eds) (2001) Climate change 2001: the scientific basis. Contribution of WGI to the Third Assessment Rep of the IPCC. Cambridge Univ Press, Cambridge, 881 pp

Hrol VP (1993) Energetic balance of the north polar area. Hydrometeo Press, St Petersburg, 168 pp (in Russian)

Hublaryan MG (1995) The Caspian Sea phenomenon. Herald Russ Acad Sci 65(7):616–630 (in Russian)
Hurell A, Woods N (eds) Inequality, globalization, and World politics. Oxford Univ Press, Oxford, 240 pp
Inozemzev VL (1999) Broken civilization. Academia-Science Press, Moscow, 724 pp (in Russian)
Ivanov BV, Makshtas AP (1990) Quasi-stationary zero-measure model of arctic ices. Proc Arctic and Antarctic Institute, Gidrometeopress, Leningrad, 420:18–31 (in Russian)
Ivashov LG (2000) Russia and the world in a new millennium. Paleya-Mishin Press, Moscow 336 pp (in Russian)
Kabanov MV, Matvienko GG, Tikhomirov AA, Soldatkin NP, Klyuev VV, Filinov VN, Ketkovitch AA (2000) Optical ecological monitoring. In: Klyuev VV (ed) Ecological diagnostics. Znanie, Moscow, pp 152–247 (in Russian)
Kalinkevitch AA, Kutuza BG (2001) Possibility of forest SAR image using for the estimate of the ecological situation. Earth Res Space 3:77–81 (in Russian)
Kates RW, Clark C, Corell R et al (2001) Sustainability. Science 292(5517):641–642
Kawasaki Y, Kono T (1993) Water exchange between the Okhotsk Sea and Pacific Ocean through the middle of Kuril islands. Proc 8th Int Symp Okhotsk Sea and Sea Ice and ISY/Polar Ice Extend Workshop, 1–5 Feb 1993, Mombetsu, Japan, pp 60–63
Keeling CD, Bacastow RB (1977) Impact of industrial gases on climate. In: Revelle R and Munk W (eds) Energy and climate. National Academy of Sciences, Geophysics Research Board, Geophysics Study Committee, Washington, pp 72–95
Kelley JJ, Gosink T (1992) The arctic environment-air-sea-land exchange of trace gases. University of Alaska Fairbanks, Rep CP 92-7, Fairbanks, 29 pp
Kelley JJ, Krapivin VF, Vilkova LP (1992a) A conception of model for the estimation of arctic ecosystems role in global carbon budget. Proc Int Symp on Problems of Ecoinformatics, 12–18 Dec 1992, Zvenigorod, pp 19–20
Kelley JJ, Rochon GL, Novoselova OA, Krapivin VF, Mkrtchyan FA (1992b) To the global geo-eco-information monitoring. Proc Int Symp on Problems of Ecoinformatics, 12–18 Dec 1992, Zvenigorod, pp 3–7
Kelley JJ, Krapivin VF, Popovich PR (1999) The problems of Arctic environment monitoring. Prob Environ Nat Res 6:32–40 (in Russian)
Khor M (2001) Rethinking globalization. Critical issues and policy choices. Zed Books, London, 160 pp
Kiefer DA, Mitchel BG (1983) A simple, steady state description of phytoplankton growth based on absorption cross section and quantum efficiency. Limnol Oceanogr 28:770–776
Klotzi S (1994) The water and soil crisis in Central Asia – a source for future conflicts? ENCOP Occasional Paper no 11. Center for Security and Conflict Research/Swiss Peace Foundation Bern, Zurich/Bern, May 1994, pp 124–131
Klyuev NN (2001) Ecological results of Russia's reforms. Herald Russ Acad Sci 71(3):233–239 (in Russian)
Klyuev VV (ed) (2000) Ecological diagnostics. Znanie, Moscow, 495 pp
Kogan FN (2001) Operational space technology for global vegetation assessment. Bull Am Meteorol Soc 82(8):1949–1964
Kohlmaier GH, Brohl H, Sire EO, Plochl M, Revell R (1987) Modelling simulations of plants and ecosystem response to present level of excess atmospheric CO_2. Tellus 39B:155–170
Kondratyev KY (1990) Key problems of the global ecology. VINITI Press, Moscow, issue no 9, 454 pp (in Russian)
Kondratyev KY (1992) Global climate. Science Press, St Petersburg, 359 pp (in Russian)
Kondratyev KY (1993) System of global climate observations. Earth Res Space 6:104–115 (in Russian)
Kondratyev KY (1997) Key issues of global change at the end of the second millennium. Newslett Eur Geophys Soc 63:4–8
Kondratyev KY (1998a) Environmental risk: real and hypothetical. Proc Russ Geogr Soc 130(3):13–24 (in Russian)
Kondratyev KY (1998b) Multidimensional global change. Wiley/PRAXIS, Chichester, 761 pp

Kondratyev KY (1999a) Climatic effects of aerosols and clouds. Springer/PRAXIS Publ, Chichester, 264 pp
Kondratyev KY (1999b) Current global ecodynamics. Earth Obs Rena Sens 15:823–852
Kondratyev KY (1999c) Ecodynamics and Geopolitics: from global to local scales. Ecology 1:3–8 (in Russian)
Kondratyev KY (1999d) Ecodynamics and geopolitics, part 1: global problems. St Petersburg Sci Center Press, St Petersburg, 1040 pp (in Russian and English)
Kondratyev KY (1999e) Turning-point: end of the growth paradigm. Proc Russ Geogr Comm 131(2):1–14
Kondratyev KY (2000a) Earth researches from space: scientific plane of the EOS system. Earth Res Space (Moscow) 3:82–91 (in Russian)
Kondratyev KY (2000b) Global environmental and social changes on the border of the two centuries. Ann Russ Geogr Soc 5:3–19 (in Russian)
Kondratyev KY (2001a) Environmental changes in Europe. Herald Russ Acad Sci 71(6):494–502 (in Russian)
Kondratyev KY (2001b) The effect NATO's bombardment on the Yugoslavia's environment. Proc Russ Geogr Soc 133(1):63–67 (in Russian)
Kondratyev KY (2002) Global climate change: reality, hypotheses and fantasies. Earth Res Space 1:3–14 (in Russian)
Kondratyev KY, Cracknell AP (1998) Observing global climate change. Francis and Taylor, London, 592 pp
Kondratyev KY, Demirchian KS (2000) Global climate change and carbon cycle. Ann Russ Geogr Soc 132(4):1–20 (in Russian)
Kondratyev KY, Demirchian KS (2001) Global climate and Kyoto Protocol. Herald Russ Acad Sci 71(11):452–458 (in Russian)
Kondratyev KY, Fedchenko PP (2001) Estimation of crop and soil states from remote sensing data using Harrington's "desirability" function. Earth Res Space 3:56–60 (in Russian)
Kondratyev KY, Filatov N (eds) (1999) Limnology and remote sensing. Springer/PRAXIS, Chichester, 406 pp
Kondratyev KY, Galindo I (2001) Global change situation: today and tomorrow. Universidad de Colima, Colima, Mexico, 164 pp
Kondratyev KY, Johannessen OM (1993) Arctic and climate. PROPO Press, St Petersburg, 139 pp (in Russian)
Kondratyev KY, Krapivin VF (2001a) Biocomplexity and global geoinformation monitoring. Earth Res Space 1:3–10 (in Russian)
Kondratyev KYa, Krapivin VF (2001b) Global dynamics of basic terrestrial ecosystems. Earth Res Space 4:3–12 (in Russian)
Kondratyev KY, Varotsos CA (2000) Atmospheric ozone variability: Implications for climate change, human health, and ecosystems. Springer/PRAXIS, Chichester, 758 pp
Kondratyev KY, Buznikov AA, Pokrovski OM (1996a) Global change and remote sensing. Wiley/PRAXIS, Chichester, 284 pp
Kondratyev KY, Donchenko VK, Losev KS, Frolov AK (1996b) Ecology-economics-politics. St Petersburg Sci Centre of Russian Acad Sci, St Petersburg, 828 pp (in Russian)
Kondratyev KY, Pena FM, Galindo I (1997) Sustainable development and population dynamics. Universidad de Colima Press, Colima, Mexico, 125 pp
Kondratyev KY, Krapivin VF, Pshenin ES (2000) Concept of the regional geoinformation monitoring. Earth Res Space 5:1–8 (in Russian)
Kondratyev KY, Demirchian KS, Balunas S, Adamenko VN, Bohmer-Christiansen S, Idso SB, Postmentier ES, Soon V (2001a) Global climate change: conceptual aspects. St Petersburg Sci Center of the Russian Acad Sci Press, St Petersburg, 125 pp (in Russian)
Kondratyev KY, Losev KS, Ananicheva MD, Chesnokova IV (2001b) Some problems of landscape and ecology in the context of biotical regulation. Proc Russ Geogr Soc 133(5):22–29 (in Russian)
Koptug VA, Matrosov VM, Levashov VK (eds) (2000) A new paradigm of Russia's development (complex studies on sustainable development problems). RITS GP "Oblinformpress", Irkutsk, 460 pp (in Russian)

Korgenevsky AV, Krapivin VF, Cherepenin VA (1989) Modelling the global processes of magnetosphere. In: Novitchikhin EP (ed) Methods of informatics in radiophysical investigations of environment. Nauka, Moscow, pp 25–43 (in Russian)

Kostitsyn VA (1937) Biologie mathématique. Armand Colin, Paris, 138 pp

Kotlyikov VM (2000) Anatomy of crises. Nauka, Moscow, 239 pp (in Russian)

Kovalev VI, Rukavishnikov AI, Perov PI, Rossukanyi NM (1999) Elaboration of optical methods and equipment for the control of technology and parameters of semiconductor structures of nano- and microelectronics. Radiotech Electron 44(11):1404–1408 (in Russian)

Kovel J (2002) The enemy of nature. The end of capitalism or the end of the World? Zed Books, London, 288 pp

Kozoderov VV (1998) On the development of theory for global change in biosphere by data of Earth research from space. Earth Res Space 6:25–39 (in Russian)

Kozoderov VV, Kosolapov VS (1999) A new approach to solve the inverse problem of retrieving the green phytomass volume for forest vegetation using air-space data. Earth Res Space 1:28–36 (in Russian)

Krapivin VF (1972) Theory-game methods for the synthesis of complex systems in the conflict states. Soviet Radio Press, Moscow, 192 pp (in Russian)

Krapivin VF (1978) On theory of complex systems survivability. Science Publ House, Moscow, 248 pp (in Russian)

Krapivin VF (1993) Mathematical model for global ecological investigations. Ecol Modelling 67:103–127

Krapivin VF (1995) Simulation model for the investigation of pollution dynamics in the Arctic basin. Oceanology (Moscow) 35:366–375 (in Russian)

Krapivin VF (1996) The estimation of the Peruvian current ecosystem by a mathematical model of biosphere. Ecol Model 91:1–14

Krapivin VF (1999) The greenhouse effect and global biogeochemical carbon dioxide cycle. Prob Environ Nat Res 12:2–16 (in Russian)

Krapivin VF (2000) Radiowave ecological monitoring. In: Klyuev VV (ed) Ecological diagnostics. Znanie, Moscow, pp 295–311

Krapivin VF, Klimov VV (1995) Valuation of convergence of 'physical mixture' strategies in the matrix games. Theory Control Syst 6:209–217 (in Russian)

Krapivin VF, Klimov VV (1997) Stable strategies in games with gain functions M(x-y). Methods Cybernet Inform Technol (Saratov) 2:36–45 (in Russian)

Krapivin VF, Nazaryan NA (1995) Theory games approach to the biosphere survivability. IREE Preptints, Moscow, 28 pp

Krapivin VF, Nazaryan NA (1997) Mathematical model for investigations of the global sulphur cycle. Math Model (Moscow) 9(8):36–50 (in Russian)

Krapivin VF, Kondratyev KY (2002) Global environmental change: ecoinformatics. St Petersburg Univ Press, St Petersburg, 723 pp (in Russian)

Krapivin VF, Mkrtchyan FA (1991) Applications in study of environment. Proc 8th Int Conf Control Systems and Computer Science, 22–25 May 1991, Bucharest. Polytechnical Institute Press, Bucharest, pp 49–56

Krapivin VF, Phillips GW (2001a) Application of a global model to the study of arctic basin pollution: radionuclides, heavy metals and oil hydrocarbons. Environ Model Software 16:1–17

Krapivin VF, Phillips GW (2001b) A remote sensing-based expert system to study the Aral-Caspian aquageosystem water regime. Remote Sensing Environ 75:201–215

Krapivin VF, Potapov II (2001a) Algorithms for two-dimensional image reconstruction in monitoring problems. Prob Environ Nat Res 6:16–23 (in Russian)

Krapivin VF, Potapov II (2001b) An estimation of the algorithms precision for space-temporal interpolation in geoinformation monitoring problems. Prob Environ Nat Res 6:40–45 (in Russian)

Krapivin VF, Shutko AM (1989) Observation and prognosis of the state of environmental resources, ecological and meteorological situations by geoinformational monitoring system. Proc 4th Int Symp Okhotsk Sea and Sea Ice. Mombetsu 5–7 Feb. 1989, Japan. Okhotsk Sea and Cold Ocean Res Assoc, Mombetsu, pp 1–5

Krapivin VF, Vilkova LP (1990) Model estimation of excess CO_2 distribution in biosphere structure. Ecol Model 50:57–78

Krapivin VF, Svirezhev YM, Tarko AM (1982) Mathematical modeling of the global biosphere processes. Science Publ House, Moscow, 272 pp (in Russian)

Krapivin VF, Shutko AM, Strelkov GM, Loskutov VS (1990) The modeling of ecological situation in upper sea water layer for climate conditions northern region. Proc 5th Int Symp on Okhotsk Sea and Sea Ice, 4–6 Feb 1990, Okhotsk Sea and Cold Ocean Res Assoc, Mombetsu, pp 243–247

Krapivin VF, Mkrtchyan FA, Nitu C, Petrache Gh, Tertisco M, Graur A (1991) Multidirectional data processing automatization for remote study of environment. Proc 8th Int Conf Control Systems and Computer Science, vol 2, 22–25 May, 1991, Bucharest. Polytechnical Institute Press, Bucharest, pp 57–62

Krapivin VF, Mkrtchyan FA, Graur A, Nitu CI (1994) Identification and pattern recognition in ecology. An Univ 'Stefan cel Mare' Suceava, Sect Electrica 1(1):23–27

Krapivin VF, Bui TL, Rochon GL, Hicks DR (1996a) A global simulation model as a method for estimation of the role of regional area in global change. Proc 2nd Ho Chi Minh City Conf. on Mechanics, 24–25 Sept 1996. Institute of Applied Mechanics, Ho Chi Minh City, pp 68–69

Krapivin VF, Long BT, Rochon GL, Hicks DR (1996b) A global simulation model as a method for estimation of the role of regional area in global change. Proc 2nd Ho Chi Minh City Conf Mechanics, 24–25 Sept 1996. Inst of Applied Mechanics, HCM, pp 68–69

Krapivin VF, Vilkova LP, Rochon GL, Hicks DR (1996c) Model estimation of the role of urban areas in global CO_2 dynamics. Proc Eco-Informa '96, Florida, 4–7 Nov 1996, pp 417–422

Krapivin VF, Bui TL, Dean C, Nguyen MN, Rochon GL, Hicks DR (1997a) System of survey and simulation for air pollution over large industrial regions. Proc IASTED Int Conf on Modelling, Simulation and Optimization (MSO '97), Singapore, 11–13 Aug 1997, Anaheim. IASRED/Acta Press, pp 307–311

Krapivin VF, Cherepenin VA, Nazaryan NA, Phillips GW, Tsang FY (1997b) Simulation model for radionuclides transport in the Angara-Yenisey River system. Prob Environ Nat Res 2:41–58 (in Russian)

Krapivin VF, Nazaryan NA, Potapov II (1997c) Adaptive system for decision making in ecological monitoring. Prob Environ Nat Res 8:16–29 (in Russian)

Krapivin VF, Cherepenin VA, Phillips GW, August RA, Pautkin AY, Harper MJ, Tsang FY (1998a) An application of modeling technology to the study of radionuclear pollutants and heavy metals dynamics in the Angara-Yenisey River system. Ecol Model 111:121–134

Krapivin VF, Graur A, Nitu C, Oancea D (1998b) A new approach for geo-information monitoring. Proc Int Conf D&AS, Suceava, 21–22 May 1998, pp 102–106

Krapivin VF, Klimov VV, Kovalev VI, Mkrtchyan FA (2001a) Expert system for pollution spills identify on the water surface. Ecol Syst Devices 3:21–22 (in Russian)

Krapivin VF, Mkrtchyan FA, Klimov VV, Kovalev VI, Rukavishnikov AI (2001b) Adaptive identifier – a new approach to ecological monitoring of the aquatic environment. New Technol 21st Century 5:30–32 (in Russian and English)

Kudo A, Zheng J, Tao G, Sasaki T, Sugahara M (1999) Global transport rates and future prediction of hazardous materials: Pu and Cs – from Nagasaki to Canadian Arctic. Selected Proc 3rd Int Specialised Conf Hazard Assessment and Control of Environ Contaminants (Ecohazard '99). Otsu City, Shiga, Japan, 5–8 Dec 1999, pp 163–169

Kutuza BG, Zagorin G, Hornbostel A, Schroth A (1998) Physical modeling of passive polarimetric microwave observations of the atmosphere with respect to the third Stokes parameter. Radio Sci 33(3):677–695

Kuzmin PO (1957) Hydrological investigations of land waters. Proc Int Assoc Sci Hydrol 3:468–478

Kuznezov SL, Kuznezov PG, Bolshakov BE (2000) Environment-society-humanity system. Sustainable development. Publ House "Noosphere", Moscow, 392 pp (in Russian)

Lal R, Stewart BA (eds) (1994) Soil process and water quality. CRC Press, Boca Raton, 285 pp

Lapko VV, Radchenko VI (2000) Sea of Okhotsk. Mar Pollut Bull 41(1–6):179–187

Lavrov SB (1999) Reality of globalization and illusions of sustainable development. Proc Russian Geogr Soc 131(3):1–8 (in Russian)

Lavrov SB (2000) Leo Gumilev and the fortune of his ideas. Svarog Press, Moscow, 407 pp (in Russian)

Lee N, Kirkpatrick C (eds) (2000) Sustainable development and integrated appraisal in a developing World. Edward Elgar Publ, Cheltenham, 272 pp

Legendre L, Krapivin VF (1992) Model for vertical structure of phytoplankton community in Arctic regions. Proc 7th Int Symp Okhotsk Sea and Sea Ice. 2 Febr 1992. Mombetsu, Hokkaido, Japan. Okhotsk Sea and Cold Ocean Res Assoc, Mombetsu, pp 314–316

Legendre P, Legendre L (1998) Numerical ecology. Elsevier, Amsterdam, 853 pp

Lin B-L, Sakoda A, Shibasaki R, Goto N, Suzuki M (2000) Modelling a global biogeochemical nitrogen cycle in terrestrial ecosystems. Ecol Model 135(1):89–110

Lisichkin VA, Shelepin LA (2001) Global empire of an Evil. Krymskyi bridge-9D/Forum Press, Moscow, 445 pp (in Russian)

Lonergan SC (ed) (1999) Environmental change, adaptation, and security. Kluwer, Dordrecht, 432 pp

Losev KS (2001) Environmental problems and perspectives of sustainable development in Russia in XXI century. Cosmosinform Press, Moscow, 400 pp (in Russian)

Lovelock J (1995) The ages of Gaia. A biography of our living Earth. Norton, New York, 255 pp

Luecken DJ, Berkowitz CM, Easter RC (1991) Use of a three-dimensional cloud-chemistry model to study the transatlantic transport of soluble sulfur species. J Geophys Res D 96(12): 22,477–22,490

Lvov DS (ed) (1999) A way to the XXI century (strategic problems and perspectives of Russia's economy). Economy, Moscow, 793 pp (in Russian)

Macdonald JA, Skiba U, Sheppard LJ, Ball B, Roberts JD, Smith KA, Fowler D (1997) The effect of nitrogen deposition and seasonal variability on methane oxidation and nitrous oxide emission rates in an upland spruce plantation and moorland. Atmos Environ 31(22):3693–3706

Mackay D, Fraser A (2000) Bioaccumulation of persistent organic chemicals: mechanisms and models. Environ Pollut 110(3):345–353

Mackinson S (2000) An adaptive fuzzy expert system for predicting structure, dynamics and distribution of herring shoals. Ecol Model 126(2/3):155–178

Maguire DJ (1991) An overview and definition of GIS. In: Maguire DJ, Goodchild MF, Rhind DW (eds)_Geographical information systems 1. Principles. Longman Sci and Technical, New York, pp 9–20

Maguire DJ, Goodchild MF, Rhind DW (eds) (1991) Geographical information systems, vols 1 and 2. Longman Scientific and Technical, New York, 649 and 447 pp

Malinetzky GG, Kurdumov SP (2001) Non-linear dynamics and the forecast problems. Herald Russ Acad Sci 71(3):210–224 (in Russian)

Mangum G, Winkle W (1973) Responses of aquatic invertebrates to declining oxygen conditions. Am Zool 13(12):529–541

Marhaba TF, Van D, Lippincott RL (2000) Rapid identification of dissolved organic matter in water by spectral fluorescent signatures. Water Res 34(14):3543–3550

Marsh GP (1864) Man and nature: or physical geography as modified by human action. C Scribner, New York, 120 pp

Martchuk GI, Kondratyev KY (1992) Priorities of global ecology. Science Publ House, Moscow, 262 pp (in Russian)

Martin G-P, Shumann H (2001) Western Globalization. The attack to the prosperity and democracy. Publ House "Alpina", Moscow, 335 pp (in Russian)

McCauley LL, Meier MF (eds) (1991) Arctic system science. Land/Atmosphere/Ice Interactions. ARCUS, Fairbanks, Alaska, 48 pp

Mcintyre AD (1999) The environment and the oil companies. Mar Pollut Bull 3(3):155–156

Meadows DH, Meadows DL, Ranger J, Behrens WW (1972) The limits to growth. Univ Books, New York, 208 pp

Miguel JC (2001) Environment and biology of the Kara Sea: a general view for contamination studies. Mar Pollut Bull 43(1–6):19–27

Mitnik LM (1977) Physical foundations of environmental remote sensing. Leningrad Politechnic Institute Press, Leningrad, 56 pp (in Russian)
Mohler O, Arnold F (1992) Gaseous sulfuric acid and sulfur dioxide measurements in the arctic troposphere and lower stratosphere: implications for hydroxyl radical abundances. Ber Bunsenges Phys Chem 96:280–283
Moiseyev NN (1998) Interaction of the environment and society; global problems. Herald Russian Acad Sci 68(2):167–170 (in Russian)
Moiseyev NN (1999) Humanity...will be or will be not? Znanie, Moscow, 288 pp (in Russian)
Moller D (ed) (1999) Atmospheric environmental research. Critical decisions between technological progress and preservation of nature. Springer, Berlin Heidelberg New York, 200 pp
Morgan J, Codispoti L (eds) (1995) Department of defense Arctic nuclear waste assessment program – FY's 1993–1994. Office of Naval Res., ONR 322-95-5, Arlington, VA, pp 15–30
Muller R, Peter T (1992) The numerical modeling of the sedimentation of polar stratospheric cloud particles. Ber Bunsenges Phys Chem 96:353–361
Munasinghe M, Sunkel O, De Miguel C (eds) (2001) The sustainability of long-term growth. Socioeconomic and ecological perspectives. Edward Elgar Publ, Cheltenham, 299 pp
Nakicenovic N, Grubber A, McDonald A (eds) (1998) Global energy perspectives. Cambridge Univ Press, Cambridge, 299 pp
Nesje A, Dahl SO (2000) Glaciers and environmental change. Oxford Univ Press, Oxford, 203 pp
Nielsen SN (1999) CRISP-crayfish and rice integrated system of production: 4. Simulation of nitrogen dynamics. Ecol Modelling 123(1):41–52
Nihoul JCJ, Smith J (1976) Mathematical model of an industrial river. In: Nihoul JCJ (ed) System simulation water resources Elsevier, Amsterdam, pp 333–341
Nikanorov AM, Horuzhaya TA (2000) Global environment. PRIOR Press, Moscow, 286 pp (in Russian)
Nitu C, Krapivin V, Bruno A (2000a) System modeling in ecology. Printech, Bucharest, 260 pp
Nitu C, Krapivin V, Bruno A (2000b) Intelligent techniques in ecology. Printech, Bucharest, 150 pp
Odum EP (1971) Fundamentals of ecology. Saunders, New York, 546 pp
Osterberg CL (1985) Nuclear war and the ocean. Wiley, New York, 467 pp
Palumbi SR (2001) Humans as the world's greatest evolutionary force. Science 293(5536): 1786–1790
Papakyriakou TN, McCaughey JH (1991) An evaluation of evapotranspiration for a mixed forest. Can J For Res 21(11):1622–1631
Park SU, In HJ, Lee YH (1999) Parametrization of wet deposition of sulfate by precipitation rate. Atmos Environ 33(27):4469–4475
Pauer JJ, Auer MT (2000) Nitrification in the water column and sediment of a hypereutrophic lake and adjoining river system. Water Res 34(4):1247–1254
Payne JR, McNabb GD, Clayton JR (1991) Oil-weathering behavior in Arctic environments. Polar Res 10(2):631–662
Pearce F (1995) How the soviet seas were lost. New Sci Lond 148(2003):38–42
Peng C (2000) From static biogeographical model to dynamic global vegetation model: a global perspective on modelling vegetation dynamics. Ecol Model 135(1):33–54
Petty GW (1995) The status of satellite-based rainfall estimation over land. Remote Sensing Environ 51(1):125–137
Phillips GW, August RA, Cherepenin VA, Harper MJ, King SE, Krapivin VF, Pautkin AY, Tsang FY (1997) Radionuclear pollutants in the Angara and Yenisey rivers of Siberia. Radioprotection-Colloques 32:299–304
Pimm SL (2001) The world according to Pimm. A scientist audits the earth. McGraw-Hill, New York, 303 pp
Plotnikov VV (1996) Long-term prognosis of Okhotsk Sea ice conditions with the consideration of large-scale atmospheric processes. Meteorol Hydrol 12:93–100 (in Russian)
Pokazeev KV, Medvedev AM (2000) Introduction to ecology. Moscow State Univ Press, Moscow, 200 pp (in Russian)

Porubaev VS (2000) Influence of dynamic and thermal factors on seasonal sea-ice thickness change in the Arctic Ocean. Meteorol Hydrol 11:73–79 (in Russian)
Power HC (2000) Estimating atmospheric turbidity from climate data. Ecol Model 135(1): 125–134
Precoda N (1991) Requiem for the Aral Sea. AMBIO 20(3/4):109–114
Preller RH, Cheng A (1999) Modeling the transport of radioactive contaminants in the Arctic. Mar Pollut 3(2):71–91
Preller RH, Edson R (1995) Proceedings of the ONR/NRL workshop on modeling the dispersion of nuclear contaminants in the Arctic seas. Washington, NRL/MR/7322-95-7584, 103 pp
Prykin BV (1998) Newest theoretical economics. Hyper-economics (concept of philosophy and natural science in economics). YuNITI Press, Moscow, 445 pp (in Russian)
Raven P (2001) Why we must worry. Science 293(5535):1598
Reilly J, Stone PH, Forest CE, Webster MD, Jacoby HD, Prinn RG (2001) Uncertainty and climate change assessment. Science 293(5529):430–433
Riedlinger SH, Preller RH (1991) The development of a coupled ice-ocean model for forecasting ice conditions in the Arctic. J Geogr Res 96:16,955–16,977
Robertson D, Kellow A (eds) (2001) Globalization and the environment. Risk assessment and the WNO. Edward Elgar Publ, Cheltenham, 288 pp
Rochon GL, Krapivin VF, Watson M, Fauria S, Tsang FY, Fernandez (1996) Remote characterization of the landsea interface: a case study of the Vietnam/South China Sea coastal zone. Proc 2nd Ho Chi Minh City Conf Mechanics, 24–25 Sept 1996. Institute of Applied Mechanics, Ho Chi Minh City, pp 72–74
Roden MS (1988) Digital communication systems design. Prentice-Hall, London, 519 pp
Rosen C (2000) People and ecosystems: the fraying web of life. Elsevier, Washington, 250 pp
Rosenberg GS, Mozgovoy DP, Gelashvili DP (1999) Ecology. The elements of the theoretical models of modern ecology. Samara' Science Centre of the Russian Acad Sci, Samara, 396 pp (in Russian)
Rovinsky FY, Chernogaeva GM, Paramonov SG (1995) The role of river flow and atmospheric transport in the pollution of the Russian northern seas. Meteorol Hydrol 9:22–29 (in Russian)
Rudels B, Larsson A-M, Sehistendt P-I (1991) Stratification and mater mass formation in the Arctic Ocean: some implications for the nutrient distribution. Polar Res 10(1):19–31
San Jose R (1999) Measuring and modelling investigation of environmental processes. WIT Press, Southampton, 376 pp
Sarkisyan AS (2000) Synthesis of observation data and modeling results as a perspective direction of the investigation of the oceans, seas and lakes investigation. Proc Russian Acad Sci Phys Atmos Ocean 2:202–210 (in Russian)
Sazykina TG, Alekseev VV, Kryshev AI (2000) The self-organization of the trophic structure in ecosystem models: the succession phenomena, trigger regimes and hysteresis. Ecol Model 133(1/2):83–94
Schimel DS (1995) Terrestrial biogeochemical cycles: global estimates with remote sensing. Remote Sensing Environ 51(1):49–56
Schrope M (2001) Consensus science, or consensus politics? Nature 412(6843):112–114
Schwartz DM, Deligiannis T, Homer-Dixon TF (2000) The environment and violent conflict: a response to Gleditsch's critique and some suggestions for future research. Environ. Change and Security Project Report. The Woodrow Wilson Center, Washington, DC, no 6, pp 77–92
Selivestrov YP (1998) Geography: unsolved problems or deliberate delusions. In: Kondratyev KYa (ed) Geographical Problems of the end of 20th century. Russian Geographical Community Press, St Petersburg, pp 108–128 (in Russian)
Sellers P, Meeson BW, Hall FG et al. (1995) Remote sensing of the land surface for studies of global change: models-algorithms-experiments. Remote Sensing Environ 51(1):3–26
Sellers PJ, Randall DA, Collatz GJ, Berry JA, Field CB, Dazlich DA, Zhang C, Collelo GD, Bounona L (1996) A revised land surface parametrization (SiB2) for atmospheric GCMs, part 1. Model formulation. J Climate 9(4):676–705

Shabas IN, Chikin AL (2001) A 3D sediment transport problem. Math Model (Moscow) 13(3):85–88 (in Russian)

Shah A (1998) Ecology and the crisis of overpopulation. Edward Elgar, Cheltenham, 176 pp

Shanahan P, Borchardt D, Henze M, Rauch W, Reichert P, Somlyody L, Vanrolleghem P (2001) River water quality model no.1 (RWQM1): I. Modelling approach. Water Sci Technol 43(5): 1–11

Shaw EM (1989) Engineering hydrology techniques in practice. Wiley, New York, 350 pp

Shinohara Y, Shikama N (eds) (1988) Marine climatological atlas of the Sea of Okhotsk. Tech Rep of the Meteorological Res Inst, Hokkaido, Japan, no 23, 57 pp

Shutko AM (1987) Microwave radiometry of the water surface and the ground. Science Publ House, Moscow, 204 pp (in Russian)

Shutko AM, Krapivin VF, Mkrtchyan FA, Reutov EA, Novichkhin EP, Leonidov VA, Mishanin VG, Tsankov NS (1994) Econo-ecological estimates of the effectiveness of utilizing remotely sensed data and GIS information for soil moisture and moisture related parameters determination (Geoinformation Monitoring System Approach – GIMS). Proc ICID, Varna, May 1994, pp 1–7

Shutt H (2001) A new democracy. Alternatives to a bankrupt world order. Zed Books, London, 192 pp

Sofiev MA, Galperin M (1998) The long-range transport of ammonia and ammonium in the Northern Hemisphere. Atmos Environ 32(3):373–380

Soon W, Baliunas S, Kondratyev KYa, Idso SB, Postmentier EC (2001) Calculating the climatic effects of anthropogenic CO_2 emissions: unknowns and uncertainties. Climate Res 18(3):259–275

Sorokin YI (1977) Characteristic of the Peruvian upwelling ecosystem. Proc Russ Acad Sci 236(2):497–500 (in Russian)

Soros MS (2000) Preserving the atmosphere as a global commons. Change and security project report. The Woodrow Wilson Center, Washington, DC, no 6, pp 149–155

Sportisse B (2000) Box models versus Eulerian models in air pollution modeling. Atmos Environ 35(1):173–178

Steffen W, Tyson P (eds) (2001) Global change and the earth system: a planet under pressure. IGBP Science 4, Stockholm, 32 pp

Straub CP (1989) Practical handbook of environmental control. CRC Press, Boca Raton, 537 pp

Strelkov GM (1995a) About the use of the active remote sensing method of the atmosphere in the millimetre wavelength band to define the contents of admixture gases CO and N_2O. Earth Res Space 4:3–9 (in Russian)

Strelkov GM (1995b) Active sensing of Earth's ozone layer at millimetre wavelengths. Earth Res Space 1:25–29 (in Russian)

Stroev ES (2001) Russia's self-determination and global modernization. Economics, Moscow, 352 pp (in Russian)

Stuiver M (1978) Atmospheric carbon dioxide and carbon reservoir changes. Science 199:253–258

Subetto AI (2000) The myths of Liberalisation and the destiny of Russia. Kostroma's State Univ Press, Kostroma, 143 pp (in Russian)

Sukachev VN (1945) Biogeocenology and phytocenology. Rep USSR Acad Sci 47(6):447–449 (in Russian)

Suzuki A (1992) Results of collection fishes, and trophical to temperate migrant fishes come into Okhotsk Sea coast during 1988 to 1991 in Northern Hokkaido Japan. Okhotsk Sea. Proc 7th Int Symp Okhotsk Sea and Sea Ice, 2–5 Feb 1992, Mombetsu, Hokkaido (Japan), pp 225–231

Svirezhev YM (2000) Lotka-Volterra models and the global vegetation pattern. Ecol Model 135(2/3):135–146

Swanson TM, Johnston S (2000) Global environmental problems and international environmental agreements. Edward Elgar Publ, Cheltenham, 304 pp

Terziev FS (ed) (1992) Hydrometeorology and hydrochemistry of USSR seas, vol 1. Barents Sea. Gidrometeoizdat, St Petersburg, 182 pp (in Russian)

Terziev FS, Zatuchnoy BM, Gershanovitch DE (1993) Okhotsk Sea. Gidrometeopress, St Petersburg, 167 pp (in Russian)
The Amsterdam Declaration on Global Change (2000) Amsterdam, June 2000, 2 pp
Thiessen KM, Thorne MC, Maul PR, Prohl G, Wheater HS (1999) Modelling radionuclide distribution and transport in the environment. Environ Pollut 100(1-3):151-177
Tickel C (2001a) Can development be sustainable? A lecture delivered at the Architectural and Planning Associated Parliamentary Group. House of Commons, London, 8 pp
Tickel C (2001b) Pressures for conflict: population, resources and environmental change. Harvard Distinguished Environmental Lecture Series, Cambridge, MA, 15 pp
Tikunov VS (1997) Modelling in cartography. Moscow State Univ Press, Moscow, 400 pp (in Russian)
Tol RSJ (2000) International climate policy: an assessment. IHDP Update 3:11-12
Trofimov AM, Kotliakov VM, Seliverstov YP, Panasuk MV, Rybzov VA, Pudovik EM (1999) Balanced development - sustainable state of the geosystems. Proc Russ Geogr Soc 131(3): 9-16 (in Russian)
Truskov PA, Astafiev VN, Surkov GA (1992) Problems of choice of sea ice cover parameters design criteria. Proc 7th Int Symp Okhotsk Sea and Sea Ice, 2-5 Feb 1992, Mombetsu, Japan, pp 21-25
Turner DP, Koerper G, Gucinski H, Peterson C, Dixon RK (1993) Monitoring global change: comparison of forest cover estimates using remote sensing and inventory approaches. Environ Monitor Assess 26(2/3):295-305
Ursul AD (1998) Russia's transition to sustainable development: the noosphere strategy. Noosphere, Moscow, 500 pp (in Russian)
Uvarov A (2000) Russian national self-consciousness: a modern look. ITRK Press, Moscow, 200 pp (in Russian)
Vail Regional Roundtable Report for Europe and North America (2002) 2002 summit on sustainable development, CO, 6-8 June 2001, 11 pp
Vaughan DG, Marshall GJ, Connolley W, King JC, Mulvaney R (2001) Devil in the detail. Science 293(5536):1777-1779
Vera R (1992) Demolition of a disastrous dam. New Sci (Lond), 134(1816):8
Vernadski VI (1926) Biosphere, essays, the first and second. Nauchtekhizdat, Moscow, 120 pp (in Russian)
Vernadski VI (1940) Biogeochemical essays. Nauchtekhizdat, Moscow, 130 pp (in Russian)
Vernadski VI (1944) A few words about the noosphere. Successes Mod Biol 18(2):49-93 (in Russian)
Vilkova LP, Novitchikhin EP, Santalov NP, Yakovleva GD (1998) An estimation of biogeochemical carbon cycle parameters in global and regional scales. Prob Environ Nat Res 7:24-37 (in Russian)
Vinogradov GM, Marasaeva EF, Larionov VV, Matishov GG (2001) Communities of zooplankton for the ice covered aquatories of Barents and Kara Seas in the winter-summer period of 2000. Proc Russian Acad Sci 376(6):815-817 (in Russian)
Vinogradov ME (1977) Pelagic ecosystems studies on the upwelling of the Eastern Pacific Ocean: cruise 17 of the R/V "Akademic Kurchatov". Pol Arch Hydrobiol 24:7-19
Vinogradov ME, Menshutkin VV, Shushkina EA (1972) On mathematical simulation of a pelagic ecosystem in tropical waters of the ocean. Mar Biol 16(4):261-268
Vinogradov ME, Krapivin VF, Menshutkin VV, Fleishman BS, Shushkina EA (1973) Mathematical model for the ocean ecosystem functioning in the tropical pelagic waters. Oceanology (Moscow) 5:852-866 (in Russian)
Vinogradov ME, Shushkina EA, Kukina IN (1977) Structural and functional analysis of pelagic communities in equatorial upwelling. Polar Arch Hydrobiol 24:503-524
Vital Sings 2001-2002 (2001) The trends that are shaping our future. Earthscan Publ, London, 192 pp
Vloedbeld M, Leemans R (1993) Quantifying feedback processes in the response of the terrestrial carbon cycle to global change: the modeling approach of IMAGE-2. Water Air Soil Pollut 70(1-4):615-628

Wald A (1960) Sequential analysis. Phys-Math Literature Press, Moscow, 328 pp (in Russian)
Wania F, Hoff JT, Jai CQ, Mackay D (1998) The effect of snow and ice on the environmental behaviour of hydrocarbic organic chemicals. Environ Pollut 102(1):25–41
Ward RC, Loftis JC, McBride GB (1990) Design of water quality monitoring system. Van Nostrand Reinhold, New York, 231 pp
Watson RT, Noble IR, Bolin B, Ravindranath NH, Verardo DJ, Dokken DJ (2000) Land use, land-use change, and forestry. Cambridge Univ Press, Cambridge, 377 pp
Wefford R (1997) Hijacking environmentalism. Corporate responses to sustainable development. Earthscan Publ, London, 251 pp
Wielgolaski FE (ed) (1997) Polar and alpine tundra. Elsevier, New York, 930 pp
Williamson P, Platt T (1991) Ocean biogeochemistry and air-sea CO_2 exchange. Global Change Newslett 7:3–4
Wirtz KW (2000) Second order up-scaling: theory and an exercise with a complex photosynthesis model. Ecol Model 126(1):59–71
Wyers GP, Erisman JW (1998) Ammonia exchange over a coniferous forest. Atmos Environ 32(3):441–451
Yablokov AV (2001) Radioactive waste disposal in seas adjacent to the territory of the Russian Federation. Mar Pollut Bull 43(1–6):8–18
Yakovlev OI (2001) Space radio science. Taylor and Prancis Group, London, 320 pp
Yanovsky RG (1999) Global changes and social security. Academia Press, Moscow, 358 pp (in Russian)
Yokozawa M (1998) Effects of competition mode on spatial pattern dynamics in plant communities. Ecol Model 106(1):1–16
Young R (2001) Uncertainty and the environment. Edward Elgar Publ, Cheltenham, 192 pp
Yudakhin FN (ed) (2000) The North: ecology. Ural Branch of the Russ Acad Sci Press, Ekaterinburg, 415 pp (in Russian)
Zonneveld C (1998) A cell-based model for the chlorophyll *a* to carbon ratio in phytoplankton. Ecol Model 113(1–3):55–70

Index

Location: http://www.springer.de/geosci/

You are one **click** *away*
from a **world of geoscience** *information!*

Come and visit **Springer's**

Geosciences Online Library

Books

- Search the Springer website catalogue
- Subscribe to our free alerting service for new books
- Look through the book series profiles

You want to order? Email to: orders@springer.de

Journals

- Get abstracts, ToC´s free of charge to everyone
- Use our powerful search engine LINK Search
- Subscribe to our free alerting service LINK *Alert*
- Read full-text articles (available only to subscribers of the paper version of a journal)

You want to subscribe? Email to: subscriptions@springer.de

Electronic Media

- Get more information on our software and CD-ROMs

You have a question on an electronic product? Email to: helpdesk-em@springer.de

Bookmark now:

http://**www.springer.de/geosci/**

Springer · Customer Service
Haberstr. 7 · D-69126 Heidelberg, Germany
Tel: +49 6221 345-217/218 · Fax: +49 6221 345-229
d&p · 006910_001x_1c

Springer

MIX
Papier aus verantwortungsvollen Quellen
Paper from responsible sources
FSC® C105338

If you have any concerns about our products,
you can contact us on
ProductSafety@springernature.com

In case Publisher is established outside the EU,
the EU authorized representative is:
Springer Nature Customer Service Center GmbH
Europaplatz 3, 69115 Heidelberg, Germany

Printed by Libri Plureos GmbH
in Hamburg, Germany